Dr Luis M. Chiappe is Director of the Dinosaur Institute and Curator of the Department of Vertebrate Paleontology at the Natural History Museum of Los Angeles County. Dr Chiappe has conducted extensive research on the evolution of dinosaurs, particularly the origin and early evolution of birds. His research has been published in many articles as well as books, including *Walking on Eggs* and *Mesozoic Birds: Above the Heads of Dinosaurs*. He lives in Santa Monica, California.

Glorified Dinosaurs

The Origin and Early Evolution of Birds

LUIS M. CHIAPPE

UNSW PRESS

BICENTENNIAL
1807
WILEY
2007
BICENTENNIAL

Published in Australia and New Zealand by
University of New South Wales Press Ltd
University of New South Wales
Sydney NSW 2052
AUSTRALIA
www.unswpress.com.au

Published in the rest of the world by
John Wiley & Sons, Inc.
111 River St MC 8-01
Hoboken NJ 07030-5774
USA
www.wiley.com

Library of Congress Cataloging-in-Publication Data
Chiappe, Luis M.
The glorified dinosaurs : origins and early evolution of birds / Louis M. Chiappe.
p. cm.
Includes index.
ISBN 0-471-24723-5 (cloth)
1. Birds, Fossil. 2. Birds--Evolution. 3. Paleontology--Mesozoic. I. Title.
QE871.C45 2006
568--dc22   2006029317

National Library of Australia
Cataloguing-in-Publication entry
Chiappe, Luis.
Glorified dinosaurs: the origin and early evolution of birds.
Bibliography.
Includes index.
ISBN 978 0 86840 413 4.
1. Birds - Evolution. 2. Birds, Fossil. 3. Velociraptor. 4. Paleontology. I. Title.
598.138

Design Di Quick
Cover illustration Nicholas Frankfurt
Printed in China

14792814

Contents

To my wife, Megan

Preface

What I had to do,
I did the best I could

HENRI MATISSE

With nearly 10,000 living species, birds are today's most diverse group of land animals and the heirs of a successful and fascinating story of vertebrate evolution. Ubiquitous and endowed with colorful plumages, exuberant behaviors, and an unrivaled grace for aerial propulsion, birds have always captured the interest of man. Since antiquity, philosophers, natural historians, and scientists have dedicated their efforts to studying the life of birds and understanding their place within our natural world. Still, with their feathered bodies, beaked heads, short bony tails, and other unusual features, the evolutionary origin of birds—and the search for their closest relatives—has been until recently a puzzling mystery of scientific inquiry. In the last few decades, however, our understanding of the origin and ancient divergences of these animals has advanced at an unparalleled speed. Not only has the ancestry of birds been greatly clarified but the anatomical and genealogical chasm between the first birds and the evolution of their living counterparts has been substantially filled. This book is an attempt to summarize the large volume of new evidence and provide a synthetic view of the Mesozoic saga of birds, an evolutionary odyssey narrated over 85 million years of our geologic past. This view, however, is not free of bias—my conclusions are the result of years of examining the pertinent fossil record and the interpretation of this evidence in light of specific methods for reconstructing genealogy.

Avian beginnings can be traced back to at least 150 million years ago, deep into the Mesozoic Era, a time when large dinosaurs dominated all terrestrial ecosystems, when climatic patterns were vastly different from those of today, and landmasses and oceans had a completely different configuration. But to understand the place birds have in the natural history of our planet, we need to dive far beyond, to the dawn of dinosaurs more than 230 million years before our present, for a vast amount of fossil evidence indicates that the ancestry of birds is deeply nested within these fascinating animals. Indeed, the fossils support the idea that birds evolved from ground-dwelling, predatory dinosaurs known as maniraptoran theropods. This and other hypotheses about avian origins, the development of feathers and the debate about the beginnings of flight are all related evolutionary events discussed in detail in the following pages. Because these events need to be explained through a genealogical lens, this book also presents methods and concepts used for establishing descent among organisms, setting up the *cladistic* way of thinking that will guide us throughout this book. Using these cladistic concepts, we can argue that because the ancestry of birds can be traced back to maniraptoran dinosaurs, it is logical to conclude that birds are living dinosaurs. And if birds encompass a group of dinosaurs, we could also logically reason that birds are reptiles, for all dinosaurs are classified as reptiles. Strange as it may seem, the reality (given the evidence at hand) is that birds are both avian dinosaurs and avian reptiles—they are a subset of dinosaurs and an even smaller subset of reptiles. Following these concepts, the traditional dinosaurs—from *T. rex* to *Triceratops* to *Brachiosaurus*—will be often referred to as large or nonavian dinosaurs.

The oldest preserved pages of the evolutionary history of birds are limited to a handful of fossils

of the famous *Archaeopteryx lithographica*. Few physical features set this 150-million-year-old bird—one with teeth, clawed wings, and a long skeletal tail—apart from its theropod dinosaur predecessors. In fact, new fossils and analyses have shown that many of the features commonly associated with birds—from feathers to wishbones to hollow bones—first evolved in theropod dinosaurs. Yet *Archaeopteryx* gives us paramount clues to the beginning of one of the most dramatic evolutionary events in the history of vertebrates—the development of powered flight in birds. With the Late Jurassic *Archaeopteryx* we begin to tell a story of transformation, reviewing the long history of anatomical changes that culminated in the unique aerodynamic design of modern birds. Following the sequence of early evolutionary divergences along a cladogram—a tree-like diagram of genealogical relationships—we discuss every later lineage of Mesozoic birds that played a significant role in the evolution of the group. We see how subsequent avian divergences are each a step further apart from their maniraptoran ancestors. Traveling across the cladogram beyond *Archaeopteryx*, we examine a diversity of other primitive, long-tailed birds. From here, we look at the rapid diversification of short-tailed birds. Undoubtedly, the abbreviation of the skeletal tail and the development of a number of critical aerodynamic features played a principal role in the fine-tuning of flight, an ability that birds mastered very early in their history and the recipe for their enormous success. Yet the details of this evolutionary transition are far from clear—the dawn of short-tailed birds is still shrouded in a cloud of mystery.

The 125-million-year-old *Confuciusornis sanctus* and its kin give us a picture of the earliest known short-tailed birds. These primitive-beaked birds thrived in the tropical environments that dominated eastern Asia during the Early Cretaceous—their fossils have been collected by the hundreds. The next divergence of our cladogram takes us to the most successful group of Mesozoic birds: the Enantiornithes. Like no other lineage of early birds, the history of discoveries of these fossils illustrates the remarkable burst of evidence unearthed in recent years—despite the fact that half of the known Mesozoic birds are Enantiornithes, the group itself was not recognized until two decades ago. A significant diversity of songbird-sized enantiornithines is recorded not much later than the start of the Cretaceous. The secret of such a successful and apparently rapid diversification may lie in their superior flying capabilities. Indeed, the skeletal plan of the earliest enantiornithines underwent a series of important transformations with respect to their short-tailed predecessors. It is likely that these skeletal innovations, combined with their small size, enabled them to fly nearly as well as their living relatives, and therefore to occupy a wide range of environments. The abundant fossil record of these birds, one that extends throughout the Cretaceous, provides important evidence for understanding the diversity of lifestyles, feeding behaviors, developmental strategies, and growth rates of early birds.

The next stop on our genealogical journey takes us to the earliest stages of the evolutionary radiation that culminated with the divergence of all modern birds. Here we explore the significance of the most archaic ornithuromorphs. Notorious among these primitive forebears of living birds is *Patagopteryx deferrariisi*, a Late Cretaceous animal that among other things offers clues to an early episode of flightlessness, a relatively common trend in the later evolutionary history of birds. Yet, the fossil record tells us that the diversification of ornithuromorphs is significantly more ancient, because a much older cast of these birds was unearthed from the same 125-million-year-old sediments that contain the most primitive short-tailed birds. Following the analysis of these primitive ornithuromorphs, our discussion visits other fossils that are closer to modern birds.

Remarkable among these are *Hesperornis regalis* and its allies, which with torpedo-like bodies, powerful hindlimbs, and atrophied wings, became superb foot-propelled divers of the Late Cretaceous seas.

If the Mesozoic Era witnessed a great deal of avian diversification, it was also a time of abundant extinction. Indeed, of the many lineages of birds that thrived during this long interval of our geologic past, all but one disappeared before the Mesozoic gave way to the current Cenozoic Era. Our journey through the evolutionary history of early birds ends with the humble beginnings of modern birds—the only avian lineage to survive the Mesozoic. We look at the possible closest relatives of our modern companions—toothed birds, but ones which were in many ways modern-looking. At the end of our story we also review the controversy around the earliest evolutionary divergence of modern birds, namely whether their origin is buried deep in the Cretaceous or if the first modern birds actually evolved at a significantly later time, not far from the closure of the Mesozoic.

History is often approached with two major purposes. On the one hand, we examine historical evidence as a way of easing our never-ending curiosity about what the past was like. On the other hand, we also look at history with hopes of finding clues for a better future, and as a source of information to solve our current problems. Our interest in the early history of birds is not an exception. We are naturally fascinated by birds that also had teeth, clawed wings, and long bony tails, a whole roster of unfamiliar animals that would take birding to a new dimension. We are also intrigued by the evolutionary mysteries that led to the development of the unique features of living birds and their mesmerizing aerodynamic capabilities. But by learning more about the ancient history of modern birds we also hope to reach a better understanding of the life of our familiar companions, because buried in such a history may also be clues about how to protect and manage the last descendants of a class of animals that once ruled the world.

CHAPTER ONE

Mesozoic Birds and Their World

The diversity of types, colors, songs, and behaviors of birds have always fascinated us. With approximately 10,000 living species traditionally classified in at least 27 major lineages or orders, birds are the most speciose and most conspicuous group of land vertebrates. Yet this enormous

diversity of forms is composed of just the survivors of an ancient and much larger evolutionary radiation that can be traced back to the famous 150-million-year-old

Archaeopteryx lithographica from the Late Jurassic Solnhofen Limestones of southern Germany—a toothed, crow-sized bird with powerful hand claws and a long bony tail.

The remarkable diversity of living birds constitutes a fraction of the many lineages that diverged during the long evolutionary history of the group. Much of this history was played out in the Mesozoic Era, when large dinosaurs ruled all terrestrial ecosystems. (Image: R. Urabe.)

The origin of birds, the diversity and genealogical relationships of their major lineages, the beginning and refinement of flight, the timing of divergence of extant groups, and the acquisition of their main functional and physiological specializations are just some of the evolutionary events that have captured the interest of decades of paleontological research. Because more than half of the long evolutionary saga of birds was played out in the Mesozoic Era—the geologic time interval that spans between 248 and 65 million years ago—it has long been known that the answers to these significant events remain locked in the fossil record of the early history of the group. For many decades, however, the evidence available for investigating these issues was limited to a small number of fossils, greatly separated both by time and by their physical disparity, and largely restricted to the near-shore and marine environments of the Mesozoic. Only recently has this situation changed, as increasing numbers of worldwide discoveries of Mesozoic-aged birds have begun to reveal the enormous and unexpected diversity of these early lineages and make it possible to piece together the evolutionary events that led to the familiar forms of today.

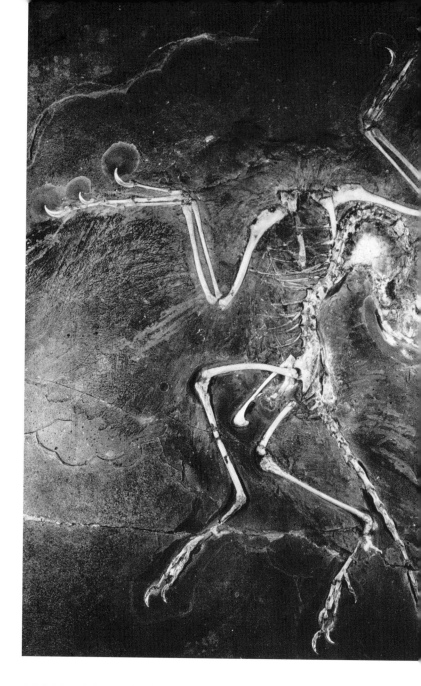

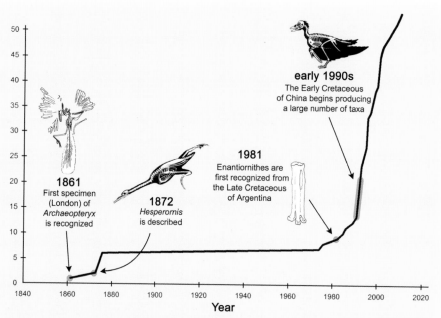

1861
First specimen (London) of *Archaeopteryx* is recognized

1872
Hesperornis is described

1981
Enantiornithes are first recognized from the Late Cretaceous of Argentina

early 1990s
The Early Cretaceous of China begins producing a large number of taxa

Year

A tremendous explosion of discoveries of Mesozoic birds has characterized the last 25 years. This trend continues unabated, with several new species of early birds being discovered every year. (Image after Chiappe and Dyke, 2002.)

The Berlin specimen of *Archaeopteryx* photographed under ultraviolet light. The 150-million-year-old *Archaeopteryx* constitutes the earliest known evidence of birds. (Image: H. Tischlinger.)

Without a doubt, the origin and early evolution of birds is today one of the most rapidly growing fields of vertebrate paleontology—the scientific discipline that studies fossils of backboned animals. Numerous fossils of dinosaurs closely related to birds and of primitive birds close to the ancestry of the group are found every year. Most of these are identified as species new to science. Dozens of scientific papers discussing their anatomy, genealogy, and physiology are published annually. These are often followed by papers presenting alternative scenarios for the evolution of complex attributes such as flight, feathers, or warm-bloodedness. This book is an attempt to summarize the enormous volume of evidence that is available on the long evolutionary history of one of the most conspicuous groups of vertebrate animals—one which draws its origins from those most fascinating creatures, the dinosaurs.

The Deep History of Birds

Birds have an enormously long history, one that is so old that it is as hard to comprehend as the vast amount of time involved in the geological processes that have shaped our planet or the astronomic events that created the universe. For good reason, scientists talk of human history as just a blink of an eye when compared to the whole history of the Earth, let alone that of the universe. Indeed, most people living in wealthy countries such as the United States or those in Europe live for only 75 or so years. The borders of most countries and their political structures have been defined within the last 200 years. Most modern religions were created during the last two millennia. Agriculture, together with the development of cities, dates back approximately 10,000 years. The magnificent Ice Age fauna of the renowned La Brea tar pits lived between 10,000 and 40,000 years ago. The earliest evidence of human-controlled fire dates back 800,000 years. Our own species evolved during this last million years and the entire human lineage can be traced back only some six or seven million years. The ice caps that cover the poles of our planet are not much older than 20 million years. Baleen whales diverged from toothed whales some 35 million years ago. Most lineages of living mammals originated during the last 60 million years. *Tyrannosaurus rex* and the last large dinosaurs died out 65 million years ago. The connection between North and South Atlantic oceans has been dated at about 100 million years old. Yet birds have an even longer past. If their earliest known records date back to 150 million years ago, the evolutionary divergence of the group is likely to be found much deeper within the Mesozoic Era, at a time when the geographic, climatic, and biotic landscapes of the world were utterly different from today.

The Mesozoic World

Indeed, the world during the Mesozoic Era was drastically different from that of the present day. Mean temperatures were much higher and their gradients were significantly lower. As a consequence, ice caps did not exist and Polar Regions housed a variety of temperate-climate plants. Seas flooded large portions of continents and latitudinal currents prevailed over other oceanic currents. Continental masses had a considerably different configuration and large dinosaurs dominated all land environments. In this radically unfamiliar world, birds and most other lineages of modern vertebrates had their evolutionary origin.

The Mesozoic is subdivided into three periods—the Triassic, the Jurassic, and the Cretaceous. In the Triassic Period (from 245 to 208 million years ago), all continents were merged into a single and enormous landmass—the Pangea. The Panthallassic Ocean, a gigantic sea occupying an entire hemisphere, washed the shores of this supercontinent.

DATING ROCKS

Dating the age of fossil-bearing rocks is usually approached in two general ways. The first one entails comparing the fossils from the rock layers to be dated with other fossils from rock layers for which the age is known—or at least better known. Usually, these are marine rock layers, which often contain microorganisms and the diversity of invertebrate fossils that are more abundant and more specific to a particular stratum than vertebrate fossils. This biostratigraphic approach has its roots in the British geological tradition of the 18th and 19th centuries, when the work of pioneer geologists such as James Hutton and Charles Lyell led the way to the geological principle of superposition—that the oldest of the rock layers in a stratigraphic column is at the bottom. This method is often aided by general stratigraphic comparisons, which look at the relative position of the strata in question with respect to strata elsewhere. This approach could provide important information about the age of a particular fossil deposit. For example, it could ascertain whether this deposit is older or younger than other rock layers, and if the ages of these rock layers are better known, a general idea of the age of the fossil deposit could be obtained.

Magnetic analyses of fossil-bearing rocks can also provide additional information about the age of strata. The positive magnetic pole has been switching from north to south and back again throughout the history of the Earth. This phenomenon seems to be caused by a series of interactions between the solid inner core of the Earth and the molten metals and rocks that form the outer layers of our planet. Occasional disturbances of the convective currents responsible for generating the magnetic field result in polarity changes. The varying polarities of the Earth's magnetic field are recorded in the orientation of iron-bearing minerals contained in many sedimentary and volcanic rocks. This occurs because particles of these minerals orient themselves according to the polarity at the time when the rocks formed. Each switch in the sequence of normal and reverse (opposite from today's) polarities documented throughout the history of Earth has been accurately dated. Rock samples from a given stratigraphic section can provide information about the particular polarity that existed when the rocks formed and the fossilized organisms were alive. This information, coupled with general stratigraphic comparisons and biostratigraphic correlations could narrow the age of a particular stratum to a relatively small time interval, at least in geological terms.

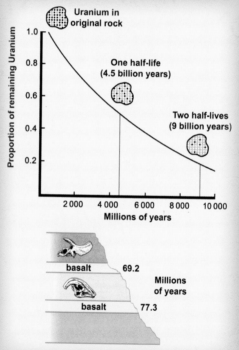

In some instances, the rocks containing fossils can be dated more precisely. Yet, to establish the time of formation of a rock layer, it is necessary to discover a kind of timepiece within the stratum in question. Several of these dating systems are known and applied today. The most common one, often known as radioisotopic dating, uses our knowledge about the radioactive nature of some chemical elements contained in volcanic rocks. Some fossils are entombed in layers of volcanic ash containing mineral crystals that were formed during an eruption. These crystals contain isotopes (forms of a chemical element, such as uranium) whose atoms decay radioactively, breaking apart into other atoms at a constant rate. The atoms that break apart are called parent atoms and those that result from the decay of the parent atoms are called daughter atoms. Certain instruments can measure the time it takes for half of the parent atoms to decay into their daughter atoms—the half-life of a radioactive isotope. The amount of parent and daughter atoms that were formed when the ancient volcano erupted can also be measured. By knowing the half-life of a radioactive element contained in a sample of ancient volcanic ash, the ratio of parent to daughter atoms that exist in the sample, and the ratio contained when it was formed, the age of the ash layer containing a fossil can be calculated.

The principle of radioisotopic dating is comparable to the familiar carbon-14 method of dating archaeological sites—except that carbon-14 is useless for dating Mesozoic rocks. Because the half-life of carbon-14 is so brief (some 5,700 years), the amount of this element left after 65 or more million years of radioactive decay is so insignificant that the resultant dating would be highly inaccurate. For Mesozoic rocks and fossils, radioactive isotopes of elements whose half-lives are billions of years long, such as uranium, need to be found if the volcanic ashes in question, and the fossils they contain, are to be dated with accuracy.

Similar dating methods can be used for other types of volcanic rocks, such as ancient lavas and basalts. These do not preserve fossils, because the scorching heat of these rocks as they form would have burned any organic remains they encountered long before cooling and consolidating. Nonetheless, fossils could be found in rocks that are stratigraphically below or above basalts, thus allowing them to be dated in relation to the age of the basalts (older or younger). (Image after Martin, 2004.)

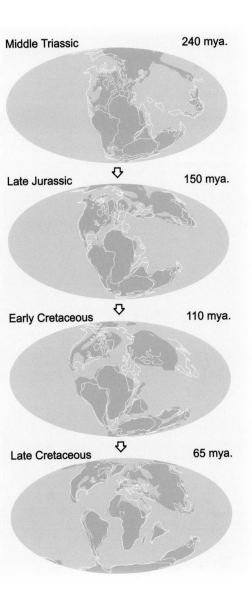

Middle Triassic — 240 mya.

Late Jurassic — 150 mya.

Early Cretaceous — 110 mya.

Late Cretaceous — 65 mya.

Drastic changes in the configuration of the continents occurred during the Mesozoic Era. Paramount among these changes was the breakdown of the Triassic supercontinent Pangea (ancient emerged landmasses in brown) into landmasses that 65 million years ago (mya) approached the modern continental configuration (white outlines). (Image: R. Urabe.)

The Triassic weather was hot and dry, and the distribution of fossil plants hints at a minimal climatic zonation. Under this warm and dry climate, vast deserts and arid regions developed and shallow seas that had earlier flooded continental areas evaporated into large accumulations of salt. The Triassic followed the largest mass extinction our planet ever endured, that of the end of the Paleozoic Era. This period of intense animal turnover witnessed a great diversification of reptiles as well as the evolution of the first mammals. The winged pterosaurs, turtles, crocodiles, and the marine ichthyosaurs all have their first fossil records and presumably their origin during the Triassic. The last 30 million years of this period also saw the emergence of dinosaurs, which progressively came to dominate the scene in all terrestrial ecosystems. Towards the end of the Triassic, the powerful tectonic processes that had created Pangea began to break this enormous supercontinent apart. These disruptions, resource and habitat competition with the newly differentiated dinosaurs, and perhaps the impact of an asteroid, drove much of the non-dinosaurian diversity of Triassic reptiles to extinction and paved the road for dinosaur supremacy.

Throughout the Jurassic Period (from 208 to 146 million years ago), the climate remained much warmer than it is today—the average annual temperatures at the beginning of this period are thought to have been close to 20°C (68°F) as opposed to the 14°C (57°F) average that prevails today. Enormous deserts continued to cover parts of South America, Africa, and southern North America, although large coal beds indicate there was a general trend to global moistening. Shallow seas teeming with life flooded large portions of continents. In this blue world, Europe became an archipelago and Asia was largely underwater. Relentless tectonic forces kept breaking Pangea apart. By the end of the Jurassic, it had been fragmented into two large continental masses, Gondwana in the south and Laurasia in the north, completely separated by an equatorial sea—the Tethys. This geographic fragmentation intensified the already ongoing evolutionary diversification of dinosaurs. In the Jurassic, stegosaurs, ankylosaurs, brachiosaurs, allosaurs, coelurosaurs, and many other familiar lineages of dinosaurs made their debut into the fossil record, and not only did dinosaurs evolve their characteristic gigantic dimensions but they also transitioned into birds. With these dramatic changes in the physical appearance of the Earth, some familiar geographical features appeared. Heralded

by intense volcanic activity, the Atlantic Ocean began to open; but it would not be until the Cretaceous (146 to 65 million years ago) that Earth began to acquire its modern continental configuration.

In this last period of the Mesozoic Era, not only did the continents acquire their basic modern features but most lineages of the organisms we know today also appeared. Gondwana began to break apart in the early part of the Cretaceous. Together, India and Madagascar split from Africa about 130 million years ago, and Africa and South America were completely separated from each other some 30 million years after that. Some believe that a narrow bridge maintained an emerged connection between Indomadagascar and Antarctica, thus allowing the exchanges of terrestrial faunas between India, Madagascar, and South America millions of years after the initial fragmentation of Gondwana. The Asian continent also began to be assembled as we know it. A series of giant slices of continental crust, which at the end of the Jurassic had separated from Gondwana, slowly drifted northwards to collide with other continental masses in the early part of the Cretaceous. This collision ended up adding today's Tibet, southern China, and Southeast Asia to the protocontinent of Asia. Slowly acquiring most of its familiar geography, Asia achieved much of its modern configuration by the end of the Cretaceous.

As in previous Mesozoic times, the Cretaceous climate was also warm. In some areas, these high temperatures hosted exuberant vegetation, which left behind vast deposits of coal. Flowering plants, which had originated early in the Cretaceous, underwent a rapid diversification, becoming the dominant flora by the end of the period. This significant event was paralleled by the evolution of many modern groups of insects, especially those of pollinating species. Cretaceous sea levels were also higher than at present—100 million years ago, the sea level was some 180 meters above today's level. For most of the last 40 million years of the Cretaceous, a large, north–south tongue of shallow sea, the Pierre Seaway (also known as the Western Interior Seaway), dissected North America. A diverse fauna of vertebrates flourished on vast coastal regions and deltas that developed on both sides of this tropical seaway. Africa and South America also had significant marine embayments, with shallow seas flooding extensive portions of these continents and unique vertebrate faunas that inhabited the lowlands that lined these extensive shores. The end of the Cretaceous was marked by the impact of a 9-kilometer-wide asteroid that left an enormous crater in the present day Yucatán Peninsula. It was also a time of active volcanism, global cooling, and falling sea levels, all factors that are likely to have contributed to a mass extinction at the end the Mesozoic Era— including the disappearance of the last large dinosaurs that ruled it.

The last 570 million years of the Earth's history are known as the Phanerozoic, a time interval subdivided into the Paleozoic, the Mesozoic, and the Cenozoic Eras. Phanerozoic fossils document the complex evolutionary history of multicellular organisms, but the timing of most major evolutionary events, including those listed here, cannot be determined with precision because of the incompleteness of the fossil record. (Image: R. Urabe.)

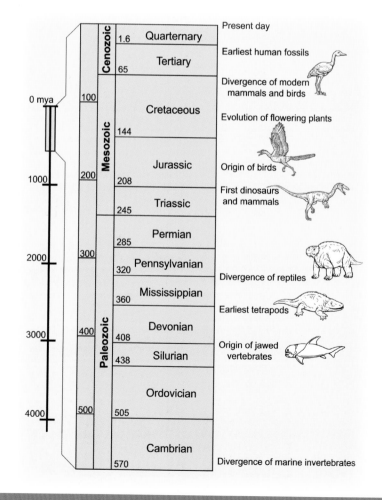

PLATE TECTONICS: UNDERSTANDING THE DYNAMIC HISTORY OF THE CONTINENTS

Plate tectonics is a universal theory that allows most geological and geographical phenomena to be explained as the result of a highly dynamic top layer of the Earth or crust, subdivided into multiple portions or plates (most containing continental masses) that over millions of years have been colliding, diving one under the other, or simply passing by each other. This theory has a precedent in the work of German climatologist Alfred Wegener, who in 1912 drew maps connecting continents based on the outline of their oceanic platforms, the types of rocks found in coastal regions, glacial evidence, and the distributions of modern and prehistoric plants and animals. Despite the large volume of evidence indicating that the continents had moved during the geologic history of our planet, Wegener's theory of continental drift was largely dismissed due to the lack of a mechanistic explanation.

Nearly fifty years later, however, when exploration of the ocean floor revealed a series of mid-oceanic mountain ridges and a symmetrical pattern of alternating normal and reversed magnetic stripes in rocks formed on both sides of the submerged ridges, a theory explaining Wegener's observations was developed. This theory suggested that the magnetic patterns observed at the bottom of the oceans were the result of new oceanic floor being formed over time at the submerged ridges. Plate tectonics also suggested that all existing oceanic floors were much younger than the continents, which passively rode on the larger tectonic plates.

Since then, a multitude of additional evidence in support of this theory has been amassed. This evidence ranges from the continuation of mountain belts and stratigraphic columns between continents that are separated by hundreds of kilometers to the simple measurement of the speed at which continents move—measured as locations change their

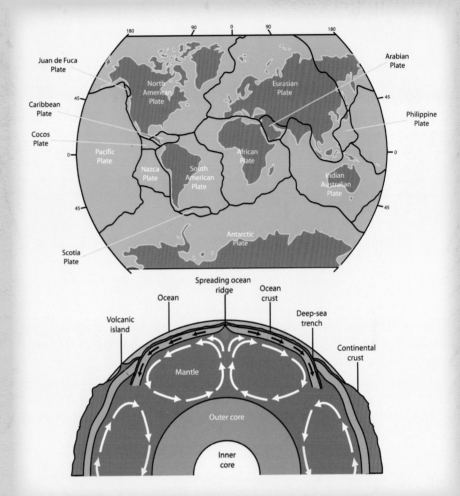

global positions (GPS positions) over time. Today, we know that on average continents move a few centimeters every year, roughly at the same speed fingernails grow. This phenomenon is a consequence of the nature of the rocks that form the continents and the floor of the oceans as well as of the characteristics of the Earth's mantle, the layer of molten rocks beneath the crust. Because crustal rocks are lighter than the molten rock of the mantle, the tectonic plates formed by continents, oceanic floor, or a combination of these crustal portions (solid lines on map) "float" on top of the molten mantle. The molten rocks of the mantle circulate by means of convecting currents (white arrows), in which the hottest rock rises in areas of upwelling and

dives deeper into the mantle in areas of downwelling. This circulation occurs because as the molten rock of the mantle sheds heat near the mantle's surface, it cools down and becomes denser; and thus, it sinks again. Such circulation is responsible for the movements of the tectonic plates, because in areas of upwelling, the molten rock of the mantle surfaces through either a spreading ocean ridge generating new ocean crust, or through volcanic islands or volcanoes associated with areas of continental rift. The ocean crust is recycled in areas of downwelling, such as deep-sea trenches, when the diving molten rock pulls down ocean crust that slides under the lighter continental crust in a process known as subduction. (Image: R. Urabe.)

Birds within the Context of Vertebrates

The end of the large dinosaurs was perhaps the last major evolutionary event that facilitated the modern supremacy of birds over land ecosystems. Yet many other breakthroughs in vertebrate evolution had to occur for birds to appear.

Vertebrates originated from small and jawless fish-like animals that lived at the beginning of the Paleozoic Era, more than 500 million years ago. These earliest vertebrates also lacked a skeleton, although it did not take long for bone to evolve. During the first part of the Paleozoic, vertebrates evolved into a great variety of aquatic lineages that, while still jawless, were covered by a bony carapace. Another momentous evolutionary event took place some 100 million years after the appearance of the first known vertebrates, with the evolution of jaws and of true predatory habits. Among these primitive jawed vertebrates were the earliest representatives of the two main groups of modern fish—sharks and other cartilaginous fish, and bony fish. For more than 150 million years since their debut in the fossil record, vertebrates remained exclusively aquatic, primarily inhabiting the marine realm. However, in the late part of the Devonian Period, somewhere between 370 and 350 million years ago, some of the early lobe-finned relatives of coelacanths underwent a series of important structural and physiological transformations that enabled their tetrapod descendants to venture into terrestrial environments.

The conquest of land is undoubtedly one of the most significant events in the evolutionary history of vertebrates. Numerous specializations had to evolve for tetrapods—four-limbed vertebrates—to function outside the aquatic realm of their ancestors. The evolution of the water-tight egg was the crucial development in becoming entirely independent of an aqueous existence.

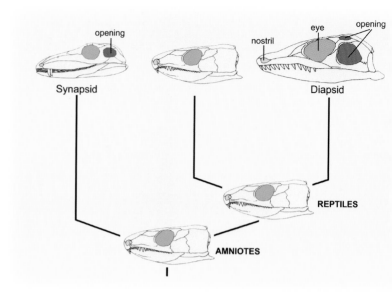

The evolutionary steps involved in the origin of this important structure are not entirely clear, but it is known that the ancestor of all amniotes—the group of tetrapods that includes mammals, turtles, lizards, crocodiles, birds, and many extinct animals—already laid water-tight eggs. The evolution of amniotes is thus framed within a terrestrial existence.

Early in their history, amniotes diverged into two main lineages that are best recognized by the characteristics of the skull portion that lies behind the orbit. Specifically, these major groups are set apart by the development and position of temporal openings, windows in the skull that most likely developed in concert with the evolution of enlarged jaw muscles. Although the position, size, and number of temporal openings differs greatly both between and within these two lineages, one of these main groups of amniotes, the synapsids, evolved a single opening placed low on the side of the rear skull. The earliest synapsids are known from the Carboniferous Period, roughly from about 300 million years ago. Following a long and complex evolutionary history, synapsids gave rise to the first mammals in the Triassic, and ultimately to us. The second major group of amniotes includes the reptiles.

Like most of their living descendants, the most primitive reptiles were probably diurnal and endowed with color vision. Very early in their history, however,

The early evolutionary history of amniotes is characterized by the development of openings in the rear portion of the skull, behind the orbit. While the group leading to mammals evolved an opening called the synapsid opening, some reptiles evolved the two openings typical of the diapsid skull. Living birds exhibit a modified version of the latter. (Image: R. Urabe.)

reptiles diverged into two main groups. Retaining the solid skull design of the most primitive amniotes, one of these lineages gave rise to the turtles. The other lineage evolved into the more lightly built diapsids—lizards, crocodiles, birds, and a variety of extinct groups. Unlike the unopened skull of turtles, diapsids

diverged into two major groups. The lepidosauromorphs evolved into a variety of now-extinct lineages, including the spectacular ichthyosaurs and plesiosaurs that ruled the Mesozoic oceans; and also into an enormous diversity of squamates, animals that include all lizards and their closest relatives. Snakes and the fearsome mosasaurs of the Late Cretaceous seas evolved from some of this primitive lizard diversity. The other major group of diapsids, the archosauromorphs, also evolved into a diverse array of animals; these are far closer to our story, since among them are the archosaurs, the group to which living crocodiles and birds belong to.

Early archosauromorphs retained the quadrupedal gait and predatory habits of their reptilian ancestors. Among these animals was the Early Triassic *Euparkeria capensis*, a small archosauromorph that played a fundamental role in the debate about bird origins and that probably resembled in many respects the ancestor of all archosaurs. Very early in their history, archosaurs diverged into two lineages: one leading to the modern crocodiles and another, the ornithodirans, which would include the flying pterosaurs as well as dinosaurs and their avian progeny. The precise reasons for the extraordinary success that dinosaurs experienced towards the end of the Triassic are not entirely understood. Yet the ancestry of dinosaurs can be safely traced back to a few ornithodirans— lightly built, bipedal animals like the Argentine *Lagosuchus talampayensis*— that lived during the Middle Triassic, some 235 million years ago.

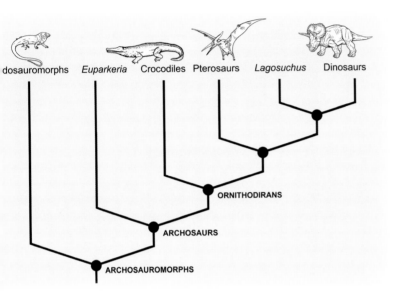

A cladogram or genealogic diagram depicting the evolutionary relationships of the main lineages of archosauromorphs. (Image: E. Heck/R. Urabe.)

The 235-million-year-old, 30-centimeter-long *Lagosuchus* is one of the best-known ornithodiran predecessors of dinosaurs. (Image after Rowe et al., 1998.)

evolved two pairs of simple holes between the bones that form the posterior part of the skull, behind the orbit. While one of these windows—the infratemporal opening—developed lower on the side of the skull, the other—the supratemporal opening—was in the top of the skull. The most primitive diapsids also evolved the predatory habits and agile gait that characterize most of their descendants, together with a larger brain and a more elaborate neurosensory system. Towards the end of the Paleozoic, more than 250 million years ago, primitive diapsids

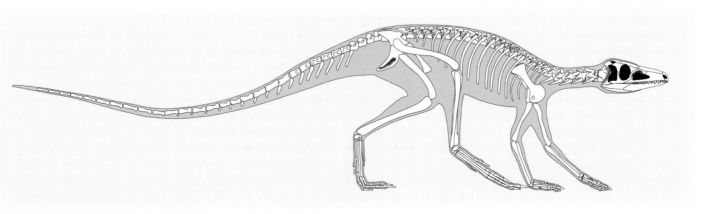

Origin and Evolution of Dinosaurs

The physical forces that shaped the face of the Earth during its 180 million years of Mesozoic history undoubtedly contributed to the remarkable diversity and success of the dinosaurs. In their early development, the breakup of Pangea, significant changes in global climate, and the disappearance of primitive lineages of archosaurs provided impetus for the divergence of these spectacular animals. Afterwards, the rapid evolution of critical anatomical, functional, physiological, and behavioral innovations equipped dinosaurs with the specializations required to rule over all Mesozoic terrestrial ecosystems, also bequeathing these to their avian descendants, which have remained ubiquitous members of today's natural and urban environments.

The earliest known dinosaurs date back to the Late Triassic; the oldest records of them are found in sedimentary rocks deposited some 230 million years ago. Yet these primitive animals already show the features characteristic of the main dinosaur groups. The small, bipedal, herbivorous *Pisanosaurus merti* from the Late Triassic of western Argentina is an early ornithischian—a group whose later and better-known members include the horned, duck-billed, armored, and dome-headed dinosaurs. The fact that *Pisanosaurus* was bipedal and a herbivore suggests that the common ancestor of all ornithischians was also a bipedal plant-eater. Furthermore, the anatomy of this early dinosaur suggests that all the members of this dinosaur group evolved from an ancestor in whom the pubis pointed backward, a feature modified from a more primitive reptilian condition—likely shared by the common ancestor of all dinosaurs—in which the pubis pointed forwards.

Among more advanced ornithischians are the thyreophorans. All thyreophorans evolved from a quadruped ancestor that had an armor of bony scutes and plates covering almost the entire surface of its body. Although fossils of this ancestral species have not been found, it is possible that the earliest thyreophorans already existed in the Late Triassic. Nonetheless, these dinosaurs make their first appearance in the fossil record of the Early Jurassic. One of the most primitive thyreophorans is the small, bipedal *Scutellosaurus lawleri* from the Early Jurassic of Arizona. Slightly more advanced thyreophorans, such as *Scelidosaurus harrisonii* from the European Early Jurassic, appear to have already developed quadrupedal habits. The most famous and well-known thyreophorans, however, are the armored stegosaurs and ankylosaurs of later Jurassic and Cretaceous times. Among these heavier dinosaurs, stegosaurs evolved small and narrow skulls, hindlimbs that were longer than the forelimbs, a characteristic pair of rows of erect plates that flanked their vertebral columns, and a series of sharp spikes that transformed their tails into formidable weapons. In the ankylosaurs, the armor evolved into a dense shield that covered much of the body, the skull incorporated extra ossifications that gave it a highly compact constitution, and the tail of the most specialized forms developed into a stiffened structure ending in an immense club.

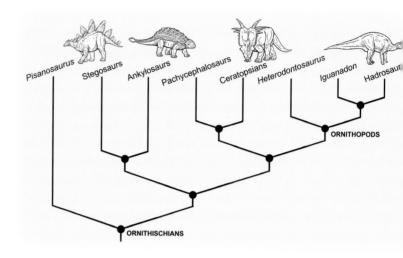

Cladogram illustrating the evolutionary relationships of the main lineages of ornithischians, the group of plant-eating dinosaurs that includes the armored stegosaurs and ankylosaurs as well as the horned and duck-billed dinosaurs. (Image: E. Heck/R. Urabe.)

Another main lineage of ornithischians includes the horned and dome-headed dinosaurs, grouped within the marginocephalians. The ceratopsians and pachycephalosaurs, as these two groups of dinosaurs are named respectively, are characterized by having a bony shelf that projects from the rear margin of the skull. All ceratopsians are united by the presence of a wedge-like bone called a rostral that covers the tip of the snout. One of the most primitive horned dinosaurs is the small, bipedal *Psittacosaurus mongoliensis*, a parrot-beaked animal that together with several other closely related species of psittacosaurids was broadly distributed in Asia during the Early Cretaceous. Another primitive ceratopsian from Asia is the small, bipedal *Chaoyangsaurus youngi* from the Early Cretaceous of

animals is the celebrated *Triceratops horridus*. In contrast, the dome-headed pachycephalosaurs remained bipedal throughout their evolutionary history. These peculiar dinosaurs are characterized by having thick bones in the skull roof. Although early forms bore flattened skulls, more advanced pachycephalosaurs developed the distinct dome-shaped head that gave them their familiar name.

The last large group of ornithischian dinosaurs, the ornithopods, appears to be more closely related to the marginocephalians than to the thyreophorans. These dinosaurs evolved sophisticated dental batteries that were highly efficient in grinding the plant matter they ate. The earliest records of ornithopods date back to the Early Jurassic, and these dinosaurs thrived during the rest of the Mesozoic Era. The primitive, Early Jurassic ornithopods, like *Heterodontosaurus tucki*, were small and agile bipeds with prominent fangs in their lower jaws. A more advanced group of ornithopods includes the iguanodontids. These dinosaurs were generally larger than the most primitive ornithopods and, although bipedal, some became facultative quadrupeds. Primitive iguanodontids include forms such as the Late Jurassic *Camptosaurus dispar* and the Early Cretaceous *Iguanodon bernissartensis*, this latter being one of the first dinosaurs to be studied. Later iguanodontids include the most famous of all ornithopods, the duck-billed dinosaurs or hadrosaurs, which while cosmopolitan in distribution, became particularly abundant in the Late Cretaceous ecosystems of Laurasia. Some hadrosaurs developed extravagant headgear, such as the spectacular crests of *Parasaurolophus walkeri* or the cassowary-like *Corythosaurus casuarius*—structures that are thought to have been important in communication and species recognition.

The saurischians, the other large group that along with the ornithischians

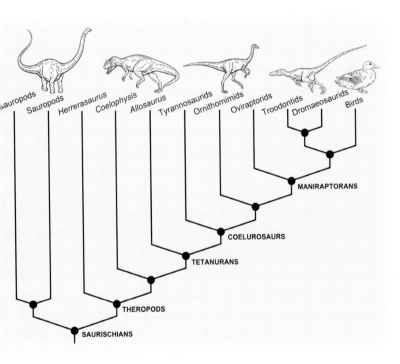

Cladogram depicting the evolutionary relationships of saurischians, the group of dinosaurs that includes the long-necked sauropods as well as all meat-eating dinosaurs (including birds). (Image: E. Heck/R. Urabe.)

northeastern China. Interestingly, the skulls of these early ceratopsians lack the characteristic horns of their later relatives. The more specialized ceratopsians evolved a great diversity of skulls that were proportionally very large for the sizes of their bodies, and they became quadrupedal. One of these

split from the common ancestor of all dinosaurs, also have their earliest records in the Late Triassic of South America. Characterized by having hands capable of grasping, the saurischians consist of the carnivorous theropods and the herbivorous sauropodomorphs. Theropods are central to this book because, as we will see later, this dinosaur group not only includes the renowned *Tyrannosaurus rex* and *Velociraptor mongoliensis* but all birds as well. Sauropodomorphs include the colossal, long-necked sauropods, the largest animals ever to walk the Earth, and a variety of smaller and more primitive dinosaurs often referred as prosauropods. Some of the earliest sauropodomorphs include the prosauropods *Saturnalia tupiniquim* and *Riojasaurus incertus* from the Late Triassic of Brazil and western Argentina, respectively. In most respects, the smaller and facultative bipedal prosauropods are less specialized than the large, quadrupedal sauropods. For example, the four limbs of sauropods are massively built, and the bones of their wrists and ankles are greatly reduced in number, a design optimal for sustaining the heavy weight these animals had. Despite their many anatomical advances, the known evolutionary history of sauropods is nearly as old as that of the prosauropods. The remains of *Isanosaurus attavipachi*, the earliest known sauropod, have been recently unearthed from Late Triassic rocks in Thailand.

Later sauropods include many familiar types of dinosaurs. *Diplodocus longus*, *Brachiosaurus altithorax*, and *Camarasaurus grandis* are all sauropods; other less familiar groups of long-necked dinosaurs include dicraeosaurs and titanosaurs. Although there is still some disagreement about the genealogical relationships among these lineages of dinosaurs, much of the available evidence supports the grouping of *Diplodocus* and its whip-tailed relatives with the dicraeosaurs of the Late Jurassic and

Early Cretaceous of Gondwana. These rather lightly built dinosaurs seem to have evolved their characteristic peg-like teeth independently from the similar teeth of the Late Cretaceous titanosaurs that flourished in Gondwana. These latter sauropods, together with *Brachiosaurus* and *Camarasaurus*, evolved from a much more robust common ancestor. Indeed, among the titanosaurs is the 90-million-year-old *Argentinosaurus huinculensis*, a gargantuan animal that may have weighed 100 tons and reached lengths of more than 35 meters.

The anatomical characteristics of both the earliest known dinosaurs and their ornithodiran relatives suggest that the common ancestor of all dinosaurs was a small bipedal predator, whose forelimbs were shorter than its hindlimbs. The fact that the earliest dinosaurs, from ornithischians to sauropodomorphs and theropods, already had the typical anatomical, locomotor, and behavioral features of their own lineages indicates that the evolutionary origin of the entire group must have predated the Late Triassic by several millions of years. Nonetheless, even though an array of Middle Triassic footprints similar to those of later dinosaurs provides circumstantial support for this notion, the fossilized bones of these first dinosaurs have yet to be discovered.

Theropod Diversification

The evolutionary history of theropods— *Tyrannosaurus* and all other carnivorous dinosaurs—also dates back to 230 million years ago, when dinosaurs first appear in the fossil record. At such an early phase of dinosaur evolution, theropods had already developed their typical predatory features—a movable joint at the center of the lower jaw, powerful clawed hands, and sharp, knife-like teeth. Undoubtedly, theropods were the most diverse dinosaurs, although they are generally less common in the fossil record than other dinosaurian groups. Throughout

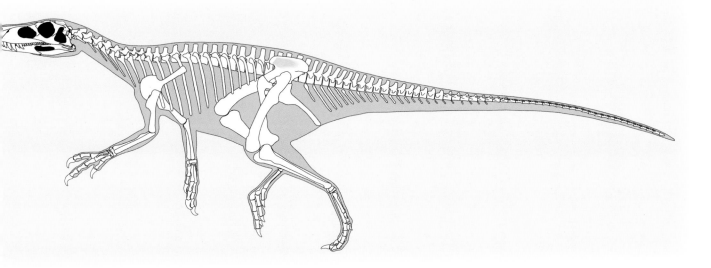

their long history, theropods developed dozens of highly specialized features, but perhaps none is more striking than feathers, which in time helped them to become warm-blooded and airborne.

Paleontologists believe that theropods originated from a bipedal dinosaur that had feet with three main toes pointing forward, of which the central one was the longest. The 230-million-year-old *Herrerasaurus ischigualastensis* from western Argentina is generally accepted as one of the most primitive theropods, although some researchers consider it to be a primitive dinosaur outside theropods. This primitive predator had already evolved the hollow limb bones, air-filled neck vertebrae, and elevated rates of growth that characterize theropods. Whether the early theropods inherited these pneumatic features from more primitive archosaurs is not clearly known, but in time their descendants would expand this system of air-filled bones, evolving air sacs that would penetrate most of the bones of their skeletons.

The subsequent early history of theropods saw the evolution of a number of primitive lineages, including the small and lightly built Late Triassic *Coelophysis bauri* and the much larger and horned *Ceratosaurus nasicornis* from the renowned Late Jurassic Morrison Formation of the American west. *Coelophysis* and *Ceratosaurus* are usually grouped with a number of other Jurassic theropods in a group called

ceratosaurians. The Late Cretaceous *Carnotaurus sastrei*, a large horned theropod found in the desolate badlands of Patagonia (Argentina), and its abelisaur relatives used to be classified within the ceratosaurians. Nonetheless, recent refinements of our understanding of the genealogical relationships of theropods indicate that *Carnotaurus*, together with all other abelisaurs, is more closely related to birds than ceratosaurians. Although abelisaur fossils have been exclusively unearthed from Cretaceous rocks, there is little doubt that this primitive lineage of theropods originated many millions of years earlier.

Another branch of primitive theropods contains a group called tetanurans. Its earliest member, the skull-crested *Zupaysaurus rougieri* of western Argentina, also dates back to the Late Triassic. This and all other tetanurans had their origins in a common ancestor that possessed a wishbone and hands with three fingers. In time, certain tetanurans differentiated into some of the most famous meat-eaters, the carnosaurs, which included large theropods such as the Late Jurassic *Allosaurus fragilis* and *Megalosaurus bucklandii* and the Cretaceous *Giganotosaurus carolinii* and *Charcharodontosaurus saharicus*.

A branch of tetanurans, the coelurosaurs, evolved from an ancestor that had longer arms and large breastbones. This first coelurosaur led to a number of small to modest-sized

BECOMING FOSSILIZED: A VERY RARE EVENT

Fossilization is a very rare event. Despite the fact that more than 99 percent of life forms are extinct, only a small fraction of this enormous diversity is known from fossils. The odds of becoming a fossil are so small because the phenomenon hinges upon a series of very infrequent events. To become a fossil, a dead animal—here illustrated by a specimen of the ornithuromorph bird *Liaoningornis longidigitrus*—needs to be among the few that are not consumed by scavengers or completely decayed. In a world where sediments form episodically, the corpse of this animal needs to be buried quickly.

While undergoing fossilization— mineral replacement of its organic components—this corpse needs to survive the pressures generated by tens or hundreds of meters of overlying rocks; and the rocks that contain it need to be preserved for millions of years without suffering metamorphosis, tectonic subduction, or erosion. These rocks also need to be uplifted to reach the surface; and as if all this were not enough, the fossil needs to be exposed by the erosion of the overlying rocks—but discovered before it is eroded away itself.

The limitations preventing us from recovering a complete record of the annals of life have been accepted since the time of Charles Darwin. Nowhere is there a complete geological column of the Earth's history. Our understanding of the changes in life forms that once inhabited our planet is based on composite stratigraphic columns derived from many different places all over the world. Collections of fossils are correlated with the amount of rock that crops out in any given region, and such exposures are limited. Sedimentary rocks—the types of rock that contain most of the available fossils—represent only a small percentage of the rocks that crop out on the surface of our planet; and this fraction is biased against certain environments and the organisms that lived in those environments.

When compared to other vertebrates, fossil birds are quite uncommon. Their small and hollowed bones, their eminently terrestrial habits, and their widespread and mobile populations, make them unlikely candidates for fossilization. It is hence remarkable that their ancient representatives have been preserved so well in the fossil record. (Image: M. Ellison.)

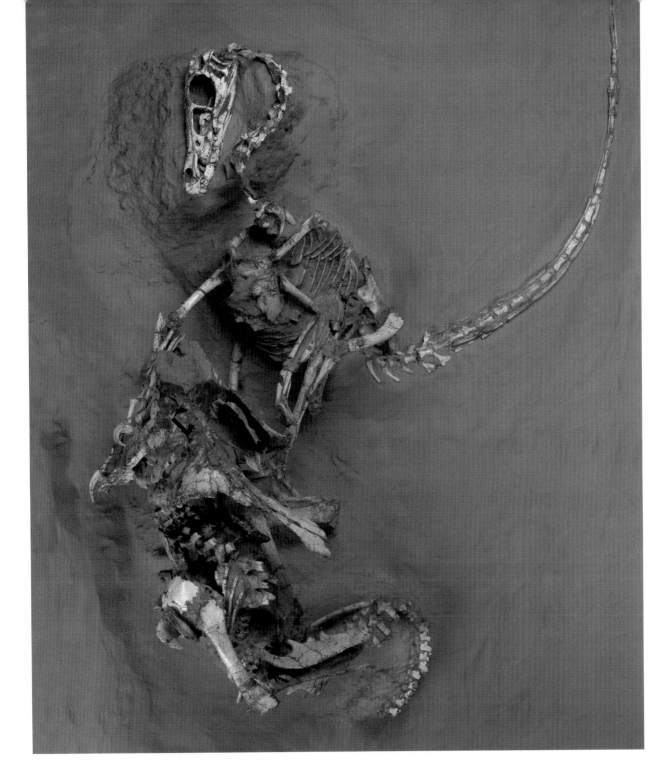

The vicious dromaeo-saurid *Velociraptor* (top), here engaged in combat with the plant-eating *Protoceratops*, is commonly considered a close relative of birds. (Image: M. Ellison.)

theropod lineages, including the bird-mimic ornithomimids and a variety of other lineages closer to the origin of birds. A notable feature of coelurosaurs is the size of the brain in relation to the body—their brains were proportionally much larger than those of any other nonavian reptile. Other coelurosaurs included the fearsome *Tyrannosaurus* and its relatives. These gigantic meat-eaters,

once believed to be close relatives of the carnosaurs, are now most commonly viewed as over-grown coelurosaurs. Definitive remains of coelurosaurs are known from the Late Jurassic, although fragmentary remains discovered in earlier rocks, and the diversity these dinosaurs exhibited 150 million years ago, indicate that the origin of the group is likely to be much older.

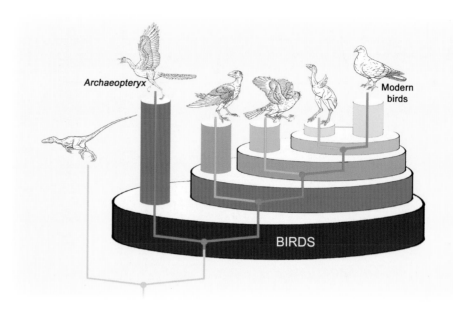

Scientists define birds, in an evolutionary context, as all species (living or extinct) that can trace their descent to the most recent ancestor of the Late Jurassic *Archaeopteryx* and modern birds. In this figure, the extinct lineages between *Archaeopteryx* and modern birds (pink branches of the cladogram) are regarded as avian since their ancestry can be traced to such a common ancestor. (Image: R. Urabe.)

Among coelurosaurs are the maniraptorans, the theropod group that contains the forerunners of birds. The parrot-headed oviraptorids, the lightly built troodontids, and the sickle-toed dromaeosaurids are the best-known lineages of maniraptorans. These theropods evolved arms even longer than their coelurosaur forerunners, and larger breastbones, together with a series of specializations of the pelvis and tail. Although these lineages originated from animals scarcely the size of a turkey, some Cretaceous maniraptorans reached lengths of over 6 meters. Despite the fact that the maniraptoran family tree is not entirely clear, mounting evidence indicates that troodontids and dromaeosaurids are more closely related to one another. These dinosaurs are also widely regarded as the closest relatives of birds—but more on this in the next chapter.

Defining Birdness

Like all other groups of organisms, birds need to be defined in biological terms. In other words, the limits to what makes an organism a bird, and not a mammal or an insect, needs to be established. All attempts at providing a definition for a particular group of organisms entail a certain dose of arbitrariness—after all, definitions are *per se* subjective. Birds have often been defined as all animals with feathers. Alternatively,

they have been defined on the basis of other physical attributes, such as skeletal characteristics that may not be found elsewhere. Nonetheless, at a time when more and more close relatives of birds are being discovered, features now thought to be unique to birds may turn up in the future in animals that have traditionally been classified as something else. Take the case of feathers, arguably the quintessential characteristic of birds. If we were to argue that all animals with feathers are birds, we

The similarity in relative position and number of phalanges (finger bones) of the thumb of a human and an early tetrapod (top figures) is an example of a homologous feature. This similarity is explained as the result of the common ancestry we share with early tetrapods. However, homologous features need not be similar, as is illustrated by the well-documented homology between some of our middle ear bones and those that formed the articulation between the lower jaw and the skull of early tetrapods (bottom figures). (Image: R. Urabe.)

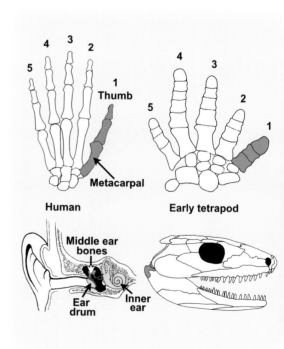

would have to classify as birds a variety of recently discovered dinosaurs whose bodies are covered with structures that are remarkably similar, if not identical, to modern feathers. These fossils, however, show a multitude of skeletal characteristics that identify them as members of other lineages of theropod dinosaurs, such as dromaeosaurids and oviraptorids. In fact, even iconographic dinosaurs such as *Tyrannosaurus* would have to be nested within birds, because the placement of this fearsome animal within the family tree of theropods suggests it also could have had plumage. As we discover more and more theropod lineages that acquired birdness, or evolved bird-like features, the line between what is a bird and what is not becomes increasingly blurred.

In this book, birds are defined in an evolutionary context, namely as the group of organisms encompassing the common ancestor of the Late Jurassic *Archaeopteryx* and extant birds, and all the descendants from this ancestor. Although this definition is, like all others, arbitrary, it allows birds to be defined on the basis of their common ancestry rather than on the presence of a physical attribute. Under this type of evolutionary definition, an organism (whether extinct or not) is regarded as a member of a group only after hypothesizing that the organism in question shares the specified genealogical relationship. To use an example germane to our story, a given Mesozoic fossil is regarded as avian only if it can be hypothesized that it has descended from the common ancestor of all birds, which, as defined above, is the most recent shared ancestor of *Archaeopteryx* and living birds. As explained in the following section, such hypotheses are reached through a method of reconstructing the evolutionary history of organisms called cladistics.

Today, it seems quite easy to distinguish between what is a bird and what it is not. When dealing with ancient fossils, however, the choice is often not that simple. Sometimes, fossils may seem quite different from our visual concept of what birds are. In the following pages, we will encounter fossils of animals with teeth, sickle-clawed feet, long bony tails, and sharp and powerfully clawed forelimbs, that although classified with birds are quite different from those that surround us today. These fossils are classified as birds because their hypothesized genealogical relationships to other fossils suggest that they evolved from the common ancestor of the Late Jurassic *Archaeopteryx* and its extant relatives.

Cladistics: Reconstructing Evolutionary Relationships

Today, most paleontologists and evolutionary biologists use cladistics to reconstruct the genealogy of plants and animals. German entomologist Willi Hennig first proposed this scientific method in the early 1950s; in the 50 years that followed, cladistics became the dominant paradigm for establishing the genealogical relationships among organisms. Its simplicity and efficiency has led scholars of other historical sciences to adopt it. For example, linguists use similar concepts to reconstruct the relationships among languages, and literature academics have used cladistics in their search for the oldest surviving copies of classic works—such as Geoffrey Chaucer's *Canterbury Tales*, of which the original manuscript has not survived.

Cladistics has advantages over other methods of inferring genealogical relationships because it sets out the fundamentals for grouping organisms clearly and allows competing hypotheses of relationships to be contrasted easily. Cladistics does not use overall similarity among organisms; rather, it establishes groups that are supported only by a special type of similarity called derived similarity or synapomorphy. In cladistics, synapomorphy is equivalent to homology—a fundamental

evolutionary concept that makes reference to similarities of corresponding characteristics in different types of organisms, whose similar nature is due to common descent. Yet it is important to bear in mind that homologies—and their equivalent synapomorphies—are statements about the common origin of characteristics shared among organisms, and that these statements are only indirectly related to similarity. Indeed, although most proposed homologies show some degree of similarity, homologous characteristics need not be similar—think of the well-documented homology between our ear ossicles and some of the bones of a fish's lower jaw. Very different homologous structures, however, may be very difficult to identify.

Synapomorphies are hypotheses of homology that are best proposed through a two-step procedure in which, first, a particular characteristic often needs to show some degree of similarity among the organisms being grouped or have a similar developmental pathway, and second, the characteristic needs to be congruent with other synapomorphies supporting the same type of grouping. Let us illustrate this two-step procedure with an example from the world of art. For centuries, artists have illustrated

Synapomorphies are hypotheses of homology that need to be congruent with other such statements supporting the same grouping of species. Although the wings of the cherubs shown in this painting of the Cuzco School are remarkably similar to those of birds—and thus, may suggest a cherub–bird grouping— the hypothesis indicating the homology between the wings of the cherubs and those of birds cannot be defended in light of many other features (arms, hair, cheeks, and so on) that do not support a close relationship between cherubs and birds. On the contrary, these features support a closer relationship between cherubs and humans, as well as the homology of their arms. (Image: Museo de Arte Hispanoamericano Isaac Fernández Blanco.)

cherubs (infant angels with chubby faces) with a pair of bird-like wings. If cherubs were to be studied from an evolutionary perspective, it would be reasonable to ask whether their wings, instead of their arms, are homologous to those of birds—that is, whether cherubs and birds inherited their wings from a winged common ancestor. Because the feathered wings of cherubs and birds are identical in every detail, it would be reasonable to state that these structures are homologous, that they indeed evolved in a hypothetical ancestor that was common to cherubs and birds only. Such a conclusion, however, would encounter problems during the second phase of the two-step procedure used to formulate hypotheses of homology. In addition to wings, cherubs have arms that look identical to our own arms. Furthermore, they have many other features (hair, fleshy ears and cheeks, a hand with five fingers, belly buttons, and five toes, among others) that make them look very human. Thus, because our initial conclusion regarding the wings of cherubs and birds is incongruent with the many similarities suggesting that the winged infants should be grouped with humans, the hypothesis of homology between the wings of cherubs and birds would have to be abandoned. Conversely, examination of the arms of these infants would reveal similarities (an upper arm bone, two forearm bones, and wrist and hand bones) with the skeletal structure of the wings of birds. Thus, the hypothesis that the wings of birds are homologous to the arms of cherubs would meet the two-step procedure well—it would be congruent with grouping cherubs and humans together, and then grouping these and birds within a larger group. This example illustrates how our interpretations about the homology of particular structures—in this case, wings and arms—depend on how congruent our interpretation is with similar interpretations about other structures.

As you might imagine, identifying homologies in the real world is far more difficult. However, in spite of the enormous diversity of living and extinct organisms, a clear pattern of synapomorphies emerges when different kinds of characteristics (anatomical, behavioral, physiological, molecular, and others) are mapped among organisms. Cladists, the scientists practicing cladistics, use such a pattern to recognize groups of organisms and arrange smaller groups within larger groups. This arrangement of groups within groups is interpreted as the pattern resulting from the process of evolution. When an ancestral species evolves into one or more species, the descendants inherit the characteristics newly evolved by the ancestral species. These characteristics, now shared by the ancestral species and all its descendants, become synapomorphies of the group formed by the ancestor and its progeny. The distribution of different synapomorphies among the diversity of extinct and extant organisms determines the order in which different groups evolved, thereby interpreting the sequence of evolutionary history. The genealogy of a particular group of organisms is illustrated through branching diagrams known as cladograms—diagrams of species or lineages (see *Cladograms and Classifications*). Instead of focusing on ancestral-descendant relationships of the species contained in the group or groups under study, cladograms focus on their common ancestry, thus establishing relationships of sister species or sister groups—the species or groups that share a most recent common ancestor. It is important to emphasize that the temporal dimension—the ages of the species or groups included in a cladogram—is irrelevant in the construction of a cladogram. The sister species or sister group relationship contained in a cladogram is established purely on the pattern of shared synapomorphies among the species or groups.

Parsimony: Choosing Among Alternative Genealogies

Like most other aspects of biology—from ecological interactions to developmental trajectories—evolutionary relationships are naturally untidy. Reconstructing the genealogy of organisms is often complicated by the fact that their characteristics did not evolve only once, nor in a linear fashion. On the contrary, these characteristics often form a much more complex pattern. Just as homologous structures between two or more organisms need not be similar, similar structures are not always homologous. Similar characteristics may have evolved in organisms that are not closely related to each other, as a result of a phenomenon called convergent evolution. Indeed, this phenomenon seems ubiquitous throughout the vast spectrum of life's diversity, from very simple organisms to very complex ones. For example, among living animals, wings evolved not just in birds but also in insects and bats; and warm-bloodedness evolved in mammals, birds, and tuna. Although it is rather easy to identify these examples as convergent evolution (after all, the wing architecture of birds, bats, and insects is as drastically different as the specifics of the warm-blooded metabolism of mammals, birds, and tuna), features that are strikingly similar and yet have evolved independently in distantly related groups are indeed abundant. The independent evolution of leaf-reduction and succulence in a variety of desert-dwelling plants is exemplary of the many well-documented instances of convergent evolution that are far less apparent than those we have mentioned before.

An important component of cladistics involves sorting out whether the similar characteristics shared by different species or groups of organisms are likely to have evolved from a single ancestor or from different ancestors. To make this decision, cladists rely on the principle of parsimony. Given alternative genealogical

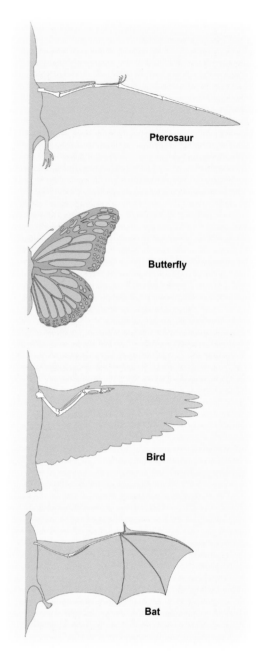

The similar wing design of an extinct pterosaur, a butterfly, a bird, and a bat is a classic example of convergent evolution. There is plenty of evidence indicating that, overall similarities notwithstanding, these lineages evolved their wings independently; but distinguishing between convergent evolution and homology is often tricky. (Image: S. Abramowicz.)

Pterosaur

Butterfly

Bird

Bat

interpretations, the parsimony principle chooses the simplest explanation that accounts for the arrangement of all species or groups on a cladogram. This fundamental principle is often referred to as Ockham's razor in reference to the 14th century English philosopher William of Ockham, whose logic led to the notion that Nature follows the simplest path; in other words, that we should not assume the existence of more things than are logically needed. Classical philosophers first drafted such a concept, and this important scientific principle has come

CLADOGRAMS AND CLASSIFICATIONS

Evolutionary biologists use branching diagrams called cladograms to illustrate the hypotheses of genealogical relationships they propose. Cladograms—like the one shown here—are the result of cladistic analyses—computational analyses of genealogical relationships based on large numbers of variables that contain specific characteristics shared by species or groups of organisms. A numerical matrix of attributes like the one shown here could contain information from bones, muscles, feathers, eggs, DNA, and other features of the organisms under study. The example illustrated here only shows 40 variables (each column in the matrix), but cladistic analyses often include hundreds or thousands of variables.

Every group of organisms or clade included in a cladogram (in our example, mammals, birds, reptiles, amniotes, tetrapods, or vertebrates) contains the most recent common ancestor of all the species within the group; in other words, all clades featured in a cladogram contain an ancestor and all the species that have descended from this ancestor. The branches of a cladogram constitute species or clades of organisms, whose names are listed at the end of the branch (in our case, a kind of fish, frog, rodent, monkey, crocodile, duck, or songbird). The branching points or nodes of the cladogram represent the ancestors; the two branches splitting from a node represent sister groups. These ancestors are considered to be hypothetical (H1 through H6 in our example), because it is impossible to be certain about their identity—after all, these branching events occurred thousands or millions of years ago. Each ancestor in a cladogram constitutes a hypothetical species that has evolved novel characteristics, inherited by its descendants, depicted on higher branches of the diagram. For example, the fish, frog, crocodile, birds, and mammals illustrated in our example all have a vertebral column composed of a multitude of individual vertebrae. All these animals, including us, belong to the group known as vertebrates. The vertebral column is thought to have evolved in the common ancestor of the entire group—with all the descendants of this ancestor subsequently inheriting

this feature. Within vertebrates, other less inclusive groups (or clades) share other characteristics. For instance, frogs, crocodiles, birds, and mammals have two pairs of limbs. As such, they belong to a subgroup of vertebrates called tetrapods. Again, four limbs originally evolved in the ancestor of all tetrapods, and all the descendants of that common ancestor inherited some version of its four limbs. Since having a backbone is more widespread among animals than having four limbs, the backbone

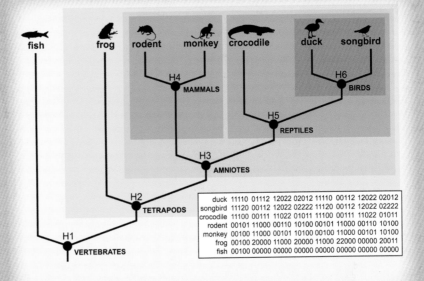

is thought to have evolved before the limbs did. Tetrapods are considered therefore to represent a subgroup within vertebrates. In this example, the fossil record appears to confirm this idea, because the first known vertebrate lived about 500 million years ago, whereas the first known tetrapod appeared about 350 million years ago. However, the fossil record does not always confirm the evolutionary sequence of characteristics that is suggested by looking at the characteristics shared by larger and smaller groups. This is not too surprising because only a small fraction of the species that once inhabited the Earth are represented in the fossil record and, as shown by the multitude of new species that are annually named on the basis of fossil remains, many of those are yet to be discovered.

Cladists see a direct correspondence between genealogy and classification. Under this framework, groups of organisms are classified by converting cladograms into hierarchies where groups are clustered within larger groups. For example, cladistic classifications depict birds (Aves) as a subgroup of reptiles (Reptilia). Cladistic studies of the evolutionary relationships among different groups of reptiles produce cladograms in which birds are nested within

theropod dinosaurs. Thus, if birds constitute a subgroup of theropod dinosaurs and these dinosaurs are a subgroup of reptiles, then birds are also a subgroup of reptiles. This simple syllogism, derived from the most parsimonious cladogram produced by studies of reptilian evolutionary relationships, provides the logic underlying the claim that birds are a flying, feathered group of reptiles. Cladistic classifications are superior to classificatory schemes used prior to the development of this method in that they best summarize the distinctive features of the groups they contain, and at the same time allow predictions about the distribution of characteristics that are unique to these groups. (Image: R. Urabe.)

down to us not just via Ockham but also through a cohort of brilliant scientists including Leonardo da Vinci and Pierre de Fermat. The parsimony principle is nothing more than a biological application of this well-accepted concept of scientific discovery, which need not be interpreted literally—that is, as if evolution always follows the simplest course—but more as an operational procedure for choosing the most preferable hypothesis of genealogical relationships. In other words, the parsimony principle identifies the evolutionary sequence of events supported by the majority of available evidence and contradicted by the fewest pieces of evidence at hand. Thus it makes cladistic hypotheses testable, and as such, scientific, because scientific hypotheses must be able to be confronted by the available data or additional data and be either rejected or supported by these data. Hypotheses of evolutionary relationships, together with all other hypotheses that belong to the realm of historical sciences, cannot be reproduced. The only procedure for testing them is to test the predictions that can be made from them.

The origin of birds is an event that cannot be reproduced. However, predictions made by the hypothesis that birds evolved from theropod dinosaurs can be tested against new discoveries. We will see how the prediction that feathers had to be found in these dinosaurs has been corroborated recently by the discovery of a variety of feathered theropods. Does this revelation prove that birds evolved from theropod dinosaurs? Certainly not, but it lends additional support to the hypothesis and indicates that this theory is still valid.

Modern studies of genealogical relationships compare large numbers of characteristics among many species or groups of organisms. Armed by preliminary hypotheses of homology based on the similarities shared by the groups under consideration, cladists assign numerical values to each

characteristic and use these values to generate a numerical matrix. This matrix is analyzed using computer algorithms that are designed to search for the combinations of similar values that correspond to the most parsimonious hypothesis—and the resulting cladogram summarizes the relationships among the organisms using all the available evidence. Cladograms resulting from these analyses not only provide a basis for classifying the organisms but also identify the instances of convergent evolution. Cladistic methodology applied in this way compares similarities among different groups of organisms, and then uses parsimony as the criterion for choosing the best genealogical

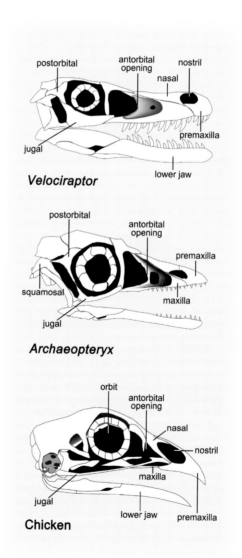

Comparison between the skulls of a dromaeosaurid theropod (*Velociraptor*), the most primitive bird (*Archaeopteryx*), and a chicken. Drawings are not to scale. (Image after Rowe et al., 1998.)

interpretation—the one that is supported by the largest number of derived similarities or synapomorphies.

Understanding the genealogical relationships between ancient organisms is crucial for reconstructing the origins of living animals and plants, as well as for understanding the evolution of their anatomical, physiological, and behavioral systems. When looking at a particular characteristic, such as the avian wing, the explanation of its evolutionary origin would be very different if we were to assume that birds originated from crocodiles instead of from theropod dinosaurs. In this book, our discussion of how birds evolved their unique physical systems is always underpinned

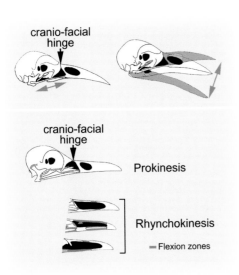

The avian skull is characterized by a number of flexible zones that allow the elevation of the snout independently from the opening or closing of the mouth. Depending on the position of these flexible zones, two major types of cranial kinesis are found among modern birds: prokinesis and rhynchokinesis. (Image after Chiappe et al., 1999.)

by the genealogical hypothesis that these animals evolved from theropod dinosaurs. But before addressing the evidence in support of this evolutionary hypothesis, it is critical that we look at the general anatomy of the modern avian skeleton, for we will use this as a reference when discussing both the origin and early evolution of birds.

The Bird Skeletal Plan

Skeletal anatomy is at the center of vertebrate paleontology because, aside from a few exceptions, bones are the only

tools paleontologists have to understand the genealogical relationships and functions of the extinct organisms they study. Given that birds are superb flying animals, it is not surprising to learn that much of their skeletal design is highly specialized for aerial locomotion. Indeed, the avian skeleton is formed by light, thin-walled, hollow bones, some of which anchor large flight muscles while others form compound bones—by the fusion of individual bones during post-hatching development—that can resist the great forces exerted on the skeleton during flying and landing.

The modern bird skull is characterized by having beaked jaws, enormous orbits, and a compact and vaulted braincase that houses a large brain. The beak, one of the most conspicuous avian trademarks, has evolved a wide range of shapes in response to a vast spectrum of feeding mechanisms; and the fusion of skull bones has provided additional strength for the strenuous activities it performs. The premaxillary, maxillary, and nasal bones form most of the snout. The maxillary bone of modern birds has become greatly reduced from the important component of the upper jaw it used to be in their primitive relatives and dinosaurian predecessors. The nostrils are usually large and distinctly bigger than another important opening of the snout—the antorbital opening. This opening, which typically forms a small triangular window between the nostrils and the orbits, houses air sacs that invade the surrounding bones—this pattern of air invasion is common for many parts of the avian skeleton, where bones have been resorbed by a process known as pneumatization. Like its upper counterpart, the lower jaw of modern birds is toothless and greatly covered by a horny beak, with sharp edges that execute the tasks of grabbing, manipulating, cutting, and crushing food items.

Another attribute of the skull bones of modern birds is their great degree of flexibility. Many of these bones form

elastic zones of thin and plywood-like bone tissue which enhance flexibility while also strengthening the skull. The existence of these bones is related to the high degree of kinetism of the modern avian skull—that is, the ability of independent motion between subunits such as the snout or the braincase. Most birds exhibit a type of kinetism called prokinesis, in which the entire snout hinges upon a flexible zone at its junction with the braincase—and movement here is independent of the movements related to opening or closing the mouth. Other birds exhibit a variety of other kinetic patterns involving the snout, in which only a portion of it moves with respect to the rest of the snout and the braincase—this broader range of kinetic movements is generally known as rhynchokinesis. In all instances, however, these movements are directed by bony struts that connect the palate (and snout) to the quadrate (a bone linking the skull and the lower jaw), which can swing back and forth with respect to its joint with the braincase. The fore–aft movement of the quadrate is instrumental for the mechanism that operates the kinesis of the upper jaw, because the forces transmitted to the snout by the bony struts of the palate produce the independent elevation or depression of all or part of the snout. The functional benefits of these systems are not well understood, but augmenting the repertoire of movements, shock-absorbing capacity, closing speed, grasping ability, and biting force of the jaws are all possible roles. The specific pathways in the evolution of the modern types of skull kinesis are also not entirely clear, but fossils of Mesozoic birds show a great deal of experimentation towards these conditions.

Earlier in this chapter, when we discussed birds as vertebrates, we highlighted the diapsid relationship of birds, indicated by the existence of two temporal openings in their skulls. Modern birds, however, have greatly modified this primitive diapsid skull configuration.

Indeed, the large orbit of living birds is confluent with the infratemporal opening of their primitive diapsid ancestors. The other temporal opening that characterizes the skulls of diapsid reptiles—the supratemporal opening—is greatly reduced and also partially surrounded by bony margins. Not surprisingly, these drastic modifications were coupled with a substantial reduction in the musculature originating on these openings (muscles primarily involved in biting) and the expansion of the brain. The relative reduction of these openings—a trend well-documented throughout the Mesozoic history of birds—is likely related to the complete reduction of teeth and the transformation of the gizzard, instead of the mouth, into the main organ for food processing.

Modern birds have a long and muscular neck, typically composed of 14 to 25 vertebrae. These bones are characterized as having saddle-shaped joints, because the front surface of each vertebra is concave from side to side and convex from top to bottom and the back side is the reverse. In addition, the neck vertebrae are arranged in three functional zones, which give the neck its characteristic S-shaped appearance. While the joints between the vertebrae of the front zone are designed for greater downward movements, those of the middle zone move mostly backwards and the ones of the rear zone move downwards—coupled with the mobility provided by the saddle-shaped joints, this zoning makes the neck of living birds a very flexible organ. Like most other attributes of these animals, this elaborate system followed a piecemeal evolutionary pace along the lengthy history of the group.

The long neck of modern birds is joined to a relatively short trunk (the combined thoracic and lumbar segments of the human backbone)—the portion of the vertebral column that is between the neck and the sacrum. Most of the ribs of the ribcage are fitted with short

bony struts called uncinate processes that interconnect one rib with the next, thus stengthening the ribcage against the compressing forces applied during flight. The ribs joined to the trunk vertebrae are composed of two segments, one joined to the vertebrae and another joined to the sternum (breastbone), which together form a hinged ribcage that can be moved up and down during respiration.

In the pelvic region, the vertebral column also forms a very solid structure, with many vertebrae (usually more than a dozen) fused into a rigid structure called a synsacrum. Fused to each ilium, the synsacrum supports the hindlimbs, and thus the entire body while standing, walking, or running, and perhaps most importantly, during landing.

Unlike those of their most primitive counterparts, the bony tail of modern birds is short and composed of five to eight individual elements, followed by a rigid structure—known as a pygostyle—made up of all the remaining vertebrae (usually four to six) fused together. The pygostyle is the main portion of the skeleton supporting the rectricial bulb, the fatty structure known as the parson's nose that anchors most tail feathers (the central pair usually attach directly to the pygostyle). A series of elaborate muscles that connect the tail, and the parson's nose, to the pelvis provide the array of critical movements that help birds control the amount of lift produced by their feathered tail—and by doing this, to brake and steer.

The large sternum of modern birds, a structure significantly larger than that of any other vertebrate animal, provides the surface necessary for anchoring the main muscles involved during flight—namely, the pectoral and supracoracoid muscles. This large surface is usually expanded by the development of a keel-like, downward projection. The development of this keel is generally correlated with the size of the muscles that power the wings, giving the sternum the required surface for their attachment, and at the same time providing strength against the compressional forces generated by these muscles. The diversity of shapes of the sternum of modern birds is remarkable, and the early evolution of the group is not an exception. Not surprisingly, the size of this bone and the development of its keel evolved in concert with the fine-tuning of flight.

Three bones—the coracoid, scapula, and clavicle—form the shoulder girdle of birds. With a pillar-like shape, the strong coracoid of modern birds acts as a strut between the sternum and the shoulder joint or glenoid. Its position within the body allows the coracoid to work as a spacer for the wings during flight and reduce the effect of compression exerted by the flight muscles against the ribcage.

The skeleton of a chicken with close-ups of its hand (top left), a top view of its pelvis (top right), sternum and ribcage (right center), and an anterior view of its foot (bottom left). (Image after Rowe et al., 1998.)

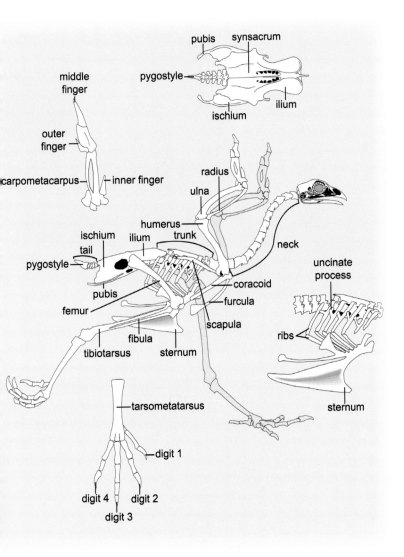

Modern birds have a scimitar-shaped scapula that is firmly attached to the ribcage, with its main body parallel to the spine. This bone and the coracoid form a wide and side-facing glenoid that allows the extensive range of wing movements required during flight. In most modern birds, both clavicles fuse together, forming a U-shaped wishbone—also called the furcula. Although the origin of the furcula can be traced back to the dinosaurian ancestors of birds, changes during the early evolution of the group have transformed it into a structure that provides an additional surface for anchoring flight muscles, storage of the elastic energy used in flight, and assistance in breathing. Another unique feature of the shoulder of modern birds involves the contribution of these three bones to the triosseal canal—a skeletal structure that plays a fundamental role in guiding the pulley-like action of the supracoracoid muscle, an important muscle responsible for providing the necessary rotation and elevation of the wing during flight.

Few other portions of the skeleton of modern birds depart as much from the ancestral reptilian pattern as the wing. The bones of the wing have the necessary stamina to withstand the forces exerted on them during flight. They also provide anchoring for the flight feathers and the skeletal framework supporting the wing's skin foils—of which the propatagium, held between the humerus and the bones of the forearm—is the most important. The humerus is typically shorter than the bones of the forearm. Its upper portion bears a large deltopectoral crest where the pectoral muscle, the main muscle responsible for depressing the wing during flight, is attached. In most cases, the humerus is pneumatized by the extension of a system of air sacs that connects to the lungs. The ulna and radius, the two bones of the forearm, are well-separated from each other. This separation and the bowed shape of the ulna provide resistance to the

bending and compression forces these bones are exposed to during flight. The secondary flight feathers are anchored to the ulna, and quill knobs are often developed at the attachment points of each of these feathers. The joint between the forearm and the humerus is such that the former portion of the wing can only be flexed and extended along its own plane. This joint, and the muscles that operate it, control the movements of the propatagium—an airfoil that is operated according to the requirements of flight. This is also the case with the joint at the wrist, which is reduced by the process of bone fusion characteristic of the whole avian skeleton to two proximal carpal bones articulating with the carpometacarpus. The latter is a compound bone formed by the fusion of several carpals and three metacarpals. The carpometacarpus forms a rigid structure that receives, together with the middle finger, the attachment of the primary flight feathers. As with secondary flight feathers, a complex system of small muscles operates with precision the subtle movements of each individual feather—movements that are critical for the nuances of flight.

The bird hand has three fingers, whose identification with respect to the pentadactyl hand of primitive tetrapods is still controversial. We will return to this issue in the next chapter, but regardless of the identification of the digits of modern birds, these are extremely reduced, likely a specialization for strengthening the hand against the stresses generated while in flight. Commonly, living birds have a single phalanx (finger bone) on the inner and outer digits, and two on the middle digit. The inner digit carries a tuft of feathers known as the "bastard wing" or alula. The relatively high mobility of this finger conveys a series of important movements to the alula, which play a critical role in flight, mostly at low speed and while landing or taking off. The other two fingers have a much more restricted

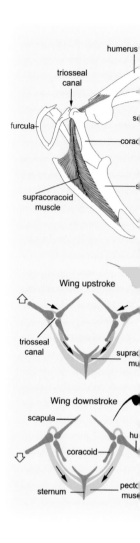

In modern birds, the supracoracoid (orange) and pectoral (yellow) muscles are largely responsible for elevating and depressing the wing, respectively, during the wingbeat cycle. Although the supracoracoid muscle is located below the point of rotation of the wing— the shoulder joint or glenoid—the pulley-like function of the triosseal canal allows this muscle to elevate the wing. (Image: S. Abramowicz.)

range of movement, being almost entirely tied to the flexion and extension movements of the hand along the plane of the wing.

The ilium, ischium, and pubis bones that are fused to one another at the level of an open hip socket or acetabulum form the pelvis of modern birds. Unlike many primitive birds and their dinosaurian ancestors, the ends of the pubis and ischium of most living birds are not joined to their counterparts. This anatomical departure of the pelvic architecture may be more convenient for animals laying large, delicate calcified eggs—a clear trend towards larger eggs is evident across the evolution from early theropods to birds.

In the hindlimb, the femur is usually short and robust, and is held nearly horizontal. This bone serves as the attachment point for a series of pelvic muscles that play important roles in posture and locomotion. Some bones of the hindlimb are formed by the same process of developmental fusion characterizing the formation of many compound bones in the adult avian skeleton. The tibiotarsus is one of them. This compound bone is formed by the fusion of the tibia to the calcaneum and the astragalus (the two upper tarsal bones of the dinosaurian ankle) and is held much more vertically than the femur. The tibia is a long bone, usually the longest in the hindlimb, whose upper end serves as the attachment point of muscles originating in the thigh and provides anchoring for other muscles attached to the foot. The second bone of the shank, the fibula, is firmly attached to the tibia. In modern birds, the fibula is reduced to a splint, lacking the articulation with the ankle—specifically, with the calcaneum— which was present in the most primitive birds and their dinosaurian predecessors. Although a relatively weak bone, the upper part of the fibula receives muscles that are critical for flexing the hindlimb, and hence this bone fulfills an important role in both posture and locomotion.

Another compound bone of the leg is the tarsometatarsus, which together with the toes forms the foot. The tarsometatarsus incorporates the three main bones of the foot (metatarsals 2, 3, and 4) as well as some ankle bones, the lower tarsals, which are all firmly fused to one another. Metatarsal 1 is much smaller than the other metatarsals and is more loosely joined by ligaments to the inner side of metatarsal 2. Because the ankle of birds is at the joint between the tibiotarsus and the tarsometatarsus, this articulation is an intertarsal one— between the upper tarsals incorporated in the tibiotarsus and the lower tarsals fused to the metatarsals. This primitive intertarsal joint, originating in the Triassic predecessors of dinosaurs, restricts the movements of the foot to the fore–aft plane. The movements of the toes are operated by a complex system of tendons attached to muscles of the upper leg. The shape and length of the tarsometatarsus and the shape, length, and position of the toes, varies across the many lineages of modern birds. The tarsometatarsus is long in waders, transversally compressed in divers and swimmers, and short in perchers and climbers. Likewise, while in most birds the three main toes point forward and the first toe points backwards, some birds have two toes pointing forwards and two pointing backwards; or all four toes facing forwards. Such an ample variation is clearly correlated with the vastly different ecological specializations that characterize the group.

Not surprisingly, the sophisticated skeletal plan of birds was acquired in piecemeal transformations over the long evolutionary history of these animals. Fortunately, many of these critical modifications are preserved in the fossil record of early birds. We will highlight these transformations throughout the following chapters; but first, let us review the controversial issue of the origin of birds.

CHAPTER TWO

The Ancestry of Birds and the Origin of Flight

The origin of birds has always been one of the greatest mysteries of biology. By combining a feathered body, clawless wings, a short bony tail, and a toothless beak, birds stand out among all living animals. Not surprisingly, given their uniqueness, deciphering the ancestry of birds has been the subject of intense scientific scrutiny since the advent of evolutionary thought. More recently, such an endeavor has captivated the public's attention like few other topics in science. After all, this evolutionary problem relates not only to how some of the most conspicuous animals on our planet evolved, but also to our perception of the dinosaurs that have long been identified as their predecessors. Were these magnificent animals feathered? Had they evolved some of the physiological adaptations and behaviors that are so typical of birds? Were they able to fly? These and other questions are intimately related to the origin of birds—or at least to identifying birds' closest relatives.

The small nonavian coelurosaur *Compsognathus*, a contemporary of the Late Jurassic *Archaeopteryx*, played a key role in 19th century discussions of the origin of birds. (Image: G. Janssen; © Bayerische Staatssammlung für Palaeontologie und Geologie, Munich.)

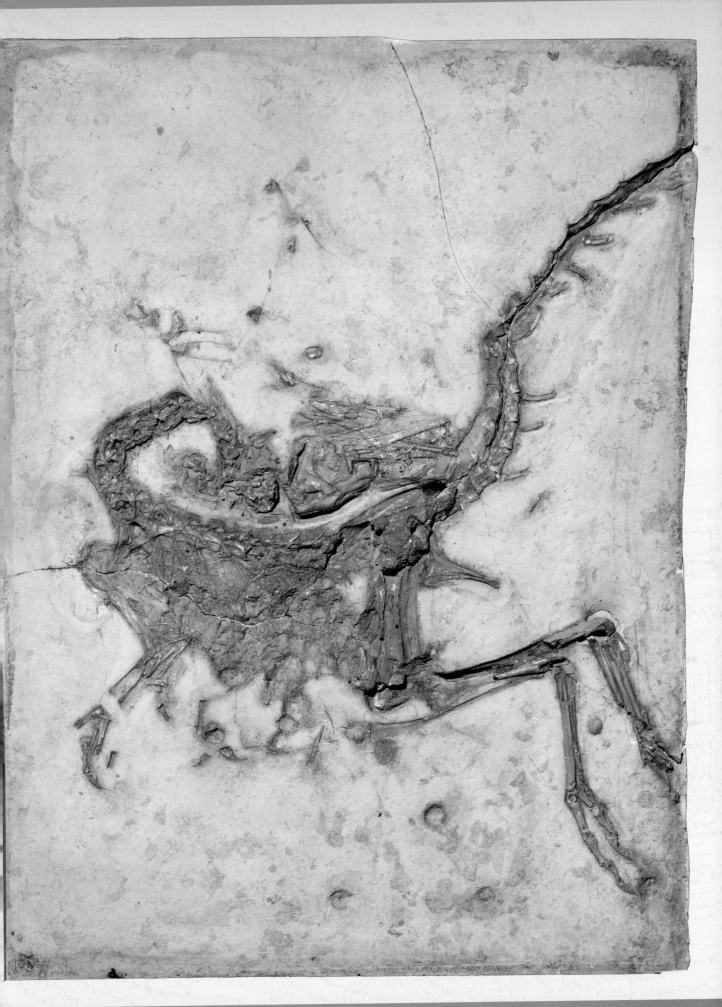

The first modern attempts at understanding the relationships of birds to other animals can be traced back to the 18th century "chains of being" used by the natural historians of the Enlightenment. In these linear and hierarchical plans of divine creation, each species had a unique position linked above and below by its closest relatives. Aquatic birds and to some extent all birds were often "linked" to flying fish. These ideas, however, were abandoned with the advent of evolutionary thinking, when a wide range of other vertebrates were postulated as either ancestral or as the closest group to birds. Early in the 19th century, the French botanist Jean Baptiste Lamarck thought that the descent of birds had to be found among turtles, which are also beaked. Proposals for the ancestry of birds multiplied after the 1859 publication of Charles Darwin's *Origin of Species*. For decades these ranged from pterosaurs (pterodactyls and other flying reptiles) to dinosaurs and from these to a variety of Triassic archosauromorphs. From time to time even theories of multiple origins—different groups of birds evolving from separate ancestors—were entertained, although these have consistently been regarded as marginal views.

Nowadays, most researchers agree that birds are the living members of theropod dinosaurs. The first suggestion that birds could have been related to dinosaurs came soon after the publication of the *Origin of Species*. In 1864, similarities in the structure of the ankle led University of Jena embryologist and vertebrate anatomist Karl Gegenbaur to place the small, Late Jurassic theropod *Compsognathus longipes* in an intermediate position between birds and other reptiles. Across the Atlantic Ocean, in Philadelphia, paleontologist Edward Drinker Cope also compared the ankle joint of the theropod *Megalosaurus* to that of an ostrich. On the basis of these comparisons and the existence of other similarities (the elongation of the neck vertebrae and the lightness of the skull), he argued for a close relationship between theropods and birds.

Thomas Henry Huxley, the British anatomist who in the 1860s became the foremost defender of the theropod origin of birds. (Image: American Museum of Natural History.)

Gegenbaur and Cope's ideas echoed in the mind of a brilliant British anatomist, who took on the problem of the origin of birds to establish a staunch defense of Darwin's concepts of evolution by means of natural selection. This man was Thomas Henry Huxley, who by 1868 was already arguing that "surely there is nothing very wild or illegitimate in the hypothesis that the *phylum* of the class Aves has its roots in the dinosaurian reptiles." In the mind of Huxley, birds became glorified dinosaurs. Like Cope, much of Huxley's initial work was done with *Megalosaurus*, a large carnivorous

The similarities in the wing of the great auk and the Snares crested penguin, and the neck of the condor and the Indian white-backed vulture, are examples of analogies—features that evolved convergently, most likely due to the adaptive success of the similar functions they perform. (Image design: R. Urabe.)

Great auk

Snares crested penguin

dinosaur (and a close relative of the better-known American theropod *Allosaurus*), discovered in the Jurassic of England. In the middle of the 19th century, dinosaur paleontology was in its infancy. Dinosaurs were first found and studied in the 1820s, by a dedicated network of fossil hunters in England. Reported in 1824, *Megalosaurus* was the first dinosaur named within the context of paleontological studies. Even though little was known about this animal, in 1869 Huxley delivered a lecture at the Royal Geological Society of London, in which he was able to recognize more than 30 skeletal similarities—primarily from the hindquarter—in support of an evolutionary connection between dinosaurs and birds. The year after, he also highlighted the similarities between birds and the small *Compsognathus*. Huxley's views, however, were not entirely well-received within the scientific community. At the same meeting of the Royal Geological Society, British paleontologist Harry Seeley wondered in response whether the similarities between *Megalosaurus* and birds were not just the result of similar functions, in particular a bipedal gait, they might have performed. As Yale University's evolutionary biologist

Jacques Gauthier so eloquently put it, "Pandora's box had been opened." The demons that had escaped were those of evolutionary convergence.

In the previous chapter, we discussed how similarities among organisms—could be divided into two groups. One type of similarity, called homologous, is interpreted as the result of common descent. In other words, similar features of two (or more) organisms—either the wings of bats and birds or the arms of humans and cherubs—are interpreted as homologous when this similarity is explained as the result of inheritance from a shared ancestor. Evolutionary biologists, however, have long recognized another type of similarity, one that is not the result of common inheritance but of parallel or convergent evolution. This type of similarity is called analogous. Convergent evolution leads to analogous similarities. Because analogous similarities are abundant in the history of life, one must consider the possibility that two organisms that look very similar may have evolved through different pathways. Nonetheless, convergent evolution should not be argued out of hand. We have seen how cladistics and its parsimony principle provide the methodological tools for distinguishing between homology and analogy: the wings of birds and cherubs are analogous because parsimony supports the homology between the bird wing and the cherub arm. In this framework, an argument that opposes a proposed homology by invoking analogy—in this case that the similarities between *Megalosaurus* and birds evolved convergently as adaptations responding to the similar functional requirements of bipedalism—could only be countered with an alternative proposal that is more parsimonious. Seeley's objections at the 1869 meeting of the Royal Geological Society, where Huxley defended his view of avian ancestry, initiated a lasting argument that still persists today. As we will see, however, this argument—

Condor

Indian white-backed vulture

the analogous similarity of nonavian theropods and birds—has never been fenced by the hand of parsimony.

The Theropod Hypothesis Loses Ground

In the last quarter of the 19th century, the idea that some theropod dinosaurs and birds had evolved from a common stock was strongly upheld by both paleontologists and embryologists. Karl Gegenbaur, Thomas Huxley, Edward Drinker Cope, and Henry Fairfield Osborn were among those who accepted this view within 20 years of the publication of Darwin's *Origin of Species*. Even Darwin, who in general had distanced himself from the debate on avian origins, endorsed this view in the 6th edition of his seminal book. Yet Seeley's arguments were also echoed by a small number of researchers, who claimed that the similarities between theropods and birds were due to convergent evolution. In most cases, these dissenters supported a close relationship between birds and pterosaurs, whose brain and flight apparatus were viewed as birdlike. However, a more serious threat to the theropod hypothesis of avian origins came in the shape of a small, primitive archosauromorph, discovered at the beginning of the 20th century in a remote corner of the world—the Karroo Desert of South Africa.

The Early Triassic *Euparkeria capensis* was described in 1913 by Robert Broom, a Scottish physician who practiced both medicine and paleontology in the Karroo, and is best known for his contributions to paleoanthropology. At the time, new discoveries of large dinosaurs—including the mighty *Tyrannosaurus*—had started to tip the scale against the theropod hypothesis. The notion that dinosaurs were too specialized to be the ancestors of birds had infiltrated the intellectual atmosphere. Slowly but surely, the demons of convergence were plucking birds off the dinosaur pantheon. In this

intellectual cauldron, Broom's claims of *Euparkeria* as the ideal ancestor of birds were quickly embraced. After all, *Euparkeria* was much less specialized than any dinosaur and this generalized primitiveness made it an excellent candidate for the ancestry of birds. Broom's ideas on avian origins greatly differed from the pterosaur–bird relationship upheld by several of those who up to then had denied the theropod hypothesis. His studies of the Triassic faunas, as well as the development of the avian embryo, led him to believe that the ancestor of birds and theropod dinosaurs had to be a bipedal animal—presumably

Robert Broom, the Scottish physician who discovered *Euparkeria* and claimed a close relationship between this Triassic archosauromorph and birds. (Image: American Museum of Natural History.)

a Triassic archosauromorph—and to the idea that pterosaurs had sprung off earlier from an even more primitive, quadrupedal archosauromorph. It is possible that Broom's ideas of a parallel origin of birds and theropods from bipedal archosaurs may have been influenced by his embryological studies of birds and his observations of a five-toed foot in the ostrich, which may have hinted at a deeper ancestry than from animals with four toes such as theropod dinosaurs. At the time, the advent of birds was generally regarded as a very ancient evolutionary event. Even Huxley's views on bird origins were framed at the end of the Paleozoic Era, more than 250 million years ago, from a hypothetical dinosaur that was much more primitive than the Late Jurassic *Compsognathus*. Regardless of the reasons, Broom's final interpretation

In the 19th century, some scientists supported a close relationship between pterosaurs, like the Late Jurassic pterodactyl shown here, and birds. (Image: L. Chiappe.)

Gerhard Heilmann, Danish artist and author of the influential *The Origin of Birds*. Despite the cool reception of his work on early birds by the scientific establishment of his country, the 1926 publication of his book virtually closed the debate on bird origins. For decades afterwards, birds were assumed to be the descendants of primitive archosauromorphs. (Image: C. Jacob Ries.)

of *Euparkeria* had a tremendous impact on the issue of bird origins. As highlighted by Lawrence M. Witmer of Ohio University (Athens) in a review of bird origins, *Euparkeria* deserves to share a place with *Archaeopteryx* as the most influential fossil discovery in the history of the debate on avian origins.

If the discovery of *Euparkeria* switched dinosaurs for less specialized archosauromorphs as the focus of the discussions on bird origins, the work that buried the dinosaur argument was the 1926 publication of *The Origin of Birds* by Danish artist Gerhard Heilmann. A dynamic artist who became interested in paleontology at an early age, Heilmann is primarily remembered as a gifted painter at the Royal Porcelain Works and a bird illustrator. In the scientific community, however, Heilmann is known for his influential volume, first published as

a series of articles in Danish between 1913 and 1916. In *The Origin of Birds*, Heilmann examined the anatomy of birds and their putative relatives in extreme detail, concluding that theropod dinosaurs had more similarities with birds than did any other group. A significant difference, however, was also noted—dinosaurs lacked clavicles or collarbones. One of the most distinctive features of birds is the presence of wishbones, formed by the fusion of the clavicles during embryonic development. During Heilmann's times, many biologists believed that if a structure had been lost during the course of evolution, it could not be regained by the descendants of the forms that had lost it. This guiding principle, known as Dollo's Law of Irreversibility—in reference to Belgian paleontologist Louis Dollo, who proposed it in 1893—led Heilmann to conclude that dinosaurs could not have been the ancestors of birds. Heilmann reasoned that even though theropods were the most similar animals to birds, they could not be the ancestors of birds because these dinosaurs had lost their clavicles. In the end, he conceded to the less compromised argument of *Euparkeria* and other generalized Triassic archosauromorphs, for which clavicles were known, as the most suitable candidates for the origin of birds.

Ironically, in the ten years between Heilmann's Danish articles about the ancestry of birds and the publication of *The Origin of Birds*, collarbones were found in the theropod dinosaur *Oviraptor philoceratops*, discovered for the first time by the renowned expeditions of Roy Chapman Andrews to the Cretaceous deposits of the Gobi Desert. These collarbones, however, were at the time misinterpreted as other bones. Like Seeley had done decades before, Heilmann concluded that the remarkable similarity between theropod dinosaurs and birds had to be the result of adaptations related to their bipedal stance and gait. The discovery of clavicles in *Oviraptor* was soon followed by the detection of the same bones in another theropod dinosaur from the Early Jurassic of Arizona—the ceratosaurian *Segisaurus halli*. Yet the enormous weight of *The Origin of Birds* held sway through much of the 20th century. Dinosaurs were viewed as too specialized and their similarities with birds were deemed as evolutionary convergences. The notion that birds had evolved many millions of years before *Archaeopteryx* from generalized archosauromorph reptiles, and independently from the origin of other archosaurs (including dinosaurs), became firmly established.

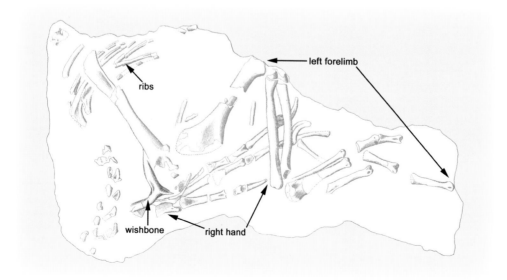

When in 1924 paleontologist Henry Fairfield Osborn described *Oviraptor*, a then newly discovered theropod dinosaur from the Gobi Desert, he confused its wishbone for another bone. The existence of wishbones in theropods was overlooked for many decades, and their assumed absence was the key argument against the theropod origin of birds. (Image after Osborn, 1924.)

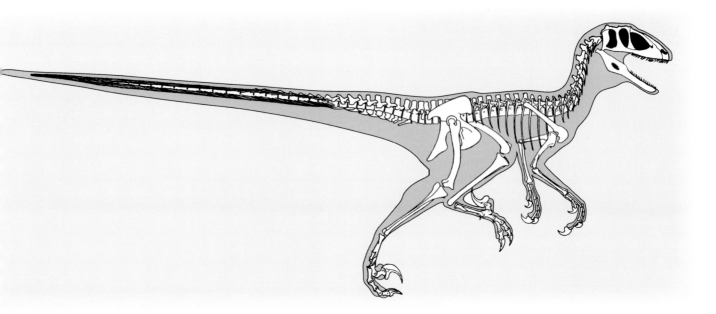

The New Age of Avian Ancestry: *Deinonychus* Comes to Life

Little research on the origin of birds was conducted after Heilmann's influential book. It seemed as if the answer had been found. However, a series of discoveries, both in the field and within museum collections, would come to strongly contest this view, and would also provide impetus for a dramatic transformation in the scientific perception of dinosaurs.

These discoveries came about in the early 1960s, when Yale University paleontologist John Ostrom undertook a careful study of a theropod specimen found three decades earlier by one of the most famous and flamboyant of dinosaur hunters—Barnum Brown of New York's American Museum of Natural History. Brown had collected the specimen from Early Cretaceous rocks in Montana and, though he had recognized it as a new, moderate-sized theropod species, he never reported on it. In Ostrom's hands, the dinosaur discovered by Brown upset the long-held notion of a basal archosauromorph origin of birds and heralded a revolution in our understanding of dinosaur biology. The "dinosaur Renaissance," as this interpretive revolution would come to be known, changed the way dinosaur researchers perceived their subjects of study forever. The traditional impression of dinosaurs as sluggish, lethargic, and stupid was replaced by the modern interpretation of them as active, agile, and gifted with sophisticated social and behavioral patterns.

The significance of the specimen Brown had discovered led Ostrom to revisit the fossil grounds where this animal had been unearthed. In the mid-1960s, he led a series of successful expeditions to Montana, where additional specimens of this theropod were collected. In 1969, a report naming this animal as *Deinonychus antirhopus* was followed by an extensive monograph on its anatomy. Ostrom noticed that the relatively small size (about 2.4 meters long), hollow bones, and stiffened tail of *Deinonychus* made it an agile, lightly built, and remarkably bird-like predator. Ostrom went on to revive Huxley's notions of a theropod ancestry of birds in a series of controversial papers published during the first years of the 1970s. Just like Huxley a century earlier, Ostrom compared the skeleton of the lightly built theropod *Compsognathus* from the Late Jurassic Solnhofen Limestones with its contemporary *Archaeopteryx*. Unlike Huxley, however, Ostrom was armed with the remarkably bird-like appearance of *Deinonychus*. Ostrom's evidence was so solid that before long much of the paleontological community rejected the decades-long view of a basal archosauromorph ancestry of birds.

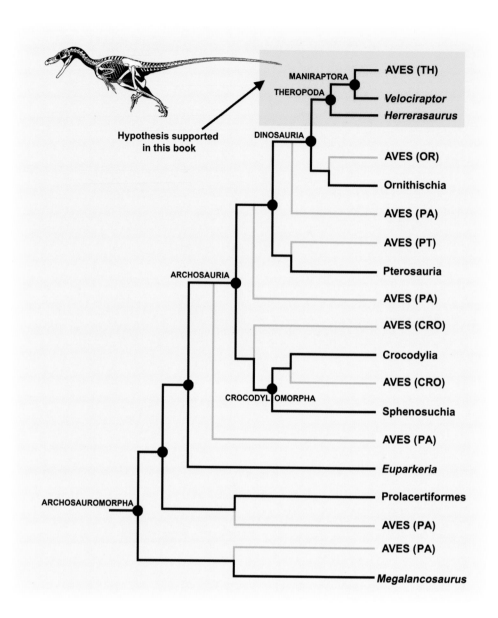

MANIRAPTORA
THEROPODA
— AVES (TH)
— *Velociraptor*
— *Herrerasaurus*

DINOSAURIA
— AVES (OR)
— Ornithischia
— AVES (PA)
— AVES (PT)
— Pterosauria
— AVES (PA)
— AVES (CRO)
— Crocodylia
— AVES (CRO)
— Sphenosuchia
— AVES (PA)
— *Euparkeria*
— Prolacertiformes
— AVES (PA)
— AVES (PA)
— *Megalancosaurus*

ARCHOSAURIA

CROCODYLOMORPHA

ARCHOSAUROMORPHA

Hypothesis supported in this book

Modern Alternatives to the Theropod Origin of Birds

Ostrom's ideas of a theropod origin came at a time of renewed interest in the ancestry of birds, an impetus that to a great extent was in reaction to his hypothesis. Nonetheless, most of the arguments and alternatives proposed to counter the growing acceptance of the theropod hypothesis were not very different from the ones that had been proposed half a century earlier. On the one hand, there were those who proposed that birds had evolved from a Triassic archosauromorph that had either predated the diversification of archosaurs—the

reptilian group encompassing birds and crocodiles—or that was a very primitive archosaur. This argument has often been referred to as the "thecodont" hypothesis, because primitive Triassic archosauromorphs used to be called "thecodonts." "Thecodonts," however, comprise an artificial group used to include all archosaurian reptiles that were neither dinosaurs, crocodiles, nor pterosaurs, and so the group has no characteristics of its own. The other hypothesis arguing that theropods were too specialized to be avian ancestors proposed the novel idea that birds had evolved from either primitive relatives of crocodiles, from within animals

called sphenosuchians, or that birds had shared a common ancestor with living crocodiles. Both these views are often referred to as the crocodylomorph hypothesis.

Most of these alternatives to Ostrom's revival of the theropod ancestry of birds have come down untouched to the present, and none of them carry enough empirical support to knock his idea off its perch. Defenders of the "thecodont" hypothesis have cited similarities in the structure of the ear region and braincase, as well as the shape of teeth and their replacement patterns in birds and Triassic archosauromorphs such as *Euparkeria*. Nonetheless, this alternative view continues to be as vague as it was at the time of Broom and Heilmann. For example, Samuel Tarsitano and the late

Max Hecht, of the Southwest Texas State University and City University of New York, respectively, have argued in favor of an origin within a group lacking armor and having a mesotarsal ankle—the type of intertarsal ankle characteristic of dinosaurs (including birds) and their immediate ancestors. They envisioned this ancestor to be genealogically intermediate between *Euparkeria* and *Lagosuchus*—a Triassic ornithodiran very close to the origin of dinosaurs—thus leaving a multitude of groups as potential alternatives. In other cases, arguments such as the fact that *Deinonychus*—Ostrom's Trojan Horse—was millions of years younger than *Archaeopteryx* were deemed sufficient to dismiss the theropod hypothesis and to call for an older origin, among Triassic "thecodonts."

An array of hypotheses about a close relationship between birds and some small Triassic reptiles is also often grouped under the "thecodont" hypothesis, even if these fossils are further removed from the ancestry of archosaurs than *Euparkeria* and have not been included in the traditional classifications of "thecodonts." The small teeth and triangular skulls of these fossils give them a superficial bird-like appearance common among juveniles of a great number of tetrapods, including theropod dinosaurs. One of these fossils is the tiny, long-limbed *Scleromochlus taylori* from the Late Triassic of Scotland. The fragmentary nature of the fossil material of *Scleromochlus*—whose specimens are primarily known from the impressions that the bones left on coarse sandstone deposits—complicates the interpretation of the anatomy and genealogical relationships of this animal. Some have also argued that all known specimens of this reptile are juveniles, further complicating the efforts of comparative anatomists. Nonetheless, several studies have identified *Scleromochlus* as either the closest relative to pterosaurs, the flying reptiles that flourished during much of the Mesozoic Era, or as an animal slightly removed from the ancestry of pterosaurs and dinosaurs.

Another poorly known fossil that was claimed to be near the origin of birds is the small *Cosesaurus aviceps* from the Middle Triassic of Spain. *Cosesaurus* not only has a triangular skull bearing tiny teeth but also a shoulder that superficially resembles that of a bird and an element that vaguely looks like a wishbone. In addition, this tiny and obscured fossil bears poorly preserved, elongated scales that were interpreted as traces of feathers. This interpretation was accepted by some and used to advance the idea that animals with avian-like features had evolved long before theropod dinosaurs. All subsequent studies of *Cosesaurus*, however, identify this tiny animal as a member of the prolacertiforms, a group of amphibious reptiles abundant in the Triassic fossil record of North America

and Europe, which is only distantly related to the ancestry of archosaurs.

Yet another of these small Early Mesozoic fossils claimed to be near the avian ancestry is the chameleon-bodied, tree-dwelling *Megalancosaurus preonensis*, whose fossils have been unearthed from Late Triassic rocks of northern Italy. In contrast with the other two examples, *Megalancosaurus* is known from much better preserved and more complete fossils. However, detailed studies of these fossils by Italian paleontologist Sylvio Renesto of Milan's Università degli Studi have provided support for the prolacertiform relationship of *Megalancosaurus* and dismissed any purported similarity to birds. According to Renesto and others, the features claimed to be shared between *Megalancosaurus* and birds are indeed unique to the Triassic prolacertiforms. Perhaps the most recent case in which one of these poorly understood, Early Mesozoic fossils was claimed to be at the base of the avian ancestry involves a study of the bizarre *Longisquama insignis* from the Osh region of Kyrgyzstan. Discovered by Russian paleontologists in the late 1960s, the most remarkable feature of the tiny *Longisquama* is a series of extremely long scales that fan out from its back. Most significantly, the fern-like appearance of *Longisquama*'s long scales, with a central shaft from which structures project perpendicularly, has supported the idea that these scales were the evolutionary precursors of feathers. We will return to *Longisquama* when discussing the origin of feathers, but the point we need to highlight here is that studies of this fossil have argued for a prolacertiform relationship. Thus, regardless of how feather-like one may see its long scales, the Triassic *Longisquama* seems to be a member of a group that lies outside the ancestry of archosaurs and is much more removed from the origin of birds.

An attempt to combine these small Triassic reptiles into a single preavian stock was presented in 1993 by Alan

Feduccia from the University of North Carolina and Rupert Wild from the Staatliches Museum für Naturkunde (Germany). Since then, this argument has been repeated at various other times by those opposing the theropod hypothesis of bird origins. Feduccia and Wild argued that *Megalancosaurus*, when combined with *Longisquama* and *Scleromochlus*, had the feather-like scales, wishbone, beak-like snout, quadrupedality, and other features that they interpreted as a requirement for the reptilian ancestor they believed to have given rise to birds. In modern comparative biology, however, combining features from animals with disparate evolutionary origins (prolacertiforms and a pterosaur relative) does not contribute to understanding the divergence of birds unless the idea of multiple independent origins for the group is entertained, a view that Feduccia and Wild do not seem to hold. Furthermore, the notion that feather-like scales are not feather precursors is now reinforced by molecular and developmental evidence, indicating that feathers did not evolve from reptilian scales. Finally, the "avian" nature of quadrupedality is an unwarranted assumption about the ancestral condition of birds, given that all members of the group are bipedal. Thus, attempts to align birds with a variety of small and quadrupedal Triassic reptiles have used fossils removed from the ancestry of archosaurs, unrelated to each other and lacking features identifiable as avian. In the end, claims for the origin of birds from this stock of small Triassic reptiles seems to adopt a formula more apt for the world of art than for scientific endeavor, as the identity of the object (the ancestor) is defended more by its symbolism than by its own physical attributes.

The second group of arguments alternative to the theropod hypothesis of bird origins is the crocodylomorph hypothesis. This view was first proposed by the late Alick Walker (University of Newcastle upon Tyne) in the wake of an early 1970s surge of research on bird origins. Birds have traditionally been considered relatives of crocodiles; both are archosaurs—and the only living members of this large group of reptiles. Yet Walker centered the support for his hypothesis on detailed observations of the braincase and ear region of the Late Triassic *Sphenosuchus acutus*, an animal closely related to true crocodilians although still outside this group. Walker's contribution greatly helped our understanding of the anatomy of *Sphenosuchus* and other primitive relatives of crocodiles, but the features he selected were either also discovered among theropods or found to be absent in *Archaeopteryx* and other early birds. The crocodile argument was later adopted by University of Kansas paleontologist Larry D. Martin, who proposed a modified version of this hypothesis, arguing that *Sphenosuchus* fell outside the crocodile–bird connection. Martin and his collaborators drew support from perceived similarities in the structure of the ankle and the teeth of crocodiles and birds, as well as refining the comparisons made by Walker. As in the case of *Sphenosuchus*, recent discoveries have shown that either these features are found among theropod dinosaurs or they are absent among the most archaic birds. In the end, although the crocodylomorph hypothesis has fewer analytical problems than the "thecodont" hypothesis, it still continues to rely on a handful of selective features present in crocodiles to claim ancestry from dinosaurs—and as these features are found to be present among theropod dinosaurs and absent in the most archaic birds, this view has become virtually untenable.

Evidence for the Theropod Origin Mounts Up

Since John Ostrom resurrected the theropod hypothesis of bird origins, a wealth of skeletal evidence has accumulated in support of the notion

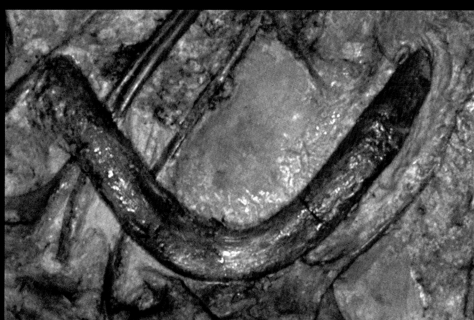

Wishbones of the
dromaeosaurids
Sinornithosaurus (top) and
Buitreraptor gonzalezorum
(middle) compared
to the wishbone of
Archaeopteryx (bottom).
(Image: L. Chiappe.)

that birds originated within often small and predominantly terrestrial animals, dinosaurs usually classified within maniraptoran theropods. Comparative study of these fossils has been greatly assisted by the new discoveries of Mesozoic-aged avian fossils, because these ancient birds possess skeletons that are slightly modified from those of their maniraptoran predecessors. This is particularly important for those structures that may have undergone substantial modification during the long evolutionary history of birds and that may consequently be drastically different from the homologous structures present in maniraptorans.

Although dinosaurs such as *Velociraptor* and *Deinonychus* are prominently featured in discussions of bird origins, the skeletal evidence for the theropod origin of birds is present in every theropod dinosaur. We have seen how Thomas Huxley—whose research was conducted decades before *Deinonychus* and *Velociraptor* were discovered—based his observations on *Megalosaurus* and the much smaller *Compsognathus*, Late Jurassic theropods far more removed from the ancestry of birds than their later famed relatives. Indeed, birds share skeletal features with even the earliest theropods. The Late Triassic *Herrerasaurus* already shows the hollow limb bones and three fully developed toes characteristic of birds.

The 75-million-year-old *Velociraptor* is a dromaeosaurid theropod and a close relative of *Deinonychus*. Dromaeosaurids are maniraptoran theropods commonly interpreted as the closest relatives of birds. (Image after Rowe et al., 1998.)

Furthermore, this animal, like all other theropods, walked only with its hindlegs. The condition of bipedalism evolved in the ancestors of all dinosaurs, but birds inherited it from their theropod predecessors. The skeleton of birds also shares with those of tetanurans—a subgroup of theropod dinosaurs—the presence of a wishbone and three-fingered hands. Today the wishbone has been reported in a variety of tetanurans, including the large *Allosaurus* and the fearsome *Tyrannosaurus* as well as the maniraptoran oviraptorids, therizinosaurids, troodontids, and dromaeosaurids. This bone has also been found in theropods more primitive than tetanurans, such as the ceratosaur *Syntarsus* from the Early Jurassic, although it is possible that this dinosaur evolved the feature independently. Within tetanurans, the coelurosaurs exhibit the long arms and large breastbones of flying birds. Doubtless these features did not evolve in the context of flight; most likely, their development occurred in connection with predatorial activities and wing-assisted running.

Not surprisingly, birds share many more features with their most immediate dinosaur relatives, the maniraptorans. Indeed, an examination of the skeleton of a dromaeosaurid, a troodontid, an oviraptorid, or any other dinosaur classified as a maniraptoran theropod would reveal similarities with birds in

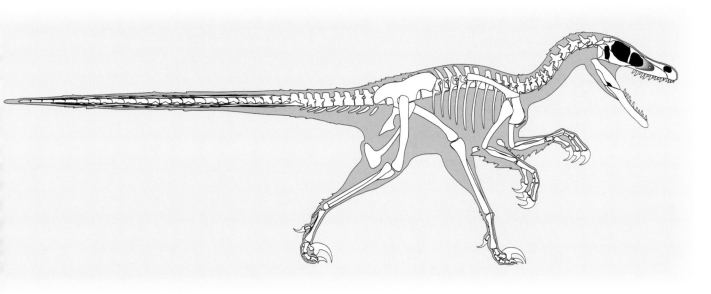

almost every bone. These similarities are particularly evident in the most primitive members of dromaeosaurids and troodontids, the two lineages of maniraptorans that are generally accepted as the closest relatives of birds. Inspection of the braincase of these dinosaurs would disclose the presence of the three air spaces that in birds (and in these dinosaurs) are connected to the ear region. At the base of the neck, the vertebral column of these theropods would reveal the same projections that in birds offer attachment to muscles that control the posture and movement of the neck. The well-preserved fossils of these maniraptorans also show that their ribcage was formed by two-part ribs, with the lower segment attached to the sternum. In birds, such a design plays a key role in breathing. Furthermore, an examination of the wrist of these dinosaurs reveals a relatively large, crescent-shaped bone identical to the one present in primitive birds and in the embryos of all birds. Such a bone allows a swivel movement of the wrists while flying or folding the wings against the body. In the pelvis, the pubic bone of some maniraptorans (like dromaeosaurids and troodontids) is either vertically oriented or angled towards the back, essentially displaying the same orientation seen in many primitive birds. These dinosaurs are also similar to a variety of early birds in other details of the pelvis, such as the shape of the boot-like expansion of the end of the pubis and the proportions between this bone and the ischium. The resemblance between maniraptorans and birds is also expressed in the hindlimbs of these animals. The attachment points for the main muscles that connect the femur to the base of the tail are greatly reduced, thus heralding the abbreviation of the tail in birds. The structure of the ankle is remarkably similar to that seen in primitive birds as well as the renowned sickle-clawed foot of dromaeosaurids and troodontids—conspicuously featured in the vicious

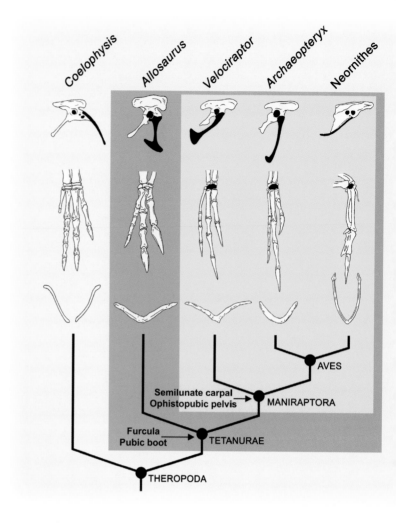

raptors of the movie *Jurassic Park*. This body of skeletal evidence represents the foundation of the theropod hypothesis and has provided the logical basis upon which birds are classified within these dinosaurs.

New Lines of Evidence Provide Even Further Support for the Theropod Hypothesis

In the last few years, a series of spectacular discoveries has provided additional evidence in support of the hypothesis that birds are evolutionarily nested within maniraptoran theropods. Discoveries of embryonic remains of maniraptorans inside their eggs have shown that the eggs of these dinosaurs share unique features with those of birds. The general appearance and microstructure of calcified eggs—whose shells are made of calcium carbonate—is specific to certain groups of extant and extinct reptiles. Until recently, the

Numerous skeletal features support the inclusion of birds within theropod dinosaurs. The presence of a furcula (wishbone) and of a terminal expansion of the pubis (pubic boot) can be traced back to at least the origin of the tetanuran theropods. The existence of a wrist with a semilunate carpal and a pubis that to some extent faces backwards (ophistopubic pelvis) can be traced back to the divergence of the maniraptoran theropods. These skeletal features—synapomorphies of tetanurans and maniraptorans—are all found in primitive birds. (Image after Padian and Chiappe, 1998.)

precise characteristics of the eggshell microstructure of theropod dinosaurs remained elusive. In the late 1980s, Russian paleontologist Konstantin Mikhailov and other specialists on dinosaur eggs identified a series of elongate Cretaceous eggs with structural similarities to those of living birds as theropod eggs. This conclusion was entirely based on the idea that birds had evolved from theropod dinosaurs, because

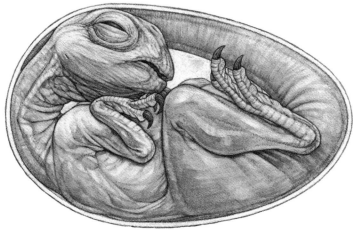

The embryonic remains (inside the egg) of the oviraptorid *Citipati*. Whether the embryos of these maniraptoran dinosaurs hatched feathered or not remains unknown. (Image: M. Ellison.)

direct evidence for the identity of those eggs—the presence of informative embryos inside them—had not yet been discovered. In 1993, a team from the American Museum of Natural History in New York City discovered a partial egg containing an embryo in the Late Cretaceous locality of Ukhaa Tolgod, in the Mongolian Gobi Desert (see *Ukhaa Tolgod: An Extraordinary Cretaceous Ecosystem*). The egg was similar to many that had been discovered in the

same region by the 1920s Central Asiatic Expeditions, a series of renowned explorations also organized by the New York museum (see *The Central Asiatic Expeditions: In Search of the Cradle of Mankind*). For decades, these eggs had been interpreted as those of the ornithischian *Protoceratops andrewsi*, a small ceratopsian dinosaur distantly related to *Triceratops horridus*, of which fossils are extremely abundant in the Late Cretaceous of the Mongolian Gobi Desert.

The contents of the 1993 egg, however, contradicted the traditional view. When the embryo inside this egg was prepared, it did not show features of its assumed ceratopsian kinship; instead, its delicate bones displayed characteristics of oviraptorids, a group of beaked, parrot-headed theropods best known from the Late Cretaceous of central Asia. Most important for the study of the origin of birds was the fact that this embryo provided the first direct evidence of the shape and shell structure of a theropod egg. As with the eggs of crocodiles and birds, the shell of this egg was composed of a variety of calcium carbonate called calcite. However, detailed comparisons between the shell microstructure of this egg and those of extant crocodiles and birds revealed features uniquely shared with birds' eggs. The most obvious of these was the fact that the calcite crystals comprising the eggshell were laid in two distinct layers, rather than the single layer they form in the eggshell of crocodiles. The crystalline structure of the innermost layer of both the oviraptorid egg and avian eggs is formed by elongate crystals that radiate from a core. This inner layer is thought to be homologous to the single layer of calcium carbonate that forms the shell of the eggs of crocodiles, sauropods, and ornithischian dinosaurs. The second layer overlaying the innermost layer of the oviraptorid egg has a contrasting manifestation—its crystals are arranged in such a way that when thin sections of the eggshell are viewed under a

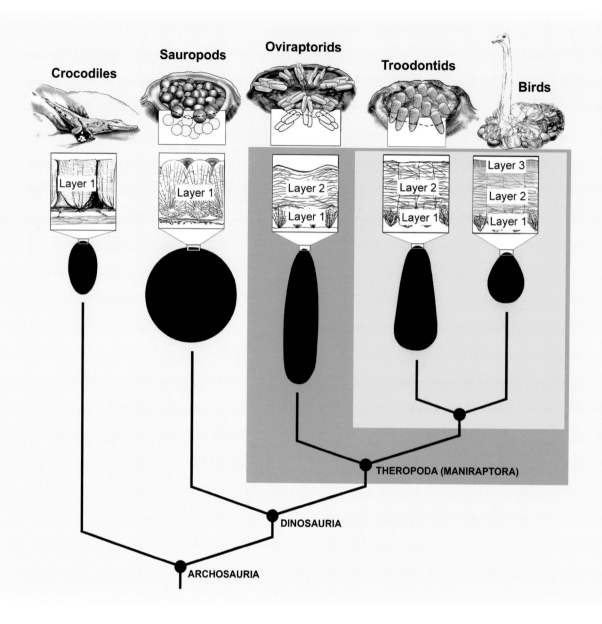

Crocodiles

Sauropods

Oviraptorids

Troodontids

Birds

THEROPODA (MANIRAPTORA)

DINOSAURIA

ARCHOSAURIA

microscope, the entire second layer has a laminated, plywood-like appearance. Avian eggs also exhibit a layer of similar appearance over their innermost layer of calcite, even if in modern birds this layer may grade into or be completely separated from a third layer.

Since the discovery of the oviraptorid egg, several other theropod embryos have been found inside their eggs. Although not all of them have been studied in detail, these embryos appear to span a good portion of the evolutionary spectrum of maniraptoran theropods, including disparate groups such as troodontids and therizinosaurids. The shells of all these eggs have two well-differentiated layers. Eggs of the lightly built troodontids also

show the asymmetric shape characteristic of the eggs of birds, in which one pole of the egg is distinctly narrower than the other. But the existence of more than one distinctive layer in the eggshell and the asymmetry exhibited by the troodontid eggs are not the only features in common between the eggs of nonavian maniraptorans and birds. These animals also evolved eggs with fewer pores (the airholes that pierce the eggshell), an elongated shape, and a proportionally bigger volume in relation to the size of the adult. All these similarities between the eggs of nonavian maniraptorans and birds suggest the dinosaur predecessors of birds had already evolved the complex cellular and glandular processes that control the

Characteristics of the eggs of several lineages of nonavian maniraptorans also support the inclusion of birds within theropod dinosaurs. The presence of at least two distinct crystallographic layers in the shells of these dinosaurs' eggs can be traced back as far as the maniraptoran divergence, and the existence of a distinctly asymmetric egg is common to troodontids and birds. (Image after Chiappe, 2004.)

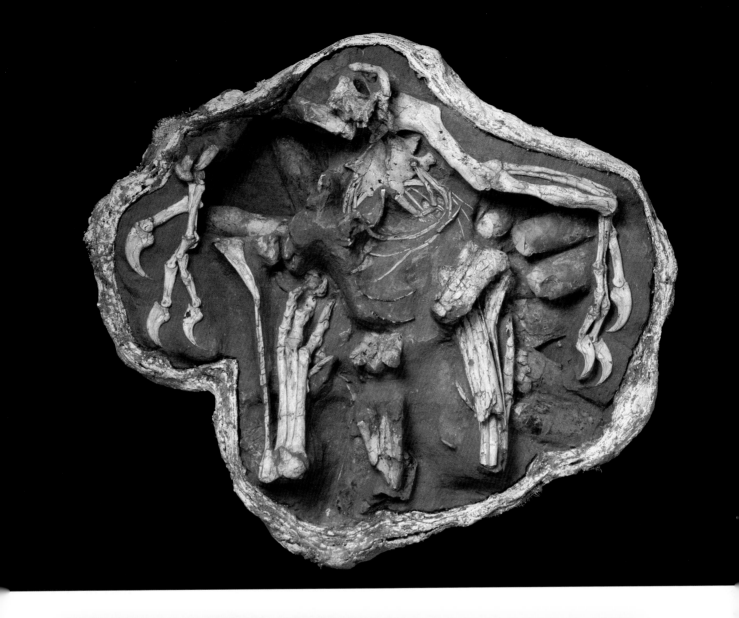

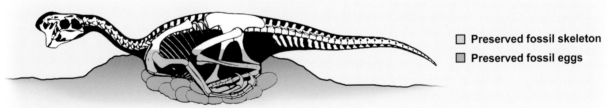

Preserved fossil skeleton
Preserved fossil eggs

A specimen of the oviraptorid *Citipati* crouching over its own clutch of eggs. Evidence from this and other exceptional fossils indicates that at least some nonavian maniraptorans had already evolved brooding behaviors comparable to those of many living birds. (Image: M. Ellison.)

unique attributes of the avian egg.

Other exceptional discoveries of the last few years involve adult specimens of maniraptorans in association with their clutches of eggs. This handful of rare fossils provides a vivid window into the furtive world of dinosaur nesting behavior. Evidence of the behavior of extinct organisms is seldom preserved in the fossil record. Unlike bones or other parts of organisms, behavior can only

be inferred from its fossilized products. Just as a fossil trackway provides evidence about the speed, posture, or other aspects of the behavioral repertoire of a trackmaker, fossilized egg-clutches and nests provide evidence of the nesting conduct of an egglaying animal. On the same day in 1993 that the above-mentioned oviraptorid embryo was recovered, paleontologist Mark Norell, co-leader of the expeditions of the American Museum

of Natural History to the Mongolian Gobi Desert, discovered the exposed large claws of an adult oviraptorid theropod. As a participant of this expedition, I joined the museum's fossil preparator Amy Davidson in excavating the claws Norell had found. To our great excitement, we soon realized that the fossil consisted of much more than the exposed claws. Indeed, the complete forelimbs, much of the trunk and ribcage, the hindlimbs, and the base of the tail were preserved almost untouched. To our enormous surprise, we also became aware that this magnificent skeleton (later identified as *Citipati osmolskae*) was crouching over a clutch of 17-centimeter-long eggs. This discovery was the first to provide direct evidence on the nesting behavior of theropod dinosaurs. Since then, the fossil-rich deposits of the Late Cretaceous Gobi Desert—both in southern Mongolia and northern China—have yielded several other adult oviraptorids in association with their egg-clutches.

In all these instances, the skeletons of these beaked dinosaurs are positioned with their hindlimbs tucked under the ribcage and their extended forelimbs demarcating the periphery of the egg-clutch. The clutches are often composed of ten or more pairs of eggs, stacked in two or three layers, and laid in a circular pattern around the ample space occupied by the adult's legs. The eggs in a pair are placed side by side, with very little space between them, and they are typically oriented towards the center. Even though none of these eggs have revealed embryonic remains, their general appearance and shell microstructure is remarkably similar to that of the Ukhaa Tolgod oviraptorid embryo. Furthermore, features of the skull of this embryo also suggest it belongs to the large oviraptorid *Citipati osmolskae*. These facts strongly support the hypothesis that this oviraptorid had already evolved brooding behavior very much like that typical of living birds.

This remarkable glimpse into the nesting behavior of some of the closest relatives of birds has received even more support from the discovery in Montana of a skeleton of the Late Cretaceous maniraptoran *Troodon formosus* in a brooding position. As in the case of the central Asian oviraptorids, the fact that similar eggs from the same deposits in Montana contained the remains of troodontid embryos (identified as *Troodon*) also suggests that the adult *Troodon* was brooding its own egg-clutch. However, the lightly built troodontids brooded their eggs by sitting directly on top of them, as opposed to within a space circumscribed by the clutch. This series of discoveries has provided unquestionable evidence that the brooding behavior typical of birds evolved prior to the origin of the group, and that this complex nesting conduct was widespread among the maniraptoran relatives of birds.

The characteristic pattern of paired eggs seen in the clutches of the oviraptorid *Citipati* has been interpreted as indicating that this dinosaur required many days to lay a clutch. Among living reptiles, such an egglaying pattern is only known for birds—the fact that avian egg-clutches do not show the distinct paired distribution seen in the egg-clutches of *Citipati* has been explained by the specialized reproductive anatomy of birds, which have only one active oviduct. In all other instances, reptiles lay all the eggs of a clutch within the span of a few hours. This more primitive behavior can also be ascribed to both sauropod and ornithischian dinosaurs, whose egg-clutches lack any kind of special arrangement. This behavior, however, cannot be ascribed to *Citipati*, because it seems highly unlikely that an animal that lays its eggs *en masse* would produce such a carefully organized clutch. The fact that a variety of fossil clutches containing elongate eggs (whose shells have the two distinct layers of theropod dinosaurs) have a similar paired arrangement and are laid in similarly designed circles around an empty space suggests that the physiological condition

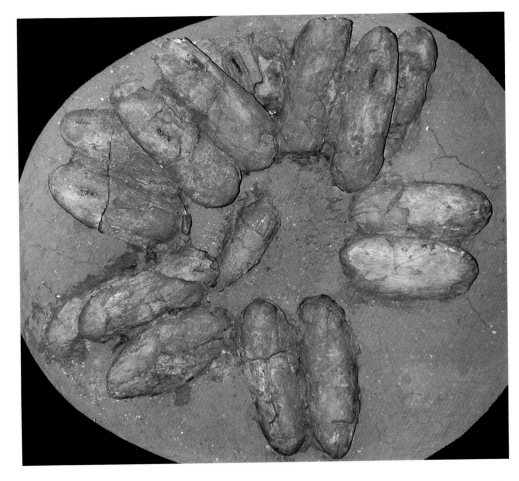

Egg-clutches like this one are common in the Late Cretaceous deposits of Mongolia and China. The general characteristics of the eggs indicate that they were laid by nonavian maniraptoran theropods, most likely oviraptorids. The paired distribution of the eggs suggests that each pair was laid sequentially and that it may have taken days for the entire clutch to be formed. (Image: L. Chiappe.)

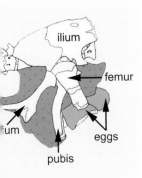

Direct evidence that the eggs of oviraptorids were shelled and laid sequentially comes from a fragmentary specimen from the Late Cretaceous of China that contains a pair of shelled eggs in what was presumably the pelvic canal of the animal. (Image after Sato et al., 2005.)

attributed to *Citipati* was widespread among theropod dinosaurs.

An extraordinary new discovery is vivid testimony to the existence of sequential egglaying among the maniraptorian ancestors of birds. The partial skeleton of an oviraptorid from the Late Cretaceous of China shows a pair of shelled eggs in the pelvic canal of the animal. This remarkable "snapshot" of dinosaur physiology provides unquestionable evidence that, as with birds, some lineages of theropods ovulated sequentially, thus requiring several egglaying stages to complete a clutch. This evidence also suggests that the anatomical specializations that characterize the reproductive system of birds evolved among their nonavian predecessors. An enormous Chinese nest containing 30-centimeter-long eggs laid in a similarly paired arrangement across a 3-meter-wide circle with an empty space at the center, indicates

that brooding and sequential egglaying were also common to theropods of large size. Based on statistical analyses of the distribution of the eggs in *Troodon* clutches, it has been suggested that this characteristically paired distribution was also common for this dinosaur. This interpretation, however, is problematic because the vertically arranged eggs of this maniraptoran do not show so clearly the paired distribution of the eggs of *Citipati* or other theropod dinosaurs. Interestingly, the lack of a paired distribution in the eggs of *Troodon* and other troodontids may indicate that one of the oviducts of this dinosaur had already atrophied. Such a conclusion, together with the particular brooding behavior of these dinosaurs (sitting directly on the egg-clutch) and characteristics of their eggs (asymmetric poles and other features) has led Gerald Grellet-Tinner—one of my former doctoral students at the Natural History Museum of Los Angeles

The fine-grained rocks of the Sihetun quarry have yielded a wealth of extraordinary fossils of Early Cretaceous age, including precious nonavian theropods and birds. These fossils, together with many others from localities spotting the western corner of Liaoning Province in northeastern China, are part of what is known as the Jehol Biota—a name derived from the site of an imperial palace of the Qing Dynasty. With thousands of exquisite fossils representing most major groups of land and freshwater vertebrates, and an enormous variety of invertebrates and plants, the Jehol Biota is the best window we have on an Early Cretaceous terrestrial ecosystem. These remarkable fossils are contained in two packages of rocks (or stratigraphic formations) that were deposited primarily on the bottom of a system of extensive and shallow lakes: the Yixian and Jiufotang formations. The thousands of meters of rocks that constitute these ancient lake systems are formed by alternating layers of multicolored shales, sandstones, conglomerates, volcanic ash, and basalt, which have been radioisotopically dated as between 128 and 120 million years old. Evidence of abundant volatile substances in the volcanic ash point at toxic emissions as the leading factor responsible for the mass mortality events inferred from the numerous fossils contained in these rocks. Most of the fossils recovered from the Jehol deposits have been collected by local farmers. Unfortunately this has resulted in critical anatomical information being lost, either during the excavation process or in subsequent handling of the fossils; but at the same time, this new way of making a living has resulted in an unprecedented number of discoveries. Today, dozens of specimens of nonavian theropods and hundreds of fossil birds have been collected from the fossil-rich deposits of western Liaoning. Although many specimens have ended up in the hands of commercial dealers and private collectors, inaccessible to most researchers, many of them have also found their way into Chinese and other public institutions. (Photo: M. Ellison.)

The small troodontid *Mei* provides evidence that some of the nonavian predecessors of birds had evolved similar resting behaviors. (Image: M. Ellison.)

corroborated predictions based on the anatomical similarities afforded by skeletal evidence alone.

As more and more fossils are uncovered, other lines of evidence are also converging to support the theropod origin of birds. The unearthing of the small troodontid *Mei long* has shown that at least some of the precursors of birds had already evolved the stereotypical resting pose so commonly seen in geese, swans, and many other birds. Several fossils of *Mei* have been discovered, all of them with the skeleton arranged such that the hindlimbs are flexed beneath the belly, the neck is turned backwards, and the head is tucked between a wing and the body. Similar poses have been found in other troodontids, such as the Early Cretaceous *Sinovenator* and *Sinornithoides*.

Other extraordinary fossils have forged yet another line of evidence supporting the theropod origin of birds— one that for many has been even more persuasive. Exquisitely preserved fossils from China have revealed the delicate structures of the plumage that covered the body of coelurosaur dinosaurs. But before turning to these exceptional fossils, let us look at the latest wave of arguments mounted against the theropod origin of birds.

County—to describe the reproductive physiology and behavior of troodontids as the most avian-like among dinosaurs.

The discovery of embryos inside eggs, complete egg-clutches, and nesting adults of theropods provided an enormous boost to the notion that birds evolved from these dinosaurs. Not only did the studies of these exceptional fossils document a remarkable similarity in the shell microstructure and general appearance of the eggs, but they also led to inferences about the nesting behavior and physiology of these dinosaurs. Furthermore, the conclusions derived from the study of these fossils

Devil's Advocates: Renewed Arguments against the Theropod Hypothesis

Despite the significant volume of evidence in support of the theropod hypothesis of bird origins, critics remain. One issue of contention has centered on the time gap that exists between the oldest known bird, the Late Jurassic *Archaeopteryx*, and the theropods typically used in discussions of bird origins—the Early Cretaceous *Deinonychus* or the Late Cretaceous *Velociraptor* and *Oviraptor*. This criticism has become known as the "temporal paradox," since it highlights the inconsistency of arguing that birds

evolved from organisms that lived many millions of years after the origin of birds took place. The "temporal paradox," however, fails to account for the fact that none of these Cretaceous dinosaurs is truly regarded as the direct ancestor of birds. In fact, the very essence of cladistic methodology opposes hypotheses that establish direct ancestor–descendant relationships. Instead, the genealogical hypotheses derived from cladistic analyses establish relationships of common ancestry. Thus, modern cladistic hypotheses supporting the theropod origin of birds postulate the existence of an ancestor common to these Cretaceous dinosaurs and *Archaeopteryx*—an ancestor that obviously existed before the differentiation of the oldest of these animals, *Archaeopteryx*. These hypotheses support the sister group relationship between *Archaeopteryx* (and birds) and those Cretaceous maniraptorans. To argue that because some of the theropods interpreted as birds' closest relatives have been discovered in rocks younger than those containing the earliest birds, the latter group could not have descended from a theropod dinosaur, has the same logical inconsistency as arguing that you cannot have an uncle younger than yourself. In other words, the "temporal paradox" is simply a misinterpretation of the cladistic hypothesis of a theropod origin of birds.

The fallacy of the "temporal paradox" has also been demonstrated from empirical grounds. Paleontologists Christopher Brochu (University of Iowa, Iowa City) and Mark Norell have statistically compared the stratigraphic record of the "thecodont," crocodylomorph, and theropod hypotheses

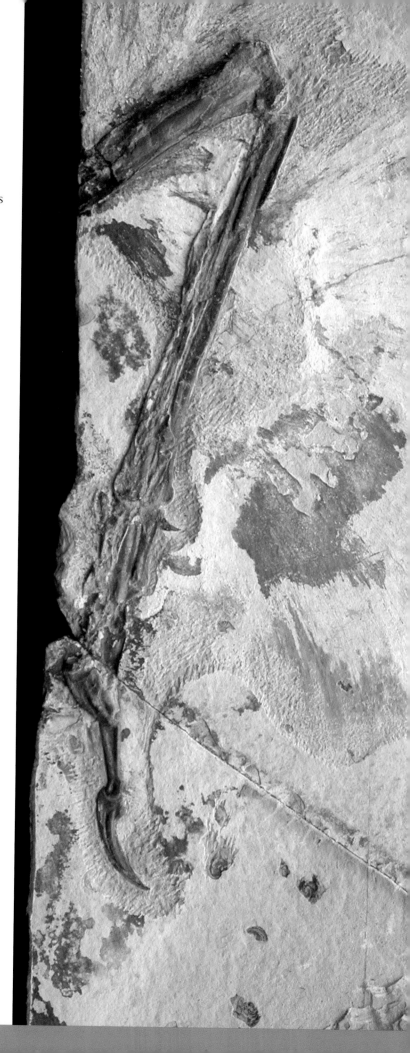

The Chinese nonavian maniraptoran *Pedopenna* is possibly a contemporary of (or older than) the Late Jurassic *Archaeopteryx*. The feet of *Pedopenna* carried long, vaned feathers. (Image: Xu Xing.)

of bird origins. These researchers have discovered that the theropod hypothesis is stratigraphically more consistent than the other two hypotheses. Indeed, if the ancestor of birds were to be found among Triassic "thecodonts" or other basal archosauromorphs, the temporal gap would be two to three times greater than the one highlighted by the "temporal paradox." In addition, advocates of the "temporal paradox" overlook evidence suggesting that maniraptorans originated before the Cretaceous, even though fragmentary Late Jurassic fossils identified as maniraptorans have been known for more than two decades. The small Chinese maniraptoran *Pedopenna daohugouensis* is one of the Late Jurassic nonavian theropod relatives of birds neglected by advocates of the "temporal paradox." More recently the therizinosaurid *Eshanosaurus degunchiianum* was unearthed from Early Jurassic rocks in southwestern China. Although only a single lower jaw was found, evidence from its overall anatomy as well as from the shape, number, and size of its teeth strongly supports the idea that *Eshanosaurus* was a therizinosaurid theropod. Information from a diversity of Cretaceous fossils indicates that this group of large, long-armed dinosaurs is

closely related to the oviraptorids and definitively nested within maniraptorans. This discovery suggests that maniraptoran theropods differentiated long before the known history of birds.

Paleontology is plagued with examples in which the oldest fossil record of two closely related groups is separated by a large temporal gap. Evolutionary biologists have little doubt that chimpanzees are the closest relatives of humans. Nonetheless, the fossil record of humans is much older than that of chimpanzees. One would have great difficulty arguing against the human–chimpanzee evolutionary connection on the basis of this human-equivalent "temporal paradox." The mismatch between the oldest fossil record of closely related lineages could stem from a number of factors. Among the most critical ones are the size of the fossils, the environments these animals inhabited, and the differential preservation of sedimentary rocks over time. It is well-known that larger animals are usually better preserved than smaller ones, so it is no surprise that small theropod dinosaurs—usually interpreted to be more closely related to birds—are rarer in the fossil record. It is also well-known

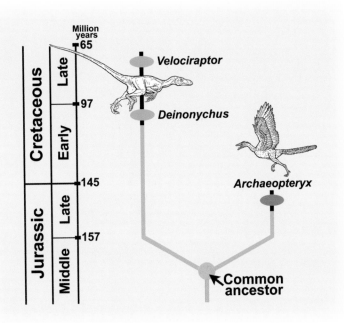

Objections to the theropod origin of birds based on the so-called "temporal paradox" can be dismissed because the current hypothesis argues that animals such as *Velociraptor* and *Deinonychus* shared a common ancestor with *Archaeopteryx* as opposed to being the direct ancestors of birds. The fact that these nonavian maniraptorans lived millions of years after *Archaeopteryx* is irrelevant in testing the validity of the theropod hypothesis of bird origins. (Image: E. Heck/R. Urabe.)

that certain environments are poorly represented in the rock record—one of these is the humid forests inhabited by chimpanzees, which may explain the virtual absence of these apes from the fossil record. Another well-known fact is that the amount of continental sediments preserved in the rock record of a given geologic period depends on the amount of emerged lands that existed at the time. As we saw earlier, the Jurassic was a period of high sea levels in which vast continental areas were flooded. In North America, the volume of Jurassic continental sediments—rocks formed mostly in riverbeds and lakebeds—represents only 25 percent of those formed during the Cretaceous. Furthermore, taking into account that the Jurassic Period only spanned about two-thirds as much time as the Cretaceous, the volume of rocks available for the preservation of the small remains of Jurassic theropods closely related to birds is only a fraction of that available in the Cretaceous.

Another criticism of the theropod hypothesis of bird origins stands on inferences about the lung structure and ventilation of nonavian theropod dinosaurs. Two types of lung ventilation exist among extant archosaurs. On the one hand, in crocodiles, the lungs are ventilated by means of muscles that primarily originate from the pubic bone and attach to a vertical wall of connective tissue—a non-muscular analogue of the mammalian diaphragm—that separates the lungs from the liver and other viscera. During inhalation, these muscles pull this vertical visceral wall and create a negative pressure that inflates the lungs. On the other hand, in birds, the lungs are ventilated by up-and-down movements of the back portion of the sternum, which are assisted by the hinge-like motions of the two-part ribs that form the ribcage. Living birds have a highly efficient type of flow-through respiratory system in which the lungs are connected to a series of air-filled sacs, where the air flows unidirectionally. The type of lung and ventilation mechanisms of extinct archosaurs, such as nonavian theropods, remains largely unknown. It should be noted that Oregon State University's John Ruben and his collaborators reconstructed the visceral cavity of these dinosaurs as having a partition between the lungs and the rest of the viscera, arguing in favor of the presence of a liver-piston mechanism for airing the lungs similar to that of crocodiles. Their argument was primarily based on a structure preserved in the type specimen or holotype (the specimen used for the description of a new species) of the Chinese feathered theropod *Sinosauropteryx* and assumed to constitute evidence of a division

The black stain within the body of this specimen of *Sinosauropteryx* (arrow) was used to infer the presence of a liver-piston mechanism for lung ventilation in this and other nonavian theropods. Subsequent studies of this and other fossils of *Sinosauropteryx* have shown this feature to be a preservational artifact. The body of *Sinosauropteryx* was covered by fuzzy integumentary structures interpreted as filamentous feathers (inset). (Image: M. Ellison.)

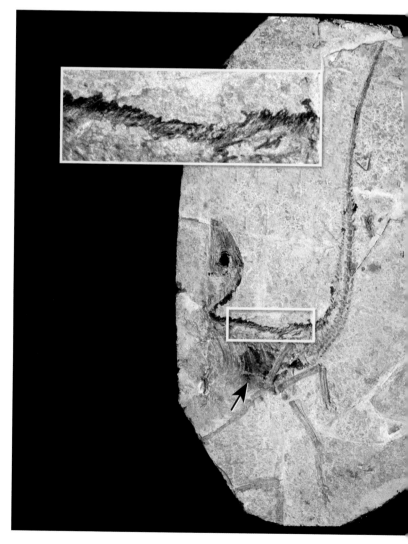

within the visceral cavity of this animal. They argued that the transition from a crocodile-like ventilation mechanism to the one characteristic of birds required the evolution of an opening in the wall dividing the visceral cavity of nonavian theropods, an event that would have compromised the breathing efficiency of the transitional forms.

Several lines of reasoning counter the arguments proposed by Ruben and his collaborators. The structure assumed to be evidence of a partition between thoracic and abdominal cavities in *Sinosauropteryx prima* has been shown to be an artifact of the preservation of the holotype. The structure in question consists of a dark stain with an anterior rounded end, while its other margins fade

away into the lighter color of the rock matrix. As noted by paleontologist Phil Currie of the University of Alberta in Edmonton, Canada, close examination of the anterior margin reveals that the stain is interrupted by a break and that its shape is clearly a consequence of the uneven split of the specimen into two slabs. None of the other known specimens of *Sinosauropteryx* shows evidence of such a structure, even though soft tissues are preserved in all of them. Currie concluded that the stain is likely to represent gut contents and ruled out the claim of a connection with the shape of the abdominal cavity. Moreover, a close look at the skeletal evidence of nonavian theropods shows that several bone correlates of the avian ventilation system were already present in these dinosaurs. For example, the general design of the ribcage and sternum of those theropods thought to be closely related to birds suggests that the coordinated movements of the ribs and sternum that ventilate the lungs of extant birds were present in their most immediate relatives. The ribs of these dinosaurs are formed by the same two bony segments that characterize the ribcage of birds—one joining the sternum and the other joining the vertebral column. These two segments are connected to one another by a movable joint that must have allowed the up and down movements that pump the lungs of extant birds. At the same time, the sternum of maniraptorans like dromaeosaurids and oviraptorids is formed by two large bony plates, a condition that also approaches what we see in birds. Thus, the skeletal evidence of nonavian theropods closely related to birds indicates that, unlike crocodiles, the ribs and sternum of these dinosaurs must have played an important role in the ventilation of the lungs. Other bone correlates of modern avian ventilation are found among nonavian theropods, even those that are not so closely related to birds. For example, small holes similar

The resemblance between the ribcage and sternum of nonavian maniraptorans and birds suggests similarities in their breathing mechanism. The dromaeosaurid *Velociraptor* shows the large sternum (sternal plates), two-segment ribs, and uncinate processes typical of birds. (Image: L. Chiappe.)

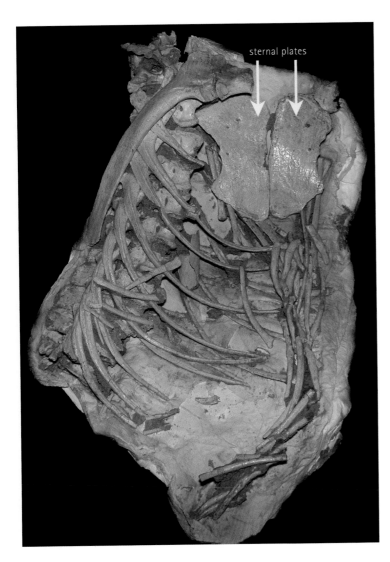

sternal plates

to the openings for the entrance of air-filled sacs connected to the lungs of modern birds pierce the vertebrae of many nonavian theropods; these holes suggest that similar air-filled sacs had already evolved among the predecessors of birds. This striking resemblance supports the idea that the respiratory system of nonavian theropods was much more similar to that of birds than to that of crocodiles—the presence in these dinosaurs of skeletal features correlated to the avian system of lung design and ventilation is undeniable.

Finally, if the respiratory system of nonavian theropods were assumed to be similar to that of crocodiles, it would be most parsimonious to reconstruct in the same fashion the breathing mechanism of a great number of archosaurian lineages, in particular those closer to the common ancestry of crocodiles and dinosaurs. Such an inference would make sense given the fact that some of the skeletal structures Ruben and associates interpreted as correlates of the crocodile type of lung ventilation (long, vertical pubis, and well-developed gastralia—a zigzag system of rod-like bones that line the belly of theropod dinosaurs and some other reptiles) are also present in these basal archosaurs. Therefore, the caveat raised by these researchers would have to also be valid for any hypothesis reconstructing the origin of birds from archosaurs closely related to crocodiles (crocodylomorph hypothesis) or from even more basal archosauromorphs ("thecodont" hypothesis). Only if birds were to have had their origin in a species whose ancestry lies far outside archosauromorphs could this caveat be bypassed. However, such a genealogical hypothesis is not sustained by any modern researcher.

Perhaps the most critical caveat to the theropod origin of birds does not come from the study of fossils or the perceived significance of the temporal gaps, but from the development of the hand of the avian embryo.

"One, Two, Three" or "Two, Three, Four": The Apparent Conflict between Paleontology and Embryology

Living birds have a highly modified hand with only three abbreviated fingers. Because birds are tetrapods, it is logical to assume that these fingers correspond to three of the five fingers of the pentadactyl hand of primitive tetrapods. Although one would think the discovery of such a correspondence an easy task, the problem of the homology of the avian fingers remains far from settled.

Embryological versus paleontological approaches to the homology of the three digits of the hand of adult birds. While the embryological evidence suggests that the fingers of modern birds correspond to digits II, III, and IV of the pentadactyl hand of primitive tetrapods, the paleontological evidence suggests a I, II, III correspondence. (Image after Larsson and Wagner, 2002.)

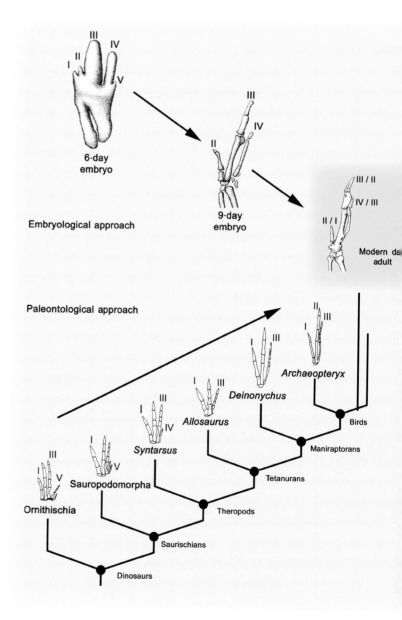

Paleontologists have approached the problem by identifying trends of digital reduction that begin with fossils whose hands have five fingers. The relative size and number of phalanges (finger bones) of each digit allow identification of the fingers of the hand of the earliest dinosaurs. In these primitive forms, the pentadactyl design of more archaic archosaurs is still traceable, even if the two most external fingers—the fourth and the fifth—are greatly reduced. In the basalmost theropods such as *Herrerasaurus* from the Late Triassic of Argentina, the fourth finger is reduced to a short, rod-like metacarpal that carries a single phalanx and the fifth finger is represented only by its small metacarpal. The well-known fossil record of theropod dinosaurs shows that these two fingers disappeared prior to the differentiation of tetanurans, whose hand bears only the first, second, and third fingers. No major changes are seen in the hand during the transition between nonavian maniraptorans and birds. The general appearance, relative size, and number of phalanges of the three fingers of oviraptorids, troodontids, and dromaeosaurids are strikingly similar to those of the hand of *Archaeopteryx*, a fact that has long supported the idea that the fingers of the earliest bird correspond to the first, second, and third digits of the five-fingered hand of primitive tetrapods. The subsequent evolutionary steps leading to the abbreviated three fingers of the hand of modern birds are also well-documented in the fossil record. In birds more advanced than *Archaeopteryx*, such as the Cretaceous enantiornithines, the innermost finger became shorter and the outermost one reached a degree of abbreviation comparable to that of extant birds. As birds evolved closer to their living counterparts, the innermost and middle fingers became even shorter, and the first phalanx of the latter developed into a flat and broad bone well-suited for anchoring the flight feathers of the tip of the wing. This stage is well-documented

in *Ichthyornis dispar*, a marine, toothed bird from the Late Cretaceous of North America. Several other Late Cretaceous fossils show that by this time, animals very close to the ancestry of all living birds had already evolved a hand virtually indistinguishable from that of their modern counterparts.

In contrast to the inferences derived from the fossil record, embryological investigations of extant birds have consistently identified the fingers of the hand as corresponding to the second, third, and fourth digits of the pentadactyl hand of early tetrapods. Until recently, embryologists defended such a view primarily on the basis of a developmental principle known as Morse's rule—in reference to the embryologist who first formulated it. This classic notion argued that when digits are reduced among amniotes, the first ones to be lost are the outermost and innermost fingers—that is, the fifth and first of the pentadactyl hand, respectively. For most of the 20th century, embryological studies documented the presence of only four condensations (aggregations of cartilaginous cells preceding the formation of a bony element) in the earliest stages of development of the avian hand. Thus, when followers of Morse established the homology of the three fingers that characterize the hand of adult birds, they did so assuming that the fifth and first digits became completely reduced during the early development of the embryo: the fifth did not even form a condensation and the first disappeared soon after its condensation was developed. Yet, because Morse's rule has its exceptions, the practice of identifying the three fingers of the adult hand as the second, third, and fourth when the embryo developed only four condensations did not convince everyone in the field. New embryological studies, however, have brought this issue closer to resolution.

These studies provide direct evidence indicating that very early in their development, extant birds possess a

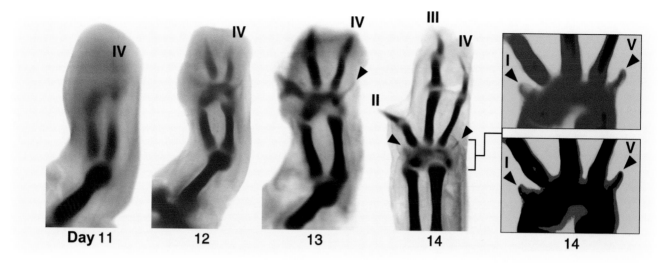

Day 11   12   13   14   14

pentadactyl framework of five digital condensations. They show that when the fate of these condensations is followed throughout the development of the embryo, the fingers that ultimately develop correspond indeed to the second, third, and fourth positions of the primitive tetrapod hand.

It is certainly tempting to claim that the discovery of five early condensations, of which the middle three develop into adult fingers, has closed the decades-old debate about the digital homology of the hand of birds, but this does not seem to be the case. New molecular evidence contradicting these recent embryological advances has emerged—contrary to expectations, it supports the correspondence between the most internal digit of the hand of living birds and the first (thumb) finger of the pentadactyl hand. The molecular studies are based on the well-documented differential expression of two regulatory genes (Hox12 and Hox13) during the late development of the hands and feet of embryonic mice and chickens. Regulatory genes are genes that control the expression of other genes. In mice and chickens, while Hox13 is expressed in the development of all digits, Hox12 is expressed in all but the first digit. And because the identification of the fingers of mice and the toes of mice and birds is uncontroversial, the discovery that the development of the hand of the chicken follows the same signature pattern of

regulatory gene expression implies that the innermost finger of the hand of birds—whose development depends only on the expression of Hox13—is indeed homologous to the first finger of the pentadactyl hand. Even more surprising is the fact that these molecular conclusions also suggest that birds could have the genetic make-up of a sixth digit, an idea already entertained by some molecular researchers.

While these developmental studies have moved beyond decades of controversy regarding the correspondence of the digits of the hand of extant birds, they have not unraveled the correspondence of the fingers of the most primitive birds. Even though some of the developmental studies have also extrapolated their conclusions to more primitive birds (including *Archaeopteryx*), there are neither empirical nor logical bases for such extrapolation. Not only

Development of the hand of an embryonic ostrich between the 11th and 14th days of incubation (right hand in dorsal view). Arrows point at condensations identified as digits I and V. Insets show high-contrast details of the wrist of the embryo on its 14th day of incubation. (Image: A. Feduccia.)

Differential expression of the regulatory genes Hox12 and Hox13 in the development of the anteriormost digit of the hand of an embryonic chicken. The arrow points at the positive expression (identified by the blue marker) of Hox13—this pattern suggests that the anteriormost finger of the chicken corresponds to the first digit of the pentadactyl hand of primitive tetrapods. (Image: A. Vargas.)

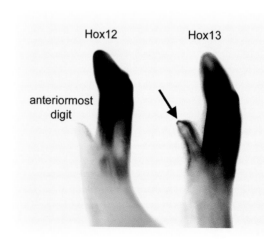

Hox12   Hox13

anteriormost digit

do we know nothing of the development of the hand of early birds, but also it is quite possible that the developmental patterns of the hand could have changed during the long evolutionary history of the group. Thus, the problem of whether the fingers of extant birds correspond to those of the hand of dromaeosaurids and other nonavian maniraptorans is the same as that of whether the fingers of *Archaeopteryx* and modern birds have the same correspondence to the pentadactyl hand. The answer to such a conundrum is not readily available because the hand of modern birds is highly transformed and embryological information is unavailable

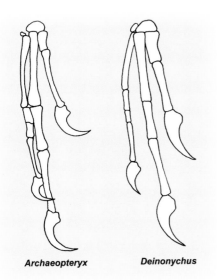

Archaeopteryx   **Deinonychus**

The remarkable anatomical resemblance between the hands of *Archaeopteryx* and the dromaeosaurid *Deinonychus* supports the homology of the fingers of these animals. Given the absence of embryological data for any of these species, there is no reason not to assume the same correspondence to the pentadactyl hand of primitive tetrapods for both *Archaeopteryx* and *Deinonychus*. (Image: S. Abramowicz.)

for *Archaeopteryx*. A myriad of examples in developmental biology has shown how developmental pathways could evolve in concert with the evolution of a particular group of organisms. Therefore, the discrepancy between the evolutionary correspondences of the three fingers of modern birds with those inferred for the hand of their theropod predecessors may well be explained by developmental evolution. If so, the embryological research on the hand of extant birds is irrelevant to the understanding of bird origins; it bears only on their later evolution, because the extrapolation of the developmental pattern observed in modern birds to all birds (including *Archaeopteryx*) is unwarranted.

Despite the fact that extrapolating the pattern of finger development in living birds to their 150-million-year-old relatives is unjustified, how could it be that with all the disparate lines of evidence favoring the theropod origin of birds, the correspondence of the fingers of living birds does not match that inferred for their maniraptoran forerunners? Is there an explanation for a developmental pattern so seemingly different? Luckily, yes. One way in which this difference has been explained centers upon a developmental pattern known as a "frame shift." Within this theoretical framework, the development of the hand of living birds transforms during evolution, in such a way that the three functional fingers that initially developed from the first, second, and third condensations, now, in extant birds, develop from the second, third, and fourth condensations. Regardless of how unintuitive this may sound, studies on the development of vertebrates indicate that such transformations, or developmental frame shifts, are perfectly possible. Indeed, a diversity of frame shifts has been identified in the development of animals, from changes in the regional identity of vertebrae to shifts in the teeth of the human lineage. An illustrative example involves the development of the hand of the two-toed earless skink. In this small lizard from Australia, the digits morphologically corresponding— meaning that they share the same number of phalanges and overall appearance—to the first and second fingers of its most closely related species actually develop from the second and third condensations. The similarity between the fingers of the two-toed earless skink and those of its closest relative is such that without embryological information, it would be impossible to tell that the hands of these two lizards have different developmental histories. Similarly, molecular and embryological manipulation of embryos has shown that the identity of condensations is not a fixed attribute of

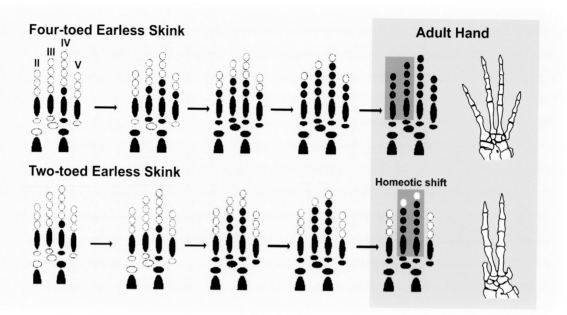

Four-toed Earless Skink

Adult Hand

Two-toed Earless Skink

Homeotic shift

the digits, and this has furthered our understanding of developmental frame shifts.

The debate about the homology of the fingers of the avian hand highlights the methodological chasm between researchers endorsing the cladistic notion of homology, which is based on how congruent an interpretation is with similar interpretations about other structures, and those using a different concept of homology, based on the similarity of developmental pathways. In other words, while cladistic approaches view homology as the result of congruence and parsimony—the congruent signal given by the study of bones, integument, eggs, and inferred behaviors of nonavian theropods and birds—other approaches view it only in light of developmental pathways, as if these were frozen in time since they first appeared. Yet developmental pathways are far from being evolutionary time capsules. In the end, the apparent conflict between the paleontological and embryological understandings of the correspondence of the avian hand also highlights the fact that for nearly the entire history of the controversy about bird origins, the debate has been less about the evidence than about the methods used to interpret it.

Feathered Dinosaurs! The Ultimate Evidence

Feathers are a remarkable evolutionary novelty and the quintessential bird feature. They are beta-keratin appendages produced by the outermost layer of the skin, which develop and grow as the result of a series of complex cellular transformations. Beta-keratin is a variant of the structural protein keratin, which is present only in reptiles. This beta-keratin is much more resistant than the softer alpha-keratin that forms the skin of all vertebrates, including our hair and fingernails. Feathers grow as skin projections from surface invaginations called follicles (see *Feathers for Tyrannosaurus rex*). Although they exhibit a wide array of shapes and functions, most feathers are multi-branched structures composed of a central shaft or rachis, numerous pairs of barbs branching from the rachis, and tiny filament-like barbules that diverge from the barbs. Feathers are attached to the skin by the quill or calamus, a hollow, tubular expansion of the rachis, and their movements are controlled by a complex network of delicate muscles. The vanes of typical feathers—flight and body feathers—are formed by the planar arrangement of barbs on both sides of the rachis. The

An example of a "frame shift" in the development of the hand of two closely related scincid lizards from Australia. In the Four-toed Earless Skink (*Hemiergis perioni*), condensations (white circles) of fingers II and III develop into the two anteriormost digits of the adult hand (top row, brown box), which have three and four phalanges (black circles), respectively. Adults of the Two-toed Earless Skink (*Hemiergis quadrilineata*) have only two fingers, the anteriormost of them with three phalanges and the other one with four phalanges. While in *Hemiergis perioni* the two anteriormost fingers—those with three and four phalanges—ossify from condensations II and III, in *Hemiergis quadrilineata*, these digits ossify from condensations III and IV. The anatomical similarity between the fingers of the adults of these lizards is such that the identity of the two digits of *Hemiergis quadrilineata* can only be verified through developmental studies. (Image after Shapiro, 2002.)

cohesion of vanes is provided by the interlocking of barbules, achieved by the differentiation of the barbules of adjacent barbs into hooked and grooved ones—a Velcro-like system that is usually limited to the central and basal regions of the vane.

The idea that feathers existed in the forerunners of birds is not new. Given that *Archaeopteryx* is the most primitive bird and that its feathers are already differentiated into structurally modern types, it has always been clear that feathers had to be found in the theropod relatives of birds. Few researchers, however, thought they would witness such a momentous discovery.

The first direct evidence that feathers evolved outside the context of birds appeared in 1996, when a small, short-armed and long-tailed theropod skeleton surrounded by a dark halo of filament-like structures was unearthed near the rural village of Sihetun, in the northeastern Chinese province of Liaoning. This remarkable fossil had been dug out by local farmers from a mudstone layer belonging to the Yixian Formation, a thick series of mostly sedimentary rock layers formed at the bottom of a 125-million-year-old lake. Although these beds had produced numerous fossils of invertebrates and fish since first surveyed by European and Japanese geologists in the 1920s, the theropod skeleton was the first of its kind.

Like most fossils from the Yixian deposits, the skeleton of this small, 60-centimeter-long theropod dinosaur was split in two halves along its plane of symmetry, with each half contained in a slab. Aware of the uniqueness of the discovery, and presumably of its commercial value, the local farmers approached paleontologists both in Beijing and Nanjing, and the specimen was split between the extensive collections of two major Chinese institutions (one slab went to the National Geological Museum of China in Beijing and the other to the Nanjing Institute of Paleontology). Teams from these institutions raced to share the discovery with the scientific community. The Beijing team first published a short article in Chinese, naming the specimen *Sinosauropteryx prima* and regarding it as a very primitive bird. The Nanjing team followed up with a more detailed study of their slab as well as of another, larger specimen. This second skeleton was lined with identical filament-like structures and also contained the remains of a meal (a lizard skeleton in its guts). The later discovery of an even larger specimen, with a length exceeding 1.2 meters, indicated that the first two skeletons of *Sinosauropteryx* were those of immature individuals. Most importantly, subsequent research showed that the reduced forelimbs, very long tail, and other characteristics of *Sinosauropteryx* did not support its inclusion within birds but next to *Compsognathus*—the primitive coelurosaur that Thomas Henry Huxley had used in promoting his view of a theropod ancestry of birds.

Published in the British journal *Nature*, the report of *Sinosauropteryx* by the Nanjing team made headlines all over the world and triggered an invasion of tiny Sihetun by television crews and reporters covering one of the most significant paleontological discoveries of the century. The reason was clear: although the Nanjing team interpreted *Sinosauropteryx* as a coelurosaur theropod closely allied to the Late Jurassic *Compsognathus*, the dark halo of filament-like structures surrounding its skeleton was considered to be nothing less than the remains of the animal's plumage. At last, predictions about the discovery of feathers in the forerunners of birds had been fulfilled. The quintessential avian attribute had been discovered in an animal that was neither a crocodile nor a basal archosauromorph but a theropod dinosaur. Yet skeptics thrived and the debate over *Sinosauropteryx* became vitriolic—after all, was the interpretation of the filamentous covering as feathers justifiable?

ng vane — trailing vane

— rachis

barbs—

calamus—

Flight feather

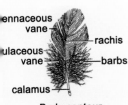

ennaceous vane

— rachis

ulaceous vane

— barbs

calamus—

Body contour

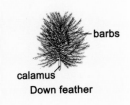

— barbs

calamus

Down feather

Anatomical components of feathers. (Image after Lucas and Stettenheim, 1972.)

The structures surrounding the skeletons of *Sinosauropteryx* are most evident as a fluffy crest on top of the head, neck, back, and both sides of the tail of these animals, clearly indicating they were some kind of integumentary coverings. These structures vary greatly in length, being about 1 centimeter long over the skull, a bit longer along the neck, and some 4 centimeters on the back and base of the tail of the second discovered specimen. The distance between the line of integument and the underlying bones also varies across the skeleton, seemingly in accordance with the expected volume of muscular and connective tissue separating the skin from the bones in different parts of the body. For example, the structures closely line the skull and pelvis (areas that in modern birds have nearly no flesh) and they are far apart from the skeleton at the base of the neck and tail. At first glance, these structures seem to lack all the features typical of modern feathers. Nonetheless, close examination reveals that like a modern calamus or quill, the bases of the main filaments are hollow. In addition, high magnification images show that in some cases, very fine filaments branched off from these calamus-like structures. In spite of their lack of vanes, the general appearance of these structures convinced many that distant theropod relatives of birds had already evolved plumage. Even so, alternative views explaining the nature of these structures were promptly formulated. One of these alternatives was proposed by a team led by John Ruben, who argued that the structures covering the body of *Sinosauropteryx* were not epidermal but instead the remains of frayed subcutaneous fibers of the protein collagen that either reinforced the skin or formed a crest along the back of these animals. This argument was soon embraced by other opponents of the theropod hypothesis of bird origins, namely Larry Martin and Alan Feduccia. Such a view, however, contradicts the observation that the line of integument is not evenly separated from the skeleton. Rather, it is in proportion to the amount of flesh that would have been present in the different parts of the animal. This observation can best be explained if the structures originated on the surface of the skin. Another explanation for the nature of these structures argued that they were ossified tendons. Yet this claim does not recognize that the structures are not made of bone and are too numerous.

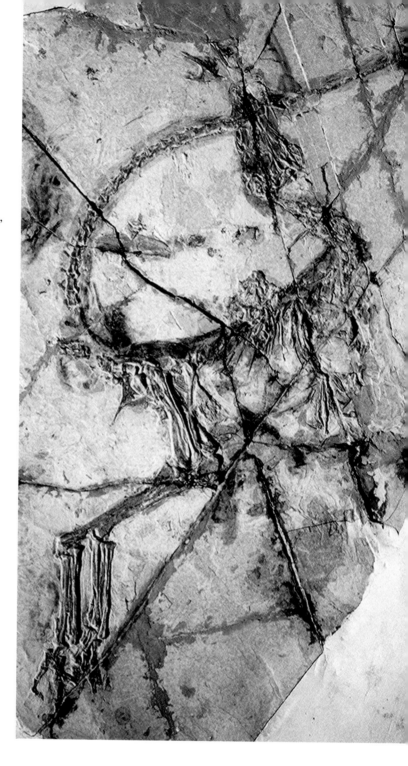

The nonavian coelurosaur *Sinosauropteryx*. The dark halo surrounding parts of the skeleton of this specimen is interpreted as part of the animal's plumage. (Image: M. Ellison.)

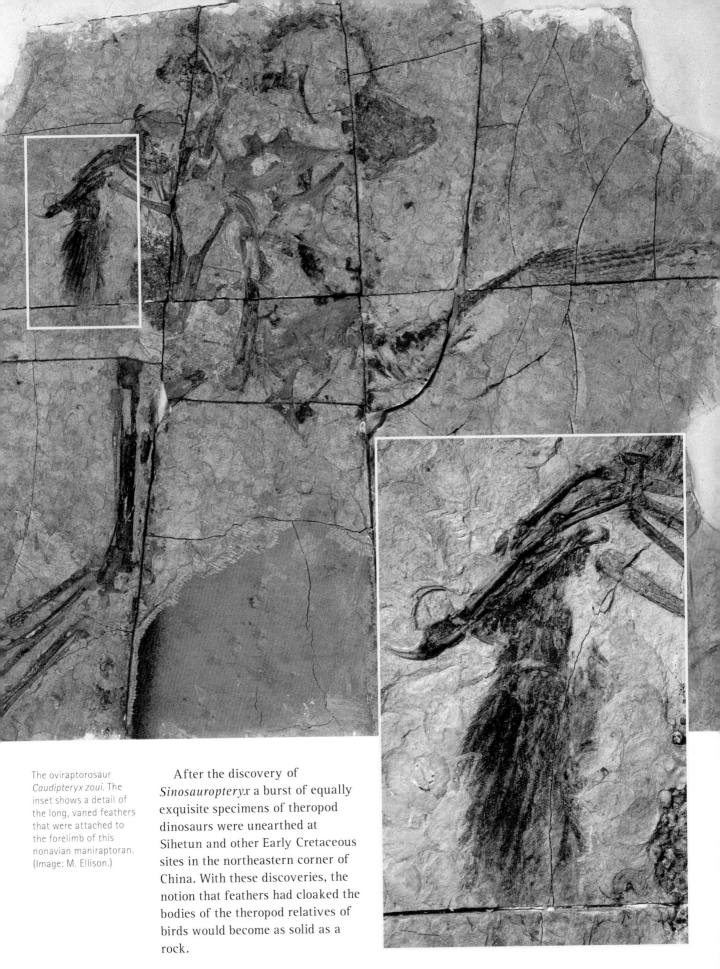

The oviraptorosaur *Caudipteryx zoui*. The inset shows a detail of the long, vaned feathers that were attached to the forelimb of this nonavian maniraptoran. (Image: M. Ellison.)

After the discovery of *Sinosauropteryx* a burst of equally exquisite specimens of theropod dinosaurs were unearthed at Sihetun and other Early Cretaceous sites in the northeastern corner of China. With these discoveries, the notion that feathers had cloaked the bodies of the theropod relatives of birds would become as solid as a rock.

The nonavian maniraptoran
Protarchaeopteryx. The
inset details a series of long,
vaned feathers that probably
attached to the tail of this
animal. (Image: M. Ellison.)

Close-up of the filament-like feathers that attached to the forelimb of the therizinosaurid *Beipiaosaurus*. (Image: L. Chiappe.)

A Feathered Cornucopia

In 1998, less than two years after the original description of *Sinosauropteryx*, two other spectacular feathered dinosaurs were reported. One of them, the turkey-sized *Caudipteryx zoui*, had long arms, a very short tail, and a tall skull bearing only a few long and procumbent teeth on the upper jaw. The body of this animal was covered by short, down-like plumulaceous feathers—but most significantly, at least 14 symmetrically vaned feathers were attached to its hand. Although proportionally much shorter than the wing feathers of *Archaeopteryx* and modern flying birds, the longest of these feathers were close to 10 centimeters in length. A tuft of similarly vaned feathers also fanned out at the end of its tail, with each of the last five or six vertebrae carrying a pair.

Initially, *Caudipteryx zoui* was thought to be very close to the ancestry of birds. Later studies, however, pointed out that a suite of characteristics in the skull, lower jaws, pelvis, tail, and other parts of the skeleton better supported its placement within the oviraptorosaurs, as a distant relative of the parrot-headed oviraptorids. Since this initial report, more than a dozen specimens of *Caudipteryx* have been discovered in several 125-million-year-old localities of Liaoning. Some are believed to represent a second species, *Caudipteryx dongi*, an animal slightly larger than *Caudipteryx zoui* but with a relatively smaller sternum and different ratios in various other bones. The discovery of these new specimens improved our understanding of what *Caudipteryx* looked like. The new finds reveal, in some specimens, vaned feathers of the forelimb reaching lengths of nearly 20 centimeters (comparable to modern birds), with the longest feathers attached to the wrist area,

FEATHERS FOR *TYRANNOSAURUS REX*

Today, nine species of feathered, nonavian theropods have already been reported. These dinosaurs encompass a wide range of shapes and sizes, and represent groups spread throughout the coelurosaur portion of the theropod genealogical tree. Undoubtedly, China's feathered dinosaurs rank among the most spectacular fossils discovered during the 20th century. With these significant fossil findings, Chinese scientists are following in the footsteps of their forerunners, who revolutionized world science with their many discoveries—gunpowder, magnetism, and paper, to mention a few. Indeed, this spectacular menagerie suggests that feathers were widespread among the dinosaurian forerunners of birds. The evidence indicates that even the colossal *Tyrannosaurus rex* (here shown attacking the horned dinosaur *Triceratops horridus*) may have been

covered with a cloak of feathers at some early stage of its life—because although for years they were regarded as close cousins of *Allosaurus* and other large predatory theropods, tyrannosaurids are now believed to be overgrown coelurosaurs with atrophied arms. The fact that these megapredators are evolutionarily

nested within a variety of feathered coelurosaurs indicates that they could have also carried plumage, and the discovery of the modest-sized *Dilong* (bottom inset shows portions of its tail feathers) has demonstrated that, indeed, some distant relatives of *T. rex* were feathered. However, most paleontologists believe that the bodies of full-grown large tyrannosaurids would not have been feathered, because the combination of this insulating covering and their enormous size might have posed a disadvantage in regulating their body temperature. A few scaly skin patches preserved in association with adult skeletons of *Albertosaurus sarcophagus*—a tyrannosaurid more primitive than *Tyrannosaurus*—seem to support this view, although these skin patches are fragmentary and do not rule out the presence of feathers over other portions of the body. (Image: R. Meier/Xu Xing.)

their size diminishing both towards the tip and the base of the forelimb. Even if the long feathers of *Caudipteryx* undoubtedly formed an airfoil, the forelimbs of these animals were proportionally too small to make these animals airborne. In fact, their long and robust hindlimbs suggest they lived an obligatory ground-dwelling existence. Numerous rounded pebbles discovered inside the bellies of these fossils suggest that *Caudipteryx* could have been a herbivore, although stomach stones are not exclusively ingested by plant-eating animals.

The second dinosaur initially reported with *Caudipteryx* was the slightly smaller but generally stouter *Protarchaeopteryx robusta*. This long-armed theropod, with fully toothed jaws and a tail longer than that of *Caudipteryx*, also exhibits feathers with strong rachises and symmetrical vanes. Like on the after-shaft of modern pennaceous feathers, the barbs at the base of the vaned feathers of *Protarchaeopteryx* are plumulaceous and the number of barbs present in the middle portions of the vanes approaches that of *Archaeopteryx*. The strikingly modern appearance of these feathers, combined with the fact that the vanes have retained their integrity, suggests that the system of interlocking hooked and grooved barbules maintaining the planar geometry of modern vanes had already developed, although these delicate structures have not been observed directly. As with *Caudipteryx*, the feathered forelimbs of *Protarchaeopteryx* may not have been large enough to propel this animal into the air. Its robust hindlimbs and the structure of its feet suggest it was also a ground-dweller. Despite the poor preservation of the only two known specimens of *Protarchaeopteryx*, the overall anatomy of this animal strongly supports its placement within maniraptoran theropods. Details of the skull and tail have led some to argue for a close relationship to *Caudipteryx*, perhaps as a primitive member of the oviraptorosaurs.

The existence of feathers has also been documented in dinosaurs substantially larger than *Sinosauropteryx*, *Caudipteryx*, and *Protarchaeopteryx*. Filament-like structures comparable to those of *Sinosauropteryx*, albeit longer, have also been found connected to the forelimbs of *Beipiaosaurus inexpectus*, an animal that exceeded 2 meters in length. These 5–7-centimeter-long filaments are arranged in parallel to one another and attached perpendicularly to the bones of the forearm. As in *Sinosauropteryx*, some of these structures show distal branching and evidence of hollowness, both characteristics of avian feathers. The structure and number of teeth, as well as similarities in the forelimb and other parts of the skeleton, indicate that *Beipiaosaurus* was a primitive therizinosaurid, a bizarre group of small-headed and often large dinosaurs, with leaf-shaped teeth, powerfully clawed hands, and robust hindlimbs. Until recently, the genealogical relationships of therizinosaurids to other dinosaurs were poorly understood. At different times, these dinosaurs have been thought to be ornithischians, prosauropods, and theropods. More recent and complete specimens, however, have greatly clarified the evolutionary relationships of this group, which is today consistently placed within the maniraptoran theropods. The discovery of the primitive *Beipiaosaurus* added support to previous studies proposing a close relationship between therizinosaurids and the parrot-headed oviraptorids.

Another relatively large feathered dinosaur was unearthed more recently. The discovery of *Dilong paradoxus*, whose adults probably reached nearly 2 meters in length, has highlighted again that feathers were not restricted to small-bodied theropods. Although little is known about its plumage, fossils of *Dilong* show that at least portions of its body were covered with branched filaments similar to those of *Beipiaosaurus* and *Sinosauropteryx*.

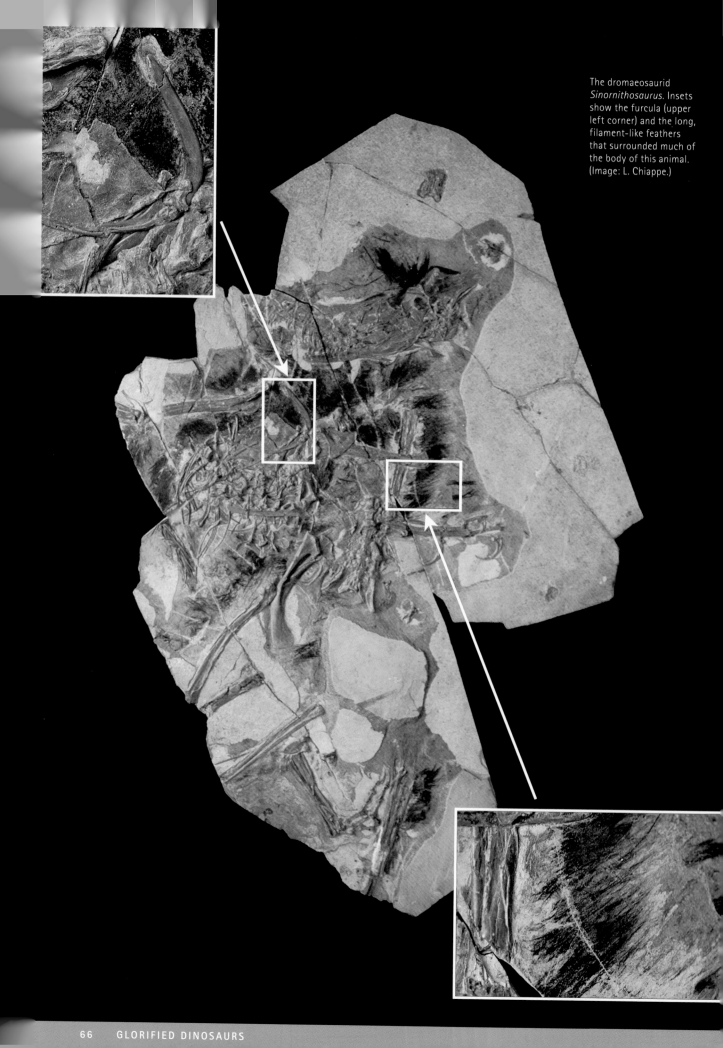

The dromaeosaurid *Sinornithosaurus*. Insets show the furcula (upper left corner) and the long, filament-like feathers that surrounded much of the body of this animal. (Image: L. Chiappe.)

However, the main significance of this newly found dinosaur relies on its genealogical relationship, because this feathered fossil has been interpreted as a long-armed, three-fingered, primitive tyrannosauroid and hence a relative of the famed *Tyrannosaurus rex*.

Not surprisingly, the reporting of these dinosaurs, in particular the exquisite fossils of *Caudipteryx*, shook the scientific community. If many had been cautious about regarding the integumentary structures of *Sinosauropteryx* as feathers, the fully vaned appendices of the maniraptorans *Caudipteryx* and *Protarchaeopteryx* left no doubt of their feather identity. Furthermore, these vaned feathers were associated in the

same specimens with structures greatly resembling those of *Sinosauropteryx*, thus lending indirect support for the feathered nature of the structures surrounding the body of the latter as well as those of *Beipiaosaurus* and *Dilong*.

The Discovery of Feathered Dromaeosaurids

Undoubtedly, the most definitive evidence in support of a theropod origin of birds comes from fossils showing vaned feathers surrounding the skeletons of animals nested within dromaeosaurids—the dinosaurs John Ostrom had presented as the closest relatives of birds. The first feathered dromaeosaurid to be discovered

Detail of the vaned feathers that also covered the body of dromaeosaurids. (Image: M. Ellison.)

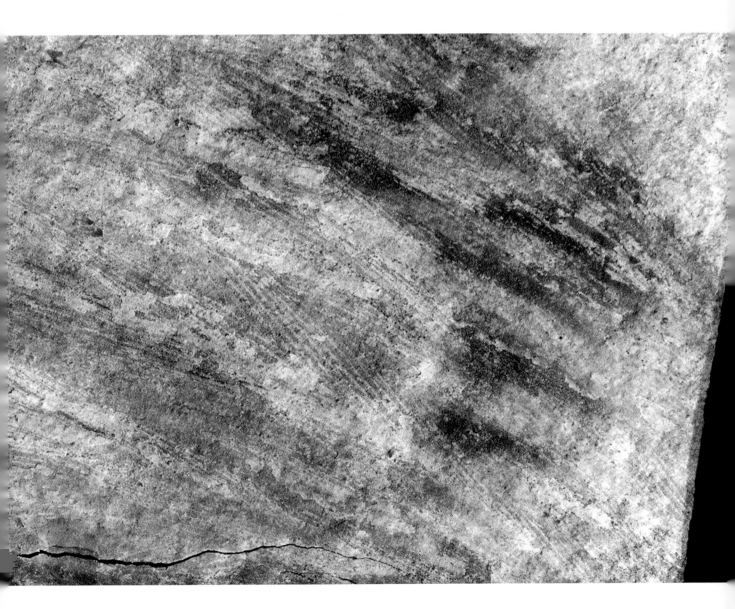

was *Sinornithosaurus millenii*, originally collected from rocks of the Yixian Formation near Sihetun. A series of features from the skull, tail, feet, and other parts of the skeleton documented *Sinornithosaurus*'s kinship with the sickle-toed *Velociraptor* and *Deinonychus*, and the 6-meter-long *Utahraptor ostrommaysorum*. Yet, being less than a meter in length, the 125-million-year-old *Sinornithosaurus* is substantially smaller than its better-known, younger relatives—and it is also more bird-like. Most of the skeleton of this fossil is covered by tufted down and 4-centimeter-long feathers with a vane-like appearance. Clearer evidence of vaned feathers was subsequently discovered in a juvenile dromaeosaurid that appears to belong to *Sinornithosaurus millenii*. The feathers in "Dave," as this specimen was nicknamed, show a distinct rachis and barbs that branch off from it, forming a pair of symmetrical vanes. The study of these important specimens has shown that the dromaeosaurid plumage exhibited a great variation in the shape and complexity of its feathers, thus demonstrating that the feather differentiation so characteristic of birds also preceded the origin of the group. Fossils of *Sinornithosaurus* appear to have been found throughout the five million years of history of the Jehol Biota (see *The Jehol Biota: The Clearest Window on a Cretaceous Ecosystem*), although this seems like a very long life for a dinosaur species and could be the result of taxonomic lumping.

Another feathered dromaeosaurid is the small *Microraptor*, whose remains are known from the youngest rocks containing the Jehol Biota, the 120-million-year-old Jiufotang Formation. The first evidence of this dinosaur was published in 2000, when Xu Xing and a Chinese team from the Institute of Vertebrate Paleontology and Paleoanthropology described *Microraptor zhaoianus*. Less than half a meter in length, *Microraptor zhaoianus* is the smallest of the feathered dinosaurs. Traces of the integument were found in the poorly preserved holotype. Long, feather-like impressions lined the forelimbs and hindlimbs. Those next to the femur showed evidence of shafts, suggesting that the animal had vaned feathers. The initial report of this fossil

Reconstruction of *Sinornithosaurus* based on "Dave." (Image: M. Ellison.)

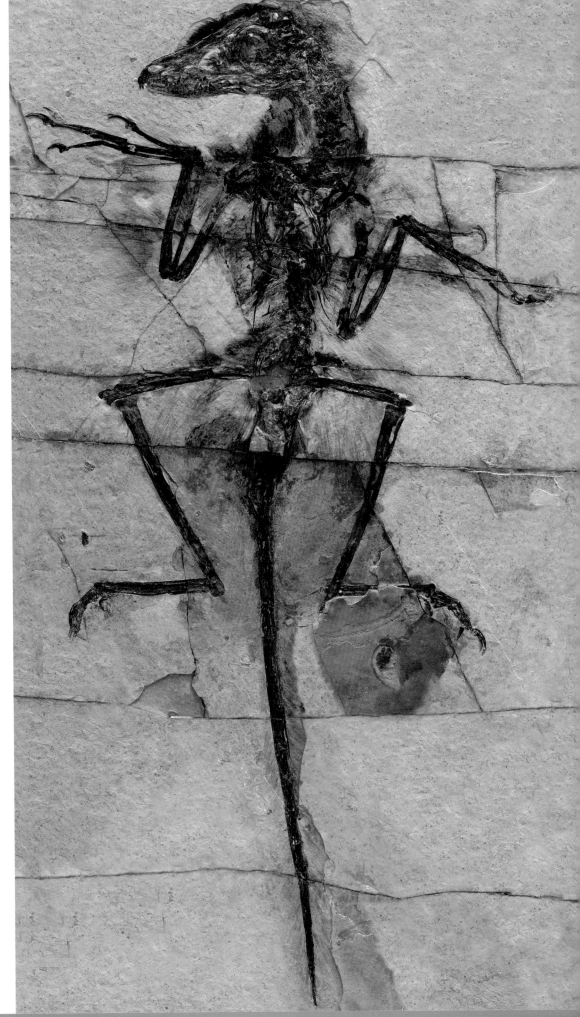

A juvenile of
Sinornithosaurus,
the specimen
nicknamed "Dave."
(Image: M. Ellison.)

was followed by the discovery of two more complete specimens of the same species. Independent studies of all these fossils have shown that *Microraptor zhaoianus* is even more primitive than *Sinornithosaurus*. Not only is the general appearance of these primitive dromaeosaurids more avian than that of the further specialized, larger and younger relatives, but the small size of *Microraptor zhaoianus* also approaches that of the earliest birds. Some have argued that the foot structure of this tiny dinosaur, with its strongly curved claws and first toes projecting farther down than in other dromaeosaurids, suggests an arboreal lifestyle. This functional interpretation, however, needs to be examined further in light of the fact that some of these features also occur in other, larger dromaeosaurids whose existence seems to be more bound to the ground. While the small size of *Microraptor zhaoianus* would clearly make it a more suitable tree-climber than its bigger dromaeosaurid relatives, nothing in the skeleton of this animal clearly indicates tree-climbing specializations.

More recently, several other specimens of tiny dromaeosaurids slightly differing from *Microraptor zhaoianus* were used to name another species of *Microraptor*, the 75-centimeter-long *Microraptor gui*. The spectacular new specimens show long and (most importantly) asymmetrical vaned feathers attached to the forelimbs, forming an extensive airfoil. Down-like feathers, nearly 3 centimeters in length, also copiously cover the body of these fossils; and the last third of the long tail carries a fanned tuft of vaned feathers. The distribution of feathers across the forelimb resembles that of birds. The feathers attached to the hand—the primaries—are longer but fewer than the secondaries, those attached to the ulna. In addition, the outermost primaries are parallel to the fingers, while the remaining wing feathers attach at an angle to the other bones of the forelimb. Remarkably, the small size, long and well-developed wings, and shoulder girdles designed to allow these airfoils to stroke up and down suggest that *Microraptor gui* could have flown. Yet *Microraptor gui* would make headlines around the world for another good reason: the lower portions of its hindlimbs also carried a set of long, vaned feathers (see Microraptor: *Mercury of the Dinosaur World*).

The dromaeosaurid *Microraptor gui*. Note the long, vaned feathers forming its wings and projecting from its hindlimbs. (Image: L. Chiappe.)

Although similar feathers also cover the hindlimbs of some living birds—particularly raptors—the much longer and lower position of the hindlimb feathers of *Microraptor gui* is unprecedented in any other feathered animal. Some researchers are skeptical about the authenticity of such feathers; after all, numerous Chinese fossils have been cleverly "enhanced" to make them more spectacular and attractive to collectors worldwide. However, the presence of these feathers in several specimens of *Microraptor gui* as well as in *Pedopenna*—a Late Jurassic fossil perhaps more closely related to birds than dromaeosaurids—makes the existence of such peculiar feathers undeniable.

At least nine species of feathered nonavian theropods have now been found. These dinosaurs encompass a wide range of shapes and sizes, and represent groups spread throughout the coelurosaur portion of the theropod genealogical tree—even the mighty *T. rex* could have been feathered (see *Feathers for* Tyrannosaurus rex)! If the feathers preserved around the bodies of these dinosaurs are important, it is equally important that the skeletal anatomy of these dinosaurs prevents their classification as birds. Rightfully, these creatures have become a source of pride and joy for China. It is an amazing experience to gaze at the entirely modern feathers of creatures which, by their skeletal characteristics, are so unquestionably dinosaurian. It feels like one has captured a significant moment in the lengthy drama of evolution.

The fact that feathers, long thought to be a characteristic unique to birds, have been found on the fossil skeletons of nonavian coelurosaurs should lay to rest any doubt that birds evolved from dinosaurs. Nonetheless, evidence of this peculiar pedigree for our familiar companions was for long time available in their bones, although incorrect assumptions about the evolutionary process and misinterpretations of the bone anatomy of nonavian theropods temporarily obscured this evolutionary transformation. More than 130 years ago, Huxley cleverly pointed out that there was nothing illegitimate in the idea that birds had evolved from dinosaurs. Since then, and especially in the last 20 years, features that for decades highlighted the uniqueness of birds, from furculae and swivel-like wrists to feathers and nesting behavior, have been discovered in maniraptoran theropods. The intersection of such disparate lines of evidence leaves little room for arguments against the theropod ancestry of birds. Within this framework, dromaeosaurids and troodontids are most commonly regarded as the closest known relatives of birds, although alternative genealogical hypotheses pointing at, among others, the parrot-headed oviraptorids as birds' closest relatives are often entertained.

Like many other scientific disciplines, paleontological inferences are always hypothetical, because no one has ever seen a fossil evolving into another. Nonetheless, the conclusion to be derived from the recently discovered menagerie of feathered fossils from northeastern China is inescapable. The only possible way to argue that the evolution of animals with skeletal features unique to specific theropod lineages and with feathers indistinguishable from those of living birds is not the result of a shared ancestry, is by using the old argument of convergence the "Pandora's box" of this scientific debate. However, evolutionary convergence—is something that needs to be demonstrated by discovering that other organisms share a greater degree of similarity with the organisms—in this case birds—whose genealogical relationships are under consideration. Granted, appearances can be deceiving, and the future may bring unexpected discoveries—but to date, no other group of organisms, living or extinct, shows greater resemblance to birds than dromaeosaurids and other groups of nonavian maniraptoran theropods.

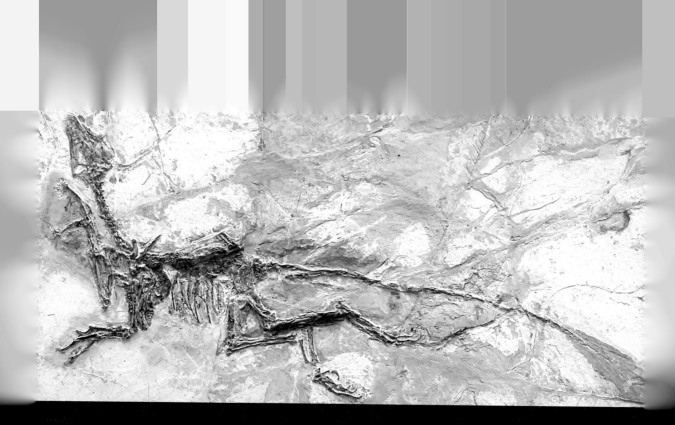

Over many decades, paleontologists have shown how the skeletons of birds, especially the most primitive ones, bear a great deal of resemblance to those of maniraptorans such as the sickle-clawed dromaeosaurids, the parrot-headed oviraptorids, and the light-legged troodontids. In the last few years, researchers have also discovered unique similarities between the eggshells of these dinosaurs and those of birds, and highlighted the fact that these animals shared aspects of their nesting behavior. The newest discoveries have shown that nonavian maniraptorans were covered by feathers, some as complex as those of living birds, and plausible explanations for their development and evolutionary origin have been devised. The large body of evidence accumulated from disparate fields, from paleontology through to developmental and molecular biology, suggests that birds evolved from maniraptoran dinosaurs and that this event probably happened sometime in the first half of the Jurassic.

The available evidence is so compelling that it has frequently been claimed that the hypothesis of the maniraptoran ancestry of birds is as strong as the argument that humans evolved from primates. Thus, if birds evolved from maniraptorans, it is logical to conclude that birds are maniraptorans—for the same reason we argue that humans are primates because we all evolved from an ancestral primate. And because maniraptorans are dinosaurs, we also need to conclude that birds are dinosaurs. This simple syllogism can also be extended to reptiles, because if dinosaurs are reptiles and birds are dinosaurs, then birds are also reptiles. Such a conclusion may puzzle those of us who learnt as we grew up that birds and reptiles were two very different classes of organisms. Yet a similar line of reasoning is used when we say that humans are primates and mammals. We argue that humans are primates because a large and diverse volume of evidence supports such a genealogical hypothesis, not because someone has seen the evolutionary transformation from a more primitive primate into one of us. At the same time, we argue that primates are mammals, because we also have a large body of evidence nesting the former group within the latter one. Clearly, there is no logical difference between arguing, on the one hand, that humans are primates and mammals, and on the other hand, that birds are dinosaurs and reptiles. There is just an ingrained perception that birds cannot be dinosaurs, because dinosaurs are reptiles. (Image: L. Chiappe/R. Urabe.)

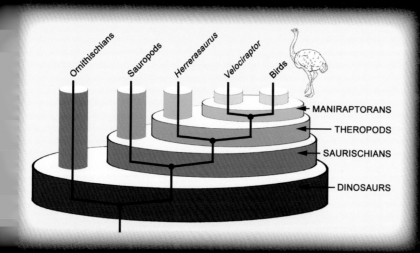

Long-scaled Reptiles: Plucking Feathered Dinosaurs

Because feathers have been long thought of as the ultimate avian attribute, the discovery of feathered nonavian dinosaurs has stimulated debates between supporters and detractors of the theropod origin of birds. The argument against the significance of the feathered nonavian dinosaurs has been structured in two ways. Some argue that feathers also existed in reptiles that are not dinosaurs and that the origin of birds could also be found elsewhere. Others assert that the feathers of these dinosaurs are irrelevant to the origin of birds because these feathered animals can in fact be interpreted as flightless birds.

The first argument centers on the tiny *Longisquama*, whose remains have been found as impressions in 220-million-year-old rocks from Kyrgyzstan. As mentioned earlier, the most salient attribute of this peculiar animal is the presence of six to eight pairs of 12-centimeter-long scale-like appendages projecting from its back. These planar structures are long and formed by a central support connected on each side to an ample, corrugated surface that gives the whole structure the appearance of a fern's frond. For 30 years, these structures have been interpreted as highly specialized scales that when deployed together could have assisted the minute *Longisquama* to parachute from a perch. In 2000, however, a team led by Terry Jones of Austin State University proposed that these structures were nonavian feathers, the primitive precursors of the feathers of birds. This claim was founded on the general vaned appearance of these structures as well as on detailed observations indicating that their basal portion was hollowed. This argument, however, was not consistent with these structures being air-filled throughout their entire length, as demonstrated by the fact that some of the concave impressions are filled with sediments that clearly invaded the hollowed structure when the dead corpse was entombed. This discovery and the corrugated nature of these structures indicate that they were built very differently from avian feathers. Furthermore, the position entertaining that these elongate scales could have been the evolutionary precursors of feathers is at odds with our understanding of the developmental origin of modern feathers. Experiments led by Chen-Ming Chuong, a leading developmental biologist from the University of Southern California in Los Angeles, have shown how barbs can be converted into a rachis by tinkering with the molecular mechanisms involved in feather development. These experiments indicate that during feather formation, barbs develop first and the rachis forms as a result of the fusion of several barbs. However, the interpretation of the long scales of *Longisquama* as the evolutionary precursors of feathers implies that the formation of barbs had to follow the formation of the rachis (both evolutionarily and developmentally)—a pattern that is not supported by the development of modern feathers.

The significance of the feathered dinosaurs from China has also been dismissed by claims that these fossils—albeit feathered—are simply another example of birds that lost their ability to fly. This view has been substantiated using comparative analysis of the hindlimb proportions and centers of gravity of *Caudipteryx*, a variety of other nonavian theropods, and extant flightless birds—namely, the work of Jones and associates, who regarded the long hindlimbs and alleged forward center of gravity of *Caudipteryx* as evidence of avian kinship. However, a close examination of this analysis reveals a number of alarming flaws. Not only are the majority of the specimens used in this study too incompletely preserved to allow replication of the given measurements, but in some cases not even the "measured" bones are preserved in the available fossil material. Equally critical is the fact that re-examination of this data within the context

The long, scale-like appendages that projected from the back (see inset illustration) of the enigmatic reptile *Longisquama*. (Image: H-D. Sues/ S. Abramowicz.)

of a more extensive set of measurements indicates that the hindlimb proportions and location of the center of gravity of *Caudipteryx* fall within the range of those of other nonavian theropods. More recently, even iconic theropod dinosaurs such as dromaeosaurids have been considered to be flightless birds. After decades of categorical dismissals of the existence of any evolutionary connection between dromaeosaurids and birds, Alan Feduccia and Larry Martin now believe that *Velociraptor* was not a nonavian dinosaur but a flightless bird!

In the end, what supporters of these feathered dinosaurs being avian seem to forget is that a great deal of evolutionary convergence would have to be explained in order to classify these animals within birds. For example, if one were to classify *Caudipteryx* within Enantiornithes—the most diverse

group of Cretaceous birds—one would have to assume that this animal re-evolved a long tail and long fingers with additional phalanges, and that numerous archaic features re-appeared in the skull, sternum, forelimb, hindlimb, and tail. Similar assumptions would have to be made if dromaeosaurids were to be placed within birds. Naturally, the most parsimonious explanation is that these animals are not birds but nonavian theropod dinosaurs with avian feathers.

But even if we were to entertain the idea that the feathered dinosaurs from Liaoning ought to be nested within birds, it would be impossible to deny their remarkable similarity to animals consistently classified as theropod dinosaurs. Is not this close to seeing the glass half full as opposed to half empty? The objective answer to this question alone should provide solid evidence for the shared evolutionary history of birds and theropod dinosaurs.

Avian feathers. These complex integumentary structures first evolved among nonavian theropod dinosaurs. (Image: R. Meier.)

Feathers: A Remarkable Evolutionary Novelty

Feathers are the most complex integumentary structures that have ever evolved among vertebrate animals, and undoubtedly figure among dinosaurs' most remarkable evolutionary novelties. The molecular basis and development of the feathers of extant birds are well-understood. Like scales and hair, feathers are keratin-structured appendages formed by the outermost layer of skin. The typical feather is a planar structure composed of a shaft bearing a series of regularly spaced branches on each side. The shaft is differentiated into a short and tubular calamus that is implanted in the skin and a longer rachis of spongy tissue that projects from the skin. During early development, the calamus has a basal opening that allows blood vessels to nourish the feather; this opening closes when the feather ceases to grow and its cells become keratinized. Feathers exhibit two general types of architecture. Plumulaceous feathers are soft and fluffy. Pennaceous feathers are stiffer and their barbs are arranged in well-defined vanes. Most feathers also exhibit three basic levels of branching: (1) either side of the shaft is subdivided into smaller branches called barbs that together form the vanes, (2) each side of a barb branches into closely spaced barbules, and (3) the barbules branch into tiny tapered projections or minute hooks. With this complex pattern of branching, it is not surprising that feathers have developed a wide range of forms that depend on variations in the shape, size, and structure of the rachis, barbs, and barbules.

The majority of feathers can be classified as either contour or down. Contour feathers cover most of the body. These include flight and tail feathers—remiges and rectrices, respectively—as well as body feathers. Remiges and rectrices are large and eminently pennaceous and the width of their vanes may show a great degree of asymmetry. As part of the wing and tail, these long pennaceous feathers provide the thrust and lift birds need to fly. The barbules of these feathers carry tiny hooks that allow them to interlock with one another, giving their vanes a close-knitted appearance and the stiffness critical for the aerodynamic functions they perform. Body feathers have a plumulaceous basal half and a pennaceous distal half—these structures are primarily responsible for water repellence. Downy feathers are small and plumulaceous, with loosely connected barbs—these feathers provide superb insulation. A wide range of even more specialized structures (filoplumes, bristles, and others that are primarily involved in sensorial functions) also exist.

Early during its development, a feather originates from a thickening on the skin's surface that grows into a peg-like projection of the skin called the feather germ. A ring of outer skin cells then folds around the feather germ, forming a moat-like structure known as the follicle, which is ultimately responsible for the formation, growth, and replacement of the feather. Proliferation of keratin cells at the base of the follicle generates a tubular, keratinic sheath, whose inner compartmentalization gives form to the barbs. In pennaceous feathers, the fusion of barbs forms the rachis; and cell differentiation of the barb surface forms the barbules. With growth, the branched feather emerges from the sheath (which is reabsorbed) and the tubular calamus is formed by the follicle.

The Origin of Feathers

The mystery of the origin of feathers has haunted evolutionary biologists for nearly as long as the problem of the ancestry of birds. Since the earliest attempts to explain the evolutionary origin of these remarkable structures, the notion that feathers are modified reptilian scales has dominated the debate.

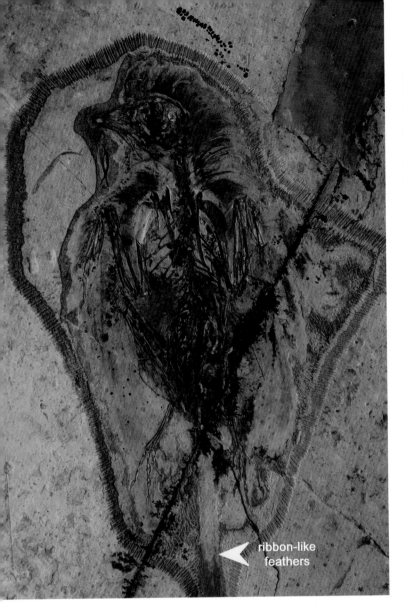

ribbon-like
feathers

The Chinese enantiornithine bird *Protopteryx fengningensis*. The long feathers of its pintail (arrow) were incorrectly interpreted as the ancestral condition of all feathers. (Image: L. Chiappe.)

pointed out that scales do not form from follicles but rather from folds of the skin, and that previous scale-to-feather generalizations have often overlooked the fact that when a chemically stimulated scale produces a feather, it does so by forming a tiny follicle (and sometimes more than one) on its surface—thus, the induced feather has essentially followed the same developmental pathway as any normal feather. These and other important differences between scales and feathers have become much clearer thanks to a series of recent molecular and developmental studies by leading feather specialists such as Chen-Ming Chuong, Alan Brush (University of Connecticut), and Richard Prum (Yale University). These studies have consolidated the view that feathers are the result of a multitude of complex evolutionary novelties, originated sequentially during the evolution of dinosaurs, most of which have neither molecular nor developmental precursors outside this group of animals.

Although the classic notion that feathers are modified scales has largely been laid to rest, the appearance and function of the ancestral feather remains elusive. Historically, theories attempting to reconstruct the physical characteristics and original function of the ancestral feather have focused on selected types of feathers assumed to be similar to the ancestral one. These selected types were often those present in lineages of birds believed to be primitive. For example, the hair-like, simplified appearance of the feathers of ostriches, rheas, kiwis, and their flightless kin—primitive birds known as ratites—were regarded as the starting point of feather evolution. Similarly, the pair of ribbon-like feathers that project from the tail of the Early Cretaceous enantiornithine bird *Protopteryx fengningensis* were recently interpreted as the ancestral condition of all feathers. Regardless of whether these feathers do or do not resemble the ancestral feather, this line of reasoning is not framed within what is known

After all, both structures are planar epidermal appendages, and the alleged transformation from one to the other was reinforced by experiments in which scales from the feet of birds were chemically stimulated to produce feathers. Recent developments, however, have emphasized the enormous differences between the molecular make-up and developmental trajectory of reptilian scales, on the one hand, and avian feathers on the other. Indeed, these studies have highlighted the fact that while most scales are composed of alpha-keratins and the smaller family of beta-keratins, feathers are exclusively composed of a particular type of beta-keratin, one whose molecule is much smaller than the rest of this family. It has also been

about the genealogy of birds. Even if ratites are among the most primitive lineages of modern birds, like all living birds, they are evolutionarily far more advanced than *Archaeopteryx*. Because *Archaeopteryx* and a host of other primitive Mesozoic birds have plumages made of contour and down feathers—indistinguishable from those of typical living birds—the plumage of ratites must be considered as a specialization derived through the simplification of more complex feathers. This is, of course, presuming that we subscribe to the widely accepted view that ratites, *Archaeopteryx*, and all other birds share a most recent common ancestor—the radical view that ratites could have originated directly from nonavian theropods has today been widely discredited. Likewise, *Protopteryx* and all other Enantiornithes are evolutionarily closer to modern birds than *Archaeopteryx* and the feathered nonavian coelurosaurs from China, which lack any evidence of having ribbon-like tail feathers. In light of this genealogical background, one must also conclude that the long, ribbon-like feathers of *Protopteryx* are specialized feathers, possibly for sexual display, with no relation to the physical appearance of the ancestral feather.

Given the compelling nature of the question, it is not surprising that theories based on the plausible functional role of the ancestral feather have remained at the heart of the discussions about the origin of feathers. In general, the architecture of these functional hypotheses includes speculations about feathers' ancestral role and the physical appearance that best fits the assumed original function. Over the years, these functional theories have ranged from those focused on aerodynamic considerations—feathers evolving in the context of flight—to those highlighting the role feathers play in heat protection, sexual display, water repellence, foraging, nesting, defense, and insulation—functions well-known for all or some living birds. These functional

hypotheses have often created scenarios in which modern feathers are envisioned as the end product of a series of adaptive steps. One common scenario pictures the ancestors of birds as arboreal animals with elongate scales that gradually became fringed, more flexible, shafted, and longer as they provided a greater advantage to a lineage of jumping or parachuting animals. By evolving these hypothetical fringed scales, these animals reduced their risk of falling and increased their ability to escape and capture prey. Clearly, these functional

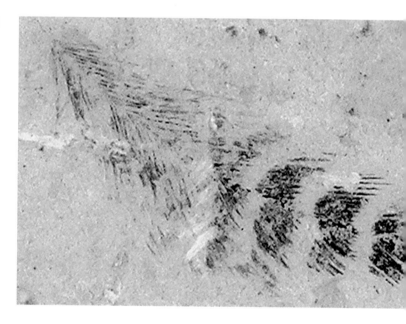

theories speculate about the role of an unknown ancestral structure and rest on substantial conjecture about the ecology of extinct organisms. As if this approach were not questionable enough, supporters often look at the fossil evidence available through the lens of their preconceptions. For example, crafters of this aerodynamic scenario reject the argument that the integumentary filaments cloaking the body of the coelurosaur *Sinosauropteryx* are feathers—because these structures lack the airfoil design envisioned as ancestral by the scenario. At the same time, however, they rely on the presence of elongate scales in *Longisquama* as supporting evidence, disregarding the data indicating that this Triassic reptile

Some fossil feathers, like this 5-centimeter-long one from the Early Cretaceous of Brazil, are preserved with color patterns intact from a time when they formed the plumage of both nonavian theropods and birds. (Image: R. Loveridge.)

is far removed from the ancestry of birds.

Another prevalent functional theory explains the origin of feathers within the context of thermoregulation: as structures evolved to insulate the bodies of animals that were becoming warm-blooded. This notion hinges upon the idea that the most primitive nonavian coelurosaurs had metabolisms that were significantly more active, and thus more able to generate body heat, than those of extant cold-blooded reptiles. From this standpoint, the ancestral plumage would have been composed of simple, hair-like structures devoid of the complex organization of typical feathers, which is often associated with aerodynamic functions. Evidence indicating that some nonavian dinosaurs, including nonavian theropods, grew more quickly than modern cold-blooded reptiles is consistent with this interpretation, because animals with high metabolic rates grow faster than those with a lower metabolism. Regardless of whether nonavian theropods were warm-blooded or on the way to becoming so, however, theories that reconstruct the ancestral appearance of feathers by envisioning the ideal design for plausible functions will remain speculative.

Many have argued that the primary function of feathers was behavioral display, either to facilitate identification of suitable mates within a species or to enhance competition for the fittest mates. We have seen that a wide range of feather forms was present in nonavian dinosaurs, and a significant record from Cretaceous deposits worldwide also reveals that diverse feathers (semiplumes, down, contour, and others) covered the bodies of the feathered animals of the Mesozoic. Isolated feathers with evidence of color patterns have been found in several Cretaceous localities. Some Early Cretaceous contour feathers show a perpendicularly barred pattern of darker and lighter bands. Others show an alternation of light and dark dots or dashes along their barbs. Likewise, bands of lighter and darker feathers are preserved in the tail of the basal coelurosaur *Sinosauropteryx*. Melanin is perhaps the most durable pigment responsible for feather color. This pigment produces dark tones such as black, grays, and browns. Carotenoids and porphyrins, lighter pigments that also tint feathers, are less durable and less resistant to wear. The fossil evidence of Cretaceous feathers suggests that in addition to the observed diversity of shapes seen, a variety of coloration patterns evolved very early in the evolution of feathers, even if the specific colors of the lighter and darker bands of these feathers cannot be ascertained. This evidence indicates that feathers had the potential to play behavioral display roles very early in their evolution—but whether their appearance as an evolutionary novelty was connected to this role or not is still uncertain.

In the end, we must keep in mind that understanding the functional advantage of an evolutionary novelty is far more complex than understanding its evolutionary origin. It is also far more conjectural, because functions are not preserved in the fossil record. To begin with, any attempt at explaining the ancestral function of feathers should rely on strong genealogical and fossil evidence. In this respect, speculations about the insulating role of feathers' original function are more consistent with what is known about the genealogy of birds and the fossil record than those arguing in favor of an aerodynamic context. As a matter of fact, it can be inferred from the feathered nonavian coelurosaurs from China that feathers did not evolve in the context of flight. With the possible exception of the small *Microraptor gui*, none of these coelurosaurs was able to take to the air, including those with the longest, vaned feathers such as *Caudipteryx* and *Protarchaeopteryx*. In these dinosaurs, both forelimbs and feathers are much shorter than in flying birds, and their

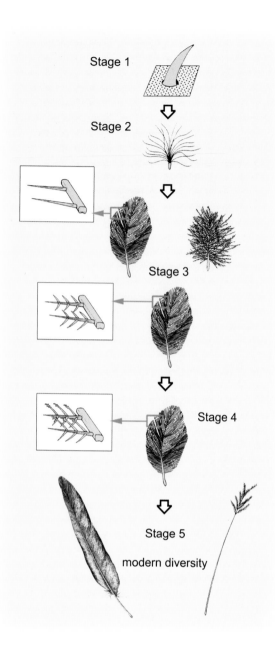

Stage 1

Stage 2

Stage 3

Stage 4

Stage 5

modern diversity

not have aerodynamically designed feathers does not mean that feathers could not have soon after become thrust generators. Research I have conducted with aerodynamicist Philip Burgers from the San Diego Natural History Museum suggests that vaned feathers may have originated in the context of thrust, evolving in running coelurosaurs that by flapping their feathered arms

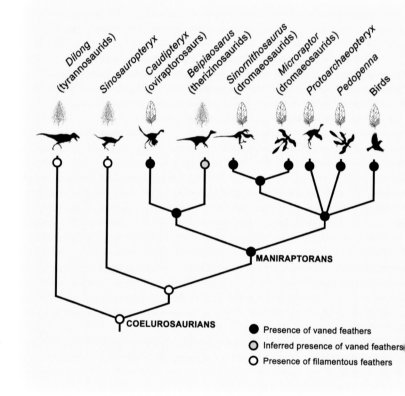

● Presence of vaned feathers
◉ Inferred presence of vaned feathers
○ Presence of filamentous feathers

Genealogical relationships of feathered nonavian theropods. Current evidence supports the hypothesis that filamentous and vaned feathers evolved with the divergence of coelurosaurs and maniraptorans, respectively. (Image: R. Urabe.)

bodies are larger. The evolution of birds and possibly flight involved a dramatic reduction in body size—the small dromaeosaurid *Microraptor gui* and the troodontid *Mei long* may be heralds of this significant evolutionary trend. The filament-like feathers of *Sinosauropteryx*, the most primitive of all known feathered dinosaurs—and far removed from the most immediate ancestry of birds—suggest that at their onset, feathers must have played a role other than flight, because these feathers could not generate significant thrust. Yet the fact that *Sinosauropteryx* does

were able to increase their running speed. It could well be, however, that vaned feathers evolved in a different context; for example, for display or egg protection—nesting oviraptorids have been found with their forelimbs circling the periphery of the egg-clutch. Although a great step has been taken by eliminating flight as one of the functional explanations for the origin of feathers, we simply do not know for certain their ancestral role. It is also within the realm of possibility that feathers had more than one original function.

Feathers' Origins: A Novel Developmental Model

A novel conceptual direction in this field has crafted a much better picture of how feathers may have evolved. This breakthrough has primarily come from the work of Richard Prum, who has proposed a developmental model for the origin of feathers that is independent from, albeit congruent with, what we know about the plumage and genealogy of the theropod predecessors of birds. This model proposes five stages of feather development that follow the evolution of the follicle—each of these stages resulting in a structure of greater complexity. In the first stage, the follicle produces a hollow and unbranched structure. The second stage hypothesizes the evolution of a tuft of barbs that are fused to the calamus or quill. Two alternatives are presented in the third stage: either the follicle first develops a feather with a shaft and barbs lacking barbules or it first develops a tuft of barbs carrying barbules. In any case, either of these hypotheses leads to a pennaceous feather with open vanes, whose barbs are not interlocked by hooked barbules. Hooked barbules appear in the fourth stage, which produces a pennaceous feather with closed vanes. From this closed, pennaceous feather, the final (fifth) stage develops not just a feather with asymmetrical vanes but also the entire spectrum of modern feather diversity. The proposed sequence of stages is justified by the developmental sequence observable in modern feathers, which makes it plausible that the ancestral follicle could have produced the designs predicted by the model. Equally important is the congruence between the proposed stages and the design of the feathers known for the different lineages of nonavian coelurosaurs. For example, the design predicted by the first and second stages occurs in *Sinosauropteryx*, the most primitive known feathered dinosaur, and closed, symmetrical feathers, like

those predicted by the fourth stage of Prum's model, occur in *Caudipteryx*, *Protarchaeopteryx*, and a variety of dromaeosaurids. Furthermore, Chen-Ming Chuong's molecular investigations have mapped the genetic pathways of feather formation and demonstrated that feather differentiation is regulated by the interplay of several proteins and regulatory genes.

Thus, recent molecular and developmental advances, coupled with the new evidence available in the spectacular fossil menagerie of Liaoning, have consolidated the view that these complex integumentary structures originated as a series of evolutionary novelties within dinosaurs. Indeed, the widespread distribution of feathers among different lineages of nonavian coelurosaurs indicates that the earliest of these evolutionary novelties had already evolved in the common ancestor of this group, which was likely a feathered animal. However, the recent discovery of a primitive coelurosaur lacking feathers in the preserved portions of its integument has highlighted the fact that the evolution of these structures could have been more complex—perhaps feathers evolved more than once or they were secondarily lost in some lineages. More fossils are needed to answer these questions and to clarify how far down the evolutionary saga of dinosaurs these remarkable structures can be traced.

Feathers' Deep History

The question of whether dinosaurs more primitive than coelurosaurs were cloaked in feathers remains far less certain. Aside from a handful of exceptional "mummified" specimens, the skin of dinosaurs other than coelurosaurs is very rarely preserved. Among ornithischians, fossils of adult duck-billed and horned dinosaurs show that the skin of these animals was formed by tubercles of various sizes and shapes. Unlike the usual scaly skin typical of lizards and snakes,

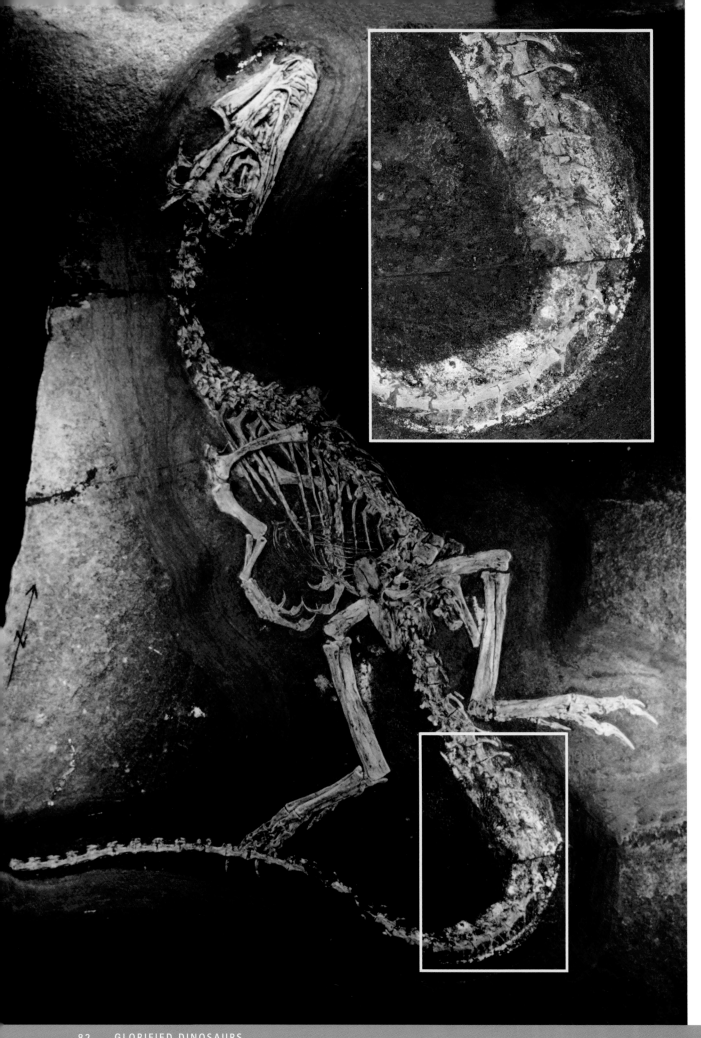

the tubercles of these dinosaurs did not overlap one another, and thus probably resembled more the beaded surface of the skin of the Gila monster. The small psittacosaurs—early relatives of horned dinosaurs with parrot-like heads—seem to be an exception among ornithischians. Much of the body surface of these animals was covered by small tubercles similar to those of other ornithischians. However, spectacular psittacosaurs from the Early Cretaceous of Liaoning show a series of long skin appendages at the end of their tails. These odd structures resemble more the long spines of a porcupine than feathers, and even if their occurrence in these fossils may point at a wider distribution of skin appendages

these theropods show evidence of plumage, but since they represent animals of large size, such as the 9-meter-long *Carnotaurus sastrei*, it is still possible that hatchlings and young immatures were feathered. The presence of feathers, albeit an advantage for animals trying to maintain a constant body temperature, could be a hindrance in animals of very large size. Because the volume of the body increases in cubic increments and its surface does so as a square function, large animals have a tendency to retain much more body heat than small ones. As a consequence, insulation could be highly disadvantageous for a large-bodied animal. This reasoning has led researchers to argue that even though *Tyrannosaurus rex* and other large

among ornithischians, nothing seems to support the presence of feathers in this large group of plant-eating dinosaurs.

Within saurischians—the other, larger group of dinosaurs—sauropods also display a tubercular type of skin, even though instances of soft-tissue preservation associated with the skeletons of these gigantic animals are extremely rare. Among theropods more primitive than the coelurosaurs, only a handful of specimens include preserved portions of skin. None of

tyrannosaurids are evolutionarily nested within coelurosaurs and are therefore expected to have evolved from a feathered ancestor, it is likely that most, if not all, of their feathered covering was shed by the time individuals of *T. rex* and their large-bodied kin became adults. The existence of feathered tyrannosauroids—remote relatives of *T. rex* and other tyrannosaurids—has been recently confirmed, but the evidence is restricted to the modest-sized *Dilong*, an animal that did not exceed 2 meters in length.

Fossil remains of baby dinosaurs are extremely rare—and rarer still are those instances where their integumentary covering is preserved. Indeed, early juvenile dinosaurs with preserved skin are exclusively known from the Late Cretaceous sauropod nesting site of Auca Mahuevo in Patagonia, Argentina. For a number of years, I have been leading the research at this spectacular locality where the discovery of thousands of sauropod eggs and clutches has clarified important aspects of the reproductive behavior of long-necked dinosaurs. Many of these eggs contain embryonic remains, which in some instances include portions of the skin that covered the bodies of the unhatched sauropods. Like in the adults of these colossal animals, the skin of baby sauropods is formed by a nodular pavement in which fields of small rounded swellings are interspersed with bigger nodules arranged in rosettes or flower-like patterns, or crossed by rows of larger lumps. None of the dozens of fossils preserving skin remnants of Auca Mahuevo's baby sauropods exhibit any evidence of plumage. The fact that neither adults nor baby sauropods show any indication of plumage suggests that these structures were absent among these dinosaurs. This also suggests that among saurischians, feathers are likely to have evolved only in theropods, even though at the moment it is difficult to say whether their origin preceded the divergence of coelurosaurs. Thus, the available evidence suggests that feathers are an evolutionary novelty of theropod dinosaurs, and the fact that no nonavian dinosaur exhibits scales that can be interpreted as transitional to feathers—elongated scales—strengthens the argument that feathers had an origin independent from scales. While these important notions are well-rooted in our empirical knowledge, the question of the original function of feathers—the *why* of feather origins—remains elusive.

Flying Dinosaurs: The Origin of Avian Flight

There is little doubt that by providing a fast mode of locomotion and an easier way to overcome physical barriers, flight is central to the evolutionary triumph of birds and their ubiquitousness in both urban and natural environments. Yet how this remarkable characteristic came about is still shrouded in mystery. This is not entirely surprising, because despite the fact that the main physical transformations that occurred during the evolution from maniraptoran theropods to modern birds are well-preserved in the fossil record, the spectacular fossils that so vividly document this evolutionary saga tell us little about how birds began to fly.

This limitation of the fossil record

Evolutionary stages proposed by the classic theory of the arboreal origin of avian flight, next to a reconstruction of a tree-climbing predecessor of birds. (Image after Heilmann, 1926.)

The pebbled skin of an embryonic sauropod dinosaur shows no evidence of feathers. These structures probably evolved only within theropod dinosaurs. (Image: L. Chiappe.)

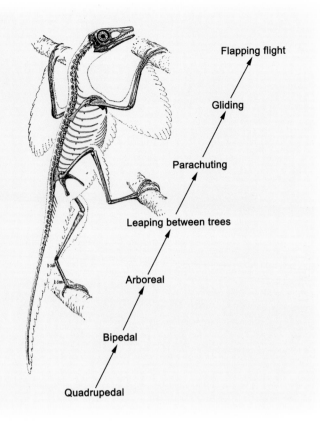

Flapping flight

Gliding

Parachuting

Leaping between trees

Arboreal

Bipedal

Quadrupedal

is not the only factor obscuring the beginnings of avian flight. The fact that the evolutionary origins of both flight and birds have often been regarded as one and the same problem has also contributed to this lack of understanding. These two problems, however, profoundly differ from each other philosophically. While approaches for understanding the origin of birds have used tangible data to resolve an intangible event, explanations of the origin of flight have used intangible data to resolve an equally intangible event. In other words, while the origin of birds can be studied by comparing the bones, eggs, and integuments of modern birds to those of other animals preserved as fossils, the origin of flight is a problem for which we have no direct data from the fossil record.

Since the late 1800s, the problem of the origin of flight in birds has been explained by means of two antagonistic and seemingly incompatible theories. On the one hand, a great number of researchers have argued that birds evolved their flight "from the trees down," assisted by gravity. Traditionally, this

view has been referred to as the "arboreal theory." On the other hand, some have favored a "ground up" origin, against the forces of gravity, from cursorial (running) animals that never developed arboreal habits. This opposing view, almost as old as the "arboreal theory," is known as the "cursorial theory." Throughout the history of these ideas, discussions have been founded on extrapolations from the anatomical, aerodynamic, behavioral, and ecological properties of living animals, by inferring similar properties in extinct organisms and creating hypothetical scenarios reconstructing the function of the physical characteristics— and especially the ecology—of the intermediate forms. Regrettably, the fact that ideas about the evolutionary origin of a function such as flight have been historically so saturated with ecological interpretations of the hypothetical ancestors has sometimes thwarted more than helped our understanding of the problem.

The notion that birds evolved their flight "from the trees down" can be traced back to O. C. Marsh, a renowned American paleontologist who in the late 1870s was involved in studying a diversity of fossil seabirds unearthed from rocks formed in the bottom of the Pierre Seaway, a shallow waterway that in the Late Cretaceous bisected North America from north to south. Perhaps influenced by Charles Darwin's plea for an arboreal origin of flight for bats, Marsh proposed that avian flight evolved among primeval birds that lived in trees and jumped from one branch to another, also parachuting to the ground. Like his contemporaries, Marsh believed that birds originated in the Paleozoic Era, long before *Archaeopteryx*. He argued that the development of even rudimentary feathers on the forelimbs of these very early birds would have endowed them with the ability to lengthen a leap or break the force of a landing. As natural selection favored generations of animals with longer feathers, these primitive

Flight is a remarkable achievement—one that humans craved for millennia until they devised its workings. When a powered flier is on its wings, two major forces are at play: lift that needs to counter gravity and thrust that needs to overcome friction and drag. Flight works because the wing—the airfoil—is constructed in such a way that it generates force when air moves relative to its surface. This phenomenon is dependent on the wing's thin, streamlined, and cambered structure, with a rounded leading edge and a sharp trailing edge. Such a shape makes air accelerate on the upper surface and decelerate on the lower surface, thus creating a differential pressure at the level of the wing that produces lift and minimizes drag—a phenomenon known as Bernoulli's Principle. The result is a series of wake vortices that force air downwards in the region behind the wing while air ascends elsewhere.

In an airplane, the wings provide the lift and the propeller generates the thrust, but in birds these two forces are supplied by a feathered wing powered by vigorous muscles. Aided by precise movements of the body and tail, the wing can adjust its geometry to meet the complex requirements of a wide range of aerodynamic performances. For this reason, unlike the motionless wings that propel a glider, the wings of an active flier must be mobile. Because gliding is a passive type of locomotion, it is energetically much cheaper than active flight. Yet this saving has its shortcomings, for gliders must expend energy climbing in order to launch themselves from a height.

The flight performance—and to some extent, the lifestyle—of a bird is determined by the size and shape of its wings and their relationship to its weight. Wing shape is often described by the "aspect ratio," a parameter that relates the span of wings to their width. High aspect ratios—found in the long-winged albatrosses, gulls, and petrels—generate a great deal of lift and minimize drag. This wing design is typical of high-speed soaring birds. At the opposite end of the spectrum, the broad and elliptical wings of many forest birds—songbirds, pigeons, and others—have low aspect ratios that enhance maneuverability. Birds that soar at lower speeds, such as vultures, storks, eagles, and hawks have intermediate aspect ratios. Another important aerodynamic parameter is "wing loading," which relates the weight of the bird to the surface area of its wings. Small birds with low wing loadings can maneuver better (although they fly more slowly) than larger birds with greater wing loadings. (Image: K. Garrett/S. Abramowicz.)

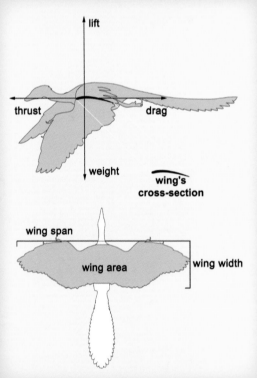

birds would have evolved gliding capabilities and, in time, powered flight. Marsh's arboreal theory was elaborated in more detail by G. Heilmann in *The Origin of Birds*. The Dutch artist and writer envisioned that flight evolved from ground-dwelling animals that became tree-climbers capable of jumping between branches and parachuting from low heights. As the range of the leaps increased, the descendants of these tree-climbers began to glide. Gliding became more and more specialized until finally the descendants of this lineage of tree-climbers evolved flapping capabilities.

Ever since Marsh and Heilmann, the idea that avian flight evolved through a series of incremental stages—from quadrupedal to bipedal ground-dwellers, to tree-climbers that learnt how to leap between branches, to parachuters and gliders that eventually evolved flapping capabilities—has persisted. Some researchers have preferred a slight variation of this view, in which the bipedal phase is excluded. The fact that gravity-aided flight seems easier to achieve than the opposite is often voiced by evolutionary biologists endorsing this traditional version of the arboreal theory of the beginnings of avian flight. Not surprisingly, this theory is also most commonly supported—but not exclusively—by the opponents of the theropod hypothesis of bird origins, because the functional and ecological requirements of the traditional arboreal hypothesis are at odds with those usually inferred for the predominantly terrestrial theropod dinosaurs. In contrast, these requirements seem a better fit for interpretations based on the small, Triassic fossils that lay at the heart of basal archosauromorph hypotheses—such as *Megalancosaurus* and *Longisquama*. Not only are these animals regarded as arboreal, but also the long scales of the latter have often been explained as specializations for parachuting. The great dissimilarity between these Triassic archosauromorphs and *Archaeopteryx*

has not stopped "arborealists" from advocating incremental evolutionary transformations from small tree-climbers to animals capable of flapping their wings. They see the arboreal theory of bird flight as the only biophysically plausible scenario. The notorious absence of fossils representing the early stages of this transition is explained by arguing that it is very unlikely for small, tree-dwelling animals to become fossilized.

Certain researchers, however, have emphasized the importance of combining the ecological and functional considerations used to infer alternatives for the origin of flight with genealogy, since any theory attempting to explain this problem must agree with our understanding of bird ancestry. Indeed, we could easily imagine how vastly different explanations for the origin of flight could be, if one were to build upon the notion that birds evolved from basal quadrupedal archosauromorphs instead of from bipedal theropods. As a result of our understanding of the origin of birds, a different and methodologically superior approach in support of an arboreal origin of flight has recently emerged. This approach argues that the features previously assumed to be predatorial specializations of theropod dinosaurs are indeed arboreal specializations, thus interweaving the arboreal theory of flight with the theropod origin of birds.

Tree-climbing Theropods: A More Plausible Arboreal Theory for the Origin of Flight

Ecological reconstructions depicting *Archaeopteryx* as a tree-dweller have long been at the heart of the idea that flight evolved from the "trees down." To a great extent, these interpretations have been based on comparisons between the shape of the claws of both the hands and feet of *Archaeopteryx* and those from the feet of living birds that climb trees. For example, in a commonly cited study, Alan Feduccia showed that

the curvature of the hand claws of *Archaeopteryx* is similar to those of the feet of trunk-climbing birds. This study also concluded that the foot claws of *Archaeopteryx* are comparable to those of perching birds, although this is more important for inferring arboreal habits than for arguing that *Archaeopteryx* had tree-climbing specializations. Feduccia's study followed an earlier investigation by Derek Yalden of the University of Manchester, who, using the shape of the claws of *Archaeopteryx* in combination with similarities in the proportions between the limbs and the backbone of tree-climbing mammals, inferred a similar lifestyle for the earliest bird. Analyses of the geometry of the claws of *Archaeopteryx*, however, appear to be inconclusive for interpreting the lifestyle of this and other very primitive birds. When in the early 1970s John Ostrom reformulated the theropod origin of birds, his comparisons between the claws of *Archaeopteryx* and those of its modern counterparts led him to favor a ground-dwelling habit for this ancient bird. Furthermore, a number of birds that spend a considerable amount of their time in trees have claws that are much less recurved than those of tree-climbers— and the degree of curvature of the latter is very similar to that of predatory birds. Equally problematic is the fact that the geometry of the claws is scaled differently depending on the animals' habits—in predatory and climbing birds their curvature increases with size, but in land-dwelling birds the claws become less hooked as the bird grows bigger. Probably because of the apparent ambiguity resulting from independent studies of claw shape, inferences about the lifestyles of early birds have more recently been interpreted on the basis of the relative lengths of toebones, excluding the claws.

Using this type of data, James Hopson of the University of Chicago argued that the feet of *Archaeopteryx* and the Chinese Early Cretaceous bird *Confuciusornis* are not so similar to those of living tree-climbing birds. Based on comparisons of nearly 200 birds, Hopson showed that the feet of these early birds are a better match with those of pigeons and land fowls, birds that spend time both in branches and on the ground but do not climb trees. He also analyzed the proportions of the bones of the hindlimb, concluding that in this respect, *Archaeopteryx* was also comparable to pigeons. An independent statistical study analyzing comparable evidence produced similar results for the short-tailed *Confuciusornis*. The researchers conducting this study, Zhou Zhonghe from the Institute of Vertebrate Paleontology and Paleoanthropology and Purdue University's James Farlow, regarded the early Chinese bird as an arboreal animal, but conceded that it could have also spent time on the ground. An important and often overlooked issue to consider, when using these results to weigh the merits of one or another alternative for the origin of flight, is that these studies tell us little about the climbing specializations of either *Archaeopteryx* or *Confuciusornis*. Arboreal habits need not imply tree-climbing capabilities. In order to argue that these birds climbed trees because they had an arboreal existence, one also needs to assume that neither of them was capable of taking off from the ground, an issue that we will examine in the next section.

In addition, the curvature of the claws of these birds is less pronounced than that present in the claws of nonavian theropods such as dromaeosaurids and oviraptorids. For example, the strongly recurved hand claws of the 3-meter-long oviraptorid *Citipati* cluster with the claws of tree-climbing birds used in Feduccia's study. The size of *Citipati* would not necessarily have prevented it from climbing trees (lions and leopards are also accomplished tree-climbers) but the presence of strongly recurved claws in large predatory animals that likely used their hands to seize prey precludes

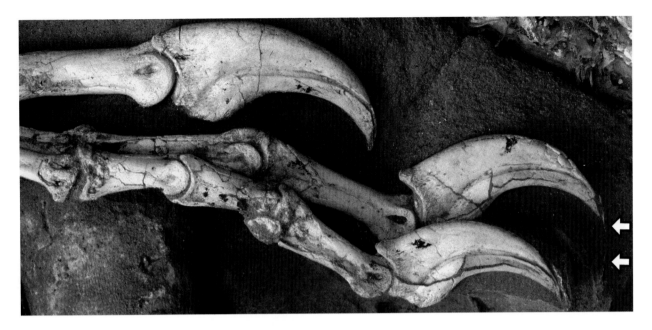

conclusive determinations about the functional significance of this feature. Likewise, the recurved claws of many early birds need not be a specialization for climbing, but could be explained simply by inheritance. It is worth noting that animals that live in trees often possess grasping specializations, from the prehensile tails and reversible ankles

Hand claws of the oviraptorid *Citipati* (photograph) with arrows pointing at the preserved horny sheaths. When these sheaths are included in the estimation of the geometry of the claws, the curvature values for *Citipati* exceed those calculated for *Archaeopteryx*. (Image after Chiappe, 1997/M. Ellison.)

Archaeopteryx

Citipati

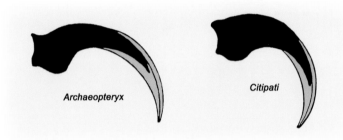

Cursorial runner → Tree-climber → Branch-leaper and parachuter → Glider

Ceratosaurus   *Ornithomimus*   *Velociraptor*   *Archaeopteryx*

The arboreal theory of flight origins framed within the theropod hypothesis of the origin of birds. (Image after Chatterjee, 1997.)

of many tree-climbing mammals and lizards to the perching feet of many arboreal birds and the hooked beaks of parrots, which are absent among the most primitive birds known.

In recent years, interpretations of tree-climbing habits in primitive birds have revived an idea first proposed at the turn of the 20th century: the possibility that theropod dinosaurs could have also been tree-climbers. Sankar Chatterjee of Texas Tech University, who first reinstated this notion, argued that the enlarged brains, elongated forelimbs, and less flexible tails of nonavian maniraptorans enabled these dinosaurs to conquer three-dimensional spaces, thus paving the road for a suite of tree-climbing specializations best evidenced among dromaeosaurid theropods. Chatterjee interpreted the long hands and sharp claws of dromaeosaurids, their swivel wrist joints, large breastbones, inflexible tails, and streamlined body shape as specializations for climbing vertical tree trunks. According to him, these features allowed dromaeosaurids to climb trees in a squirrel-like fashion, moving their forelimbs and hindlimbs alternately in unison, using their clawed wings as crampons and their rigid tail for additional support; and to use their tree-climbing skills to nest within the canopy as well as to ambush their ground-dwelling prey. Chatterjee envisioned the emergence of powered flight as the incremental evolution of parachuting to gliding to undulating flight—the alternation of flapping and gliding—to more sophisticated variants of flapping flight.

Albeit genealogically well-founded, Chatterjee's scenario leaves several important points out of consideration. The first of these revolves around the nesting behavior of theropods as inferred from a series of spectacular discoveries of brooding nonavian maniraptorans, including evidence from dromaeosaurids. Indeed, these discoveries have shown that nonavian maniraptoran theropods nested on the ground and invested a good deal of energy in parental care. If these dinosaurs were capable of climbing trees, why did not they brood their egg-clutches in the far more protected environment of the canopy? One could argue that other maniraptorans did nest in trees and that such nests (and nesters) would have been much less likely to be preserved as fossils. If this was the case, however, why do all nonavian maniraptorans (including those nesting on the ground) have essentially the same design for the features claimed to be tree-climbing specializations? Would it not be more realistic to expect substantial dissimilarity given the enormous differences in lifestyles? Furthermore, if modern birds are used in a comparative

Airfoil design of a pterosaur, the Triassic lepidosauromorph glider *Kuehneosaurus latissimus*, and a bird. The design of the airfoil of pterosaurs suggests that these animals evolved their aerial capabilities from animals that were initially gliders. In contrast, the design of the avian wing does not support a gliding origin. (Image: S. Abramowicz.)

Pterosaur

Triassic glider

Bird

The enigmatic nonavian theropod *Epidendrosaurus*. Although the long third finger of this small animal has been interpreted as a tree-climbing adaptation, the evidence supporting tree-climbing specializations in nonavian theropods is inconclusive. (Image: L. Chiappe.)

context—as they ought to be—most aerial nesters rear helpless, altricial hatchlings, but this developmental strategy contrasts with what we know about nonavian theropods, which seem to have raised a highly precocial brood. In addition, all the fossil evidence of the nesting behaviors of early birds indicates that they also nested on the ground and had precocial hatchlings, once again raising the question of why they did not inherit the supposedly more advantageous behavior of their ancestors. Spectacular new fossils from the celebrated Early Cretaceous beds of Liaoning are also at odds with this arboreal scenario for nonavian maniraptorans. Several skeletons of the troodontid *Mei long* have

shown that this tiny Chinese dinosaur slept in the stereotypical posture of many living birds—both hindlimbs flexed beneath the belly, the neck turned backwards, and the head tucked between a wing and the body. Undoubtedly, these animals died peacefully—perhaps poisoned by toxic gases from nearby volcanoes—while resting on the ground, a fact that once again raises a red flag regarding the purported arboreal habits of maniraptorans.

One other unconsidered point in Chatterjee's proposed scenario deals with the anatomy of maniraptoran theropods, because any interpretation requiring a gliding phase to explain the origin of flight is at odds with the

forelimb design of these dinosaurs. Gliding as a way of locomotion has evolved many times among vertebrates. This type of locomotion requires an extensive airfoil that is firmly attached to the body—a design that maximizes lift and reduces drag at the expense of power stroke and control during flight. Gliders have achieved this important structural design by either developing a membrane between the forelimb and the hindlimb (as in gliding frogs, some lizards, and pterosaurs) or by lengthening a membrane-bearing ribcage (as in flying lizards and some Triassic reptiles). Yet the evidence indicates that neither birds nor their most immediate maniraptoran relatives developed the kind of airfoils typical of gliding animals. In fact, throughout their evolution, the wings of these animals continued to act as mobile airfoils largely detached from the body. Likewise, during the evolution from nonavian maniraptorans to birds, the thrust-generating outer portion of the wing became expanded, thus increasing control and wingstroke power. This key aspect of the evolution of the wing—indicating that, in opposition to predictions made by the arboreal theory, the flight surface evolved from the tip of the forelimb towards its base—is also illustrated by the development of the wing in living chicks. The design of the wrist of maniraptorans (including birds) adds yet another caveat for ideas envisioning a gliding phase. As opposed to the rigid wrist of gliders, the swivel wrist of maniraptorans has always been a flexible and highly mobile joint. Not only is the structure of the maniraptoran forelimb at odds with the idea of a gliding phase, but the large muscle attachments of the sternum and humerus, and the existence of a glenoid (shoulder joint) limiting the movements of the wing to those involved in flight are also features typical of flapping birds that are absent in gliders.

Even if Chatterjee's arguments do not seem to take us beyond those previously used in favor of tree-climbing specializations in nonavian theropods, the existence of such an ability should not be dismissed so quickly. A notable finding is *Epidendrosaurus ningchengensis*, a sparrow-sized nonavian maniraptoran from Inner Mongolia, just west of the Liaoning fossil sites. Although the age of this fossil is still unclear, the significant faunal differences with respect to the better-dated Early Cretaceous rocks of Liaoning have led many Chinese paleontologists to believe that *Epidendrosaurus* lived during much earlier times, possibly the Late Jurassic. Little is known about this creature, since the only reported specimen represents an incomplete juvenile whose bones are mostly preserved as impressions. It has a toothed skull, a long tail, and forelimbs that appear longer than the hindlimbs and carry a third finger much longer than that of any other theropod. The presence of such a unique hand has led Chinese paleontologists to argue that *Epidendrosaurus* was a tree-climbing theropod. The fact that its first toe is also long and lowered, and that the remaining toes have the long penultimate phalanges typically found in grasping extremities, has also contributed to this interpretation.

Very similar to *Epidendrosaurus* is the tiny *Scansoriopteryx heilmanni*, an animal also known from a single juvenile, albeit reputedly from the Early Cretaceous of Liaoning. In light of their physical resemblance and the fact that these two fossils are likely of the same species, the temporal difference between their fossil localities is somewhat puzzling. Precise stratigraphic provenances, however, are well-documented for only a fraction of the dinosaurs unearthed from the Late Jurassic–Early Cretaceous of northern China, because most these fossils have been collected by farmers instead of paleontologists—indeed, some Chinese researchers believe that *Scansoriopteryx* is from the same locality as *Epidendrosaurus*. Most important for our present discussion is that the

MICRORAPTOR: MERCURY OF THE DINOSAUR WORLD

Most dinosaur specialists never imagined they would witness the day a feathered dromaeosaurid would be found—let alone a dinosaur with long flight feathers attached to its hindlimbs. This completely unexpected discovery came to light early in 2003, with the report of six fossils of the pheasant-sized *Microraptor gui*, all from the extraordinary Early Cretaceous deposits of Liaoning, in northeastern China.

Fossils of *Microraptor* display a number of dromaeosaurid features, including the peculiar tail of these dinosaurs—and the skeletons are dressed in full plumage. Large pennaceous feathers are attached to the forelimbs, and the distribution of these feathers resembles that of modern birds: the distal feathers (primaries) are the longest, those from the tip are parallel to the bones of the hand, and feathers increase their angle of attachment as they progress towards the base of the wing. Most remarkably, the primaries are distinctly asymmetric, with their leading vanes much thinner than the trailing vanes. Some degree of vane asymmetry is also visible in the distalmost secondaries—feathers perpendicularly attached to the ulna—although the vanes have approximately the same width in most of these feathers. Covertors of

various sizes complete the ample wing of *Microraptor*, which even shows a tuft of feathers attached to the first finger, a structure reminiscent of the bastard wing or alula of birds. Long pennaceous feathers also attach to the distal portion of the tail in a frond-like manner. Yet the most remarkable feature of *Microraptor* is on its hindlimbs, because this remarkable animal also has long pennaceous feathers projecting from its lower leg, a sort of avian version of the heel-winged Mercury of the Roman pantheon. Indeed, *Microraptor* has complicated the life of paleontologists—a plausible explanation for the function of its leg feathers is still wanting. These feathers are attached to both the metatarsals and the tibia, and are set in a pattern similar to those of the wing: the distalmost feathers are the longest and have asymmetrical vanes.

Based on its small size and the assumption that other theropods close to the ancestry of birds were tree-climbers, Xu Xing and collaborators from the Institute of Vertebrate

Paleontology and Paleoanthropology in Beijing—the Chinese team who reported the discovery—interpreted *Microraptor* as an arboreal, and presumably tree-climbing, animal. Assuming that by rotating its hindlimbs sideways this animal was able to create an additional set of "wings," these paleontologists also claimed the existence of a gliding phase in the origin of avian flight. These conclusions, however, are not straightforward. Interpretations of tree-climbing theropods are yet to provide solid anatomical evidence in support of this habit. Furthermore, the architecture of the femur of *Microraptor*—whose ball-shaped head is fitted inside the hip socket as in all other theropods—suggests that the feathered legs of this animal could have not rotated sideways by 90 degrees without dislocating.

Microraptor gui ranks among the most perplexing fossil discoveries of recent decades. The design of its hindlimbs will certainly fuel heated discussions and be the center of controversies for many years to come. For the time being, we may have to accept that we simply do not know the function of these peculiar hindlimbs, nor do we understand their possible significance in the origin of flight. (Image: U. Kikutani.)

features used to reconstruct these tiny animals as tree-climbers—the length of the third finger and the phalangeal proportions of the foot—need not be specializations for climbing tree trunks. Even though the extreme elongation of the third finger of these specimens calls for a specific function, the proportions of its phalanges are at odds with those of extremities usually assumed to be apt for grasping. In both *Epidendrosaurus* and *Scansoriopteryx* the three phalanges that precede the claw of this finger become progressively shorter towards its tip, as opposed to the penultimate phalanx being longer than the first phalanx as in hands and feet specialized for grasping. Likewise, the design of the feet of these dinosaurs cannot be distinguished from that of feet specialized for grabbing food.

Other recent fossils emblematic of this notion of tree-climbing theropods include those of *Microraptor gui*, a spectacular Liaoning animal heralded as a four-winged dinosaur (see Microraptor: *Mercury of the Dinosaur World*), and the more poorly characterized *Microraptor zhaoianus*. Ranging from the size of a crow to that of a pheasant, these very similar animals display features that suggest they are primitive dromaeosaurids. They have strongly arched claws on both hands and feet, and *Microraptor zhaoianus* seems to have a partially opposable first toe that could have been used as an incipient perching device. If, as highlighted earlier, the functional reading of these features is inconclusive, interpretations of these theropods as tree-climbers are even more open to doubt. Furthermore, the small size, well-developed wings, modern pattern of feather arrangement, and general skeletal design of *Microraptor gui* suggest that this animal (and presumably also *Microraptor zhaoianus*) may have been able to fly, thus weakening the argument that it had to climb in order to get to the treetops.

As with any other attempt at reconstructing the lifestyle of an extinct organism, the question is not whether the earliest birds or their most immediate relatives were able to climb trees but whether we have conclusive evidence in support of this interpretation. Even if the ecological milieu of early birds remains unclear, arguments contending that *Archaeopteryx* and other very primitive members of the group were tree-climbers are weak. Likewise, arguments in support of similar lifestyles for their nonavian maniraptoran relatives are not much stronger. Despite the fact that these theropods had the grasping capabilities and body sizes that could have enabled them to become tree-climbers and to have had an arboreal existence, the features used in favor of this scenario remain inconclusive, since none of these animals show traits common to animals that cling to trees. Furthermore, interpretations of the origin of avian flight need to be underpinned by a specific and well-supported hypothesis identifying birds' closest relatives, but the precise position within the theropod family tree of some of these fossils (for example, *Epidendrosaurus* and *Scansoriopteryx*) still needs to be established. We may find out that some of these purported arboreal maniraptorans are not so close to the ancestry of birds, making them less relevant in deciphering the origin of avian flight. Naturally, all this does not mean that some of the dinosaurian cousins of birds could not have ventured into trees—just that we do not have conclusive evidence of it.

Conquering Gravity: The Cursorial Origin of Bird Flight

At about the same time that O. C. Marsh was formulating his ideas on the arboreal origin of bird flight, his assistant at Yale University, S. W. Willinston, was engendering the idea that flight had evolved from running animals that, by leaping and jumping from elevated points, became airborne. This view was further expanded by the eccentric

Baron Franz von Nopcsa, a remarkable Hungarian paleontologist with a colorful life that among other things involved espionage, plotting to become the ruler of Albania, and a 5,000-kilometer-ride on a motorcycle. Nopcsa had a much better understanding of the functional significance of the wings in animals that were still ground-dwellers. In 1907, he compared the flapping of wings to the movement of a pair of oars, suggesting that by this action, a cursorial animal—a theropod dinosaur—could have substantially increased its running speed. Nopcsa argued that during their evolution, these flapping predecessors of birds increased their running speed until at a certain point they became airborne. Nopcsa defended this view throughout his prolific career, although he did not live to see his idea embraced (he shot himself in 1933). His views were criticized by a scientific community already entrenched in the notion that flight had evolved through arboreal and gliding phases, and hindered by the apparent absence of living analogs for the functions he was proposing. Besides, the theropod theory of bird origins was at the time losing ground because theropods were thought too specialized to be the ancestors of birds.

It was not until John Ostrom's revitalization of the theropod ancestry of birds, in the early 1970s, that the idea of a cursorial origin of flight gained strength. Yet instead of attempting to resurrect the already discredited notion of Nopcsa,

Ostrom proposed the idea that flight came about as the result of the incremental development of feathers and flapping in animals that used their wings to trap insects. Regarding *Archaeopteryx* as essentially a flightless cursor very close to the earliest stages of flight, Ostrom envisaged the earliest bird running and beating its wings to knock down the insects it presumably ate. This rather awkward adaptive scenario did little to promote the "ground-up theory." Few were convinced by this far-fetched idea, which among other things would have likely damaged the very feathers proto-fliers needed in order to become airborne. In time, Ostrom conceded and abandoned this view—but he did not cast off the idea that flight could have evolved from the "ground up."

With the advent of cladistic methodology in the mid-1980s, as the evolutionary relationships of birds with other reptiles grew clearer, it became more evident that key flight features were already present in the flightless maniraptoran relatives of birds. These developments were spearheaded by Jacques Gauthier of the California Academy of Sciences (who is now at Yale University) and University of California (Berkeley)'s Kevin Padian. Gauthier's seminal 1986 study of the genealogical interrelationships of archosauromorphs strongly argued for the common ancestry of birds and maniraptorans such as dromaeosaurids and troodontids. As

At the beginning of the 19th century, Hungarian paleontologist Franz von Nopcsa speculated that nonavian theropods could have increased their running speed by flapping their protowings. (Image: American Museum of Natural History.)

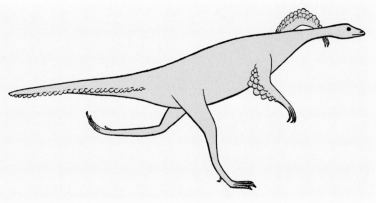

a result of this study and functional comparisons between birds and theropods, Gauthier and Padian showed that dromaeosaurids and troodontids already had the sideways-flexing wrist joint that in birds is critical to the wingbeat. The key bone that allows this movement is the semilunate carpal—a crescent-shaped bone that caps the upper ends of the metacarpals—which Ostrom had for the first time documented in the wrist of the dromaeosaurid *Deinonychus*, leading to his arguments in favor of the theropod ancestry of birds. This wrist configuration, unique to birds and other maniraptoran theropods, allows the hingelike flexion necessary for the hand to swivel, a key movement both during the upstroke–downstroke cycle of the wingbeat and when folding the wing against the body. Gauthier and Padian saw the thrust-generating movement of the forelimb of maniraptorans as a preying–seizing stroke, but they also argued that with only a slight adjustment of the angle of attack—the angle between the airfoil and the direction of movement—the predatory stroke of these theropods would have created thrust of aerodynamic significance. These researchers suggested that by performing those strokes while running and jumping, the predecessors of birds could have gradually evolved their ability to extend their leaps, eventually developing flying capabilities. Unlike the scenario proposed by the traditional arboreal theory, the anatomical and aerodynamic prerequisites for ground-up flight had now been discovered in animals identified as the closest relatives of birds.

Nopcsa Redeemed: A New Cursorial Model Comes to Light

This integration of genealogical, anatomical, and functional evidence was the starting point for a 1999 study I conducted with Philip Burgers, a privately employed aerodynamic engineer affiliated with the San Diego Natural History Museum. The initial goal of our analysis was to test whether the cursorial model of the origin of birds was aerodynamically flawed, as some had previously argued. Specifically, concerns about the plausibility of this model had focused on the apparent discrepancy between the estimated maximum running speed of *Archaeopteryx* (7 kilometers per hour for a 200-gram individual) and the estimated minimum speed this bird would need to achieve to become airborne (21 kilometers per hour). Our study was intended as a way of reconstructing the aerodynamics involved in the take-off run of *Archaeopteryx* up to the moment it lifted off the ground.

The initial assumption of our model was that *Archaeopteryx* was able to run while flapping its wings. Its hindlimb architecture, the sideways orientation of the wing socket, and the design of its wrist strongly support these interpretations—*Archaeopteryx* clearly

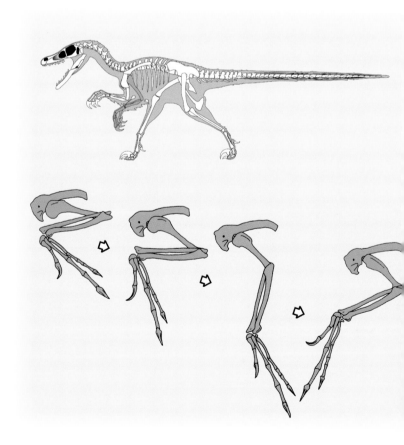

The anatomy of the shoulder and forelimb of dromaeosaurids suggests that these animals could have moved their forelimbs in a fashion similar to the movement of the wing of a bird while flying. This indicates that in the transition from nonavian dinosaurs to birds, the evolution of the "flight" stroke preceded the origin of flight. (Image after Gauthier and Padian, 1985.)

had the mechanical requisites for running and generating the downstroke–upstroke movements involved in the wingbeat cycle. Furthermore, nothing seems to suggest that its flight musculature was insufficient for flapping its wings, even if the energetic output of these muscles—as inferred from the absence of a bony sternum and the structure of the shoulder—may not have been sufficient for taking off from a standstill position. *Archaeopteryx* probably had to run in order to take off, flapping its wings much in the fashion of pelicans, flamingos, and

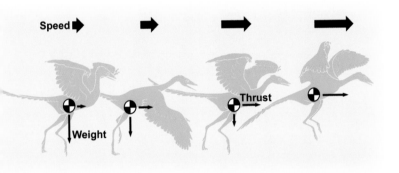

other large birds, a locomotor behavior known to increase both speed and lift.

Our model thus begins by imagining the take-off run of a 200-gram *Archaeopteryx* as it tries to lift off on a quiet morning, 150 million years ago. As the bird begins to run it also starts flapping its wings. The initial forward thrust supplied by the hindlimbs is gradually replaced by thrust generated by the flapping wings. Because the flapping wings also produce lift, even in a running animal, the action of the wings also makes its body lighter. The calculations of our model predicted that this dual force migration, the forward propulsion and the weight support migrating from the hindlimbs into the wings, would have profound implications for the maximum running speed of *Archae-opteryx*. Since the wing thrust is larger than the propulsion generated by the hindlimbs, flapping increases the bird's running speed; and since the lift produced by the flapping wings unloads the weight

of the body from the hindlimbs, our *Archaeopteryx* is able to run even faster. As its speed increases, so does the lift generated by the flapping wings. At a certain point in the run-up, this lift becomes greater than the bird's weight, and at this point, our *Archaeopteryx* takes off.

Previous estimations of the maximum running speed of *Archaeopteryx* assumed that only its hindlimbs generated propulsive force and provided support for its weight during the take-off run. The calculations and analyses of the forces involved in our reconstructed take-off scenario showed that these estimates overlook a very plausible behavior—that the bird could have run while flapping its wings. When the proposed force migrations resulting from this behavior are considered, *Archaeopteryx* can reach and even exceed the estimated minimum running speed for take-off of 21 kilometers per hour. Indeed, our calculations suggested that a 200-gram *Archaeopteryx* could have been able to run as fast as 27 kilometers per hour by means of the additional thrust of its wings. This value may seem high for such a primitive flier, but the speed it needed for taking off was significantly lower. In fact, the estimated 21 kilometers per hour a 200-gram *Archaeopteryx* required to become airborne is substantially lower than the cruising speeds of similarly sized modern birds. Furthermore, it is also possible that take-offs could have been attained at even lower speeds. Any take-off was likely assisted by an aerodynamic phenomenon known as ground effect—a thrust-generating interference between the ground and the flapping wing surfaces of any running bird that has not yet become airborne—and the minimum speed required to become airborne could have been substantially slower if early birds such as *Archaeopteryx* took off against the wind.

Even though our study focused on *Archaeopteryx*, its conclusions can be also applied to the forerunners of birds. Nonavian maniraptorans such

as *Caudipteryx*, *Sinornithosaurus*, or *Protarchaeopteryx*, show evidence of well-developed feathered airfoils attached to their forelimb bones. The shoulder sockets of several nonavian maniraptorans had already rotated sideways, thus allowing the forelimb to execute high-amplitude flapping movements, and the architecture of the wrist of these dinosaurs indicates that their hands were able to perform the coordinated movements involved during the downstroke–upstroke cycle. Because the structures necessary for wing-generated thrust were already present in the flightless ancestors of birds, we can infer that the functions necessary for generating this thrust were also in place. Even if the thrust generated by the flapping forelimbs of a running nonavian maniraptoran was smaller than that generated by the more developed wing of *Archaeopteryx*, any thrust would have been an advantage to animals either pursuing prey or avoiding predation. Our study thus argued that the vaned feathers and flight stroke of birds evolved in the context of terrestrial thrust, a force that enhanced the running speeds and control of the theropod forerunners of birds and their ability to extend their leaps. An equally important corollary of

our study was the concept that wings did not have to be fully developed to be functional (even the protowings of chicks produce thrust). Such a concept countered the adaptive arguments of arborealists critical about the selective value of proto-wings in ground-dwelling animals—organisms for which these wings had no apparent purpose.

Interestingly, a recent study on chukar partridges and other species of ground-dwelling birds by Kenneth Dial, a functional morphologist from the University of Montana in Missoula, documented a behavior similar to the one we had argued for our *Archaeopteryx* model. Using high-speed footage, Dial recorded hatchlings of these birds running up slopes of over 45 degrees while beating their tiny wings; and more mature chukars using this wing-assisted incline running to even climb vertical walls and overhanging ledges. Careful examination of the high-speed movies revealed that when climbing steep slopes these birds move their wings differently from when they fly—chukars running up an incline flap their wings in a front-to-rear direction. This behavior was further documented in a subsequent study

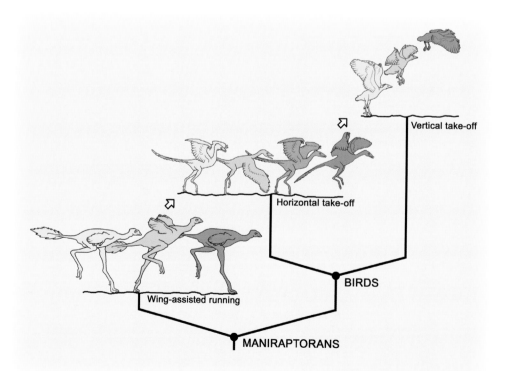

Avian flight could have originated as heavier nonavian maniraptorans capable of wing-assisted running became smaller and their feathered forelimbs increased in size. Wing-assisted running could have led to the horizontal take-off inferred for *Archaeopteryx*, and as the flight system became further refined, birds would have evolved their capacity for vertical take-off. If avian flight began in such a fashion, its evolution did not need to go through an arboreal-gliding phase. (Image: S. Abramowicz.)

Vertical take-off

Horizontal take-off

Wing-assisted running

BIRDS

MANIRAPTORANS

High-speed footage of chukar partridges has shown that these birds can climb vertical walls using the thrust and traction generated by their flapping wings. The graph shows how the contribution of the wings to forward propulsion increases (and even surpasses the amount of propulsion generated by the hindlimbs) as the bird ascends steeper inclines. (Image after Bundle and Dial, 2003.)

in which Dial and his collaborators analyzed the forces involved in wing-assisted incline running. This later study observed that during the initial phases of the downstroke, the wing generated large forces oriented upwards (lift) and in the direction of movement (thrust), and that in the later phases of the downstroke, these forces were directed towards the ground, enhancing traction. This investigation also found that as the birds engaged in wing-assisted incline running, the contribution of their forelimbs to forward propulsion

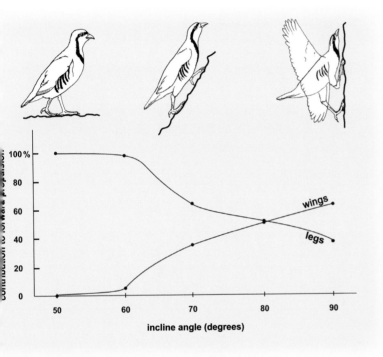

became greater, while the role of their hindlimbs was reduced significantly. Although Dial's study emphasized the enhanced traction resulting from wing-flapping, this action also generates thrust that assists the chukars in their cursorial locomotion. Dial has since documented this unexpected behavior in a variety of birds, including chickens, pheasants, turkeys, and megapods, which in the wild use wing-assisted incline running to ascend trees, rocks, and other heights. And so, a

century later, Baron Franz von Nopcsa has been redeemed. Not only has the advantage of wing-assisted running in theropods been demonstrated by aerodynamic calculations, but similar behaviors, whose apparent absence in living animals had fueled the arguments of his critics, have been discovered in a great variety of birds.

Size Reduction: A Critical Step in the Origin of Flight

If there is one point of agreement among researchers attempting to explain the origin of avian flight, it is that the evolutionary precursors of birds had to reach sizes comparable to those of *Archaeopteryx* or smaller before flight could be achieved. It is not difficult to see why size reduction must have played such a crucial role in the evolution of flight. Miniaturization of the ancestors of birds would have lightened their bodies, thus requiring substantially less energy to propel them, both on the ground and in the air. Furthermore, a decrease in size would have also reduced the value of an important flight parameter—wing loading—adjusting it within optimal aerodynamic values. Wing loading is defined as the ratio between the weight of an animal (or a plane) and its wing surface, which is calculated as the surface of the fully extended wings plus the surface of the body between them. This parameter is the most important aspect of the design of an airplane, because the relationship between the weight of the plane and the size of its wings not only determines its cruising and stall speeds but also how well the plane handles turbulence. An animal with a high wing loading (that is, one with a heavy body and small wings) is far from being the fittest flier—all the flightless birds, from the Kiwi to the Galapagos cormorant, have large values of wing loading. Lower values of wing loading can be achieved by enlarging the wings. Nonetheless,

THE CHIMERIC *ARCHAEORAPTOR*:
A PHONY MISSING LINK

A by-product of the discovery of the Jehol Biota and the prices its exquisite fossils fetch in the international market is fossil forgery. Anyone who has ever traveled to China and experienced the parade of fake art and antiques—from endless copies of Xu Beihong's horses to the chubby ladies of the Tan Dynasty—can easily imagine the new burst of paleontological counterfeit. By far, the most publicized case of recent forgery is the chimeric *Archaeoraptor*—a fossil that brought disappointment, shame, and legal battles to some professionals.

In a prominent article published by the *National Geographic* magazine in 1999, a peculiar fossil with the frontquarter of a flying bird and the long tail and raptorial foot of a small nonavian theropod was hailed as the ultimate proof for the dinosaur–bird evolutionary connection. The fossil had been purchased that same year at the Tucson Gem and Mineral Show in Arizona, but the report had the credentials of well-respected scientists and the natural endorsement of an organization long devoted to scientific discovery. Soon after its announcement, however, it became clear that this remarkable animal was in fact a hoax. The first hint was provided by the careful examination of Chinese scientists, rightfully upset at the fossil having been smuggled out of China, who discovered that the tail of *Archaeoraptor* was nothing else but the counterpart of the slab containing the holotype of *Microraptor zhaoianus*. Their claim was subsequently supported by a computerized tomography study that revealed a pattern of 88 small slabs making up the mosaic fossil. Further examination determined that the forgery had been made by combining the counterpart of the tail of *Microraptor*, the leg of another specimen of this dromaeosaurid (the two legs of *Archaeoraptor* were in *fact* the split slabs of this single leg arranged into a "natural" position) and a good portion of a fossil of the primitive ornithuromorph *Yanornis martini*, a fish-eating bird from the Jehol Biota. In the end, the *Archaeoraptor* scandal highlighted the predicament of modern paleontology in the face of a growing appetite for fossil collectibles and powerful modern technologies for paleontological counterfeiting. (Image after Rowe et al., 2001.)

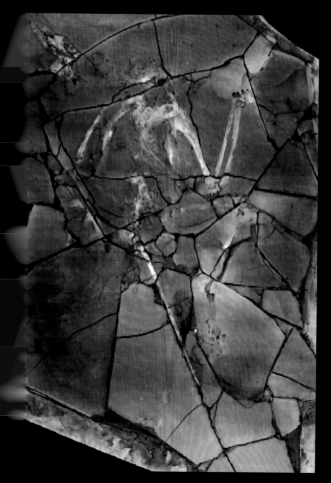

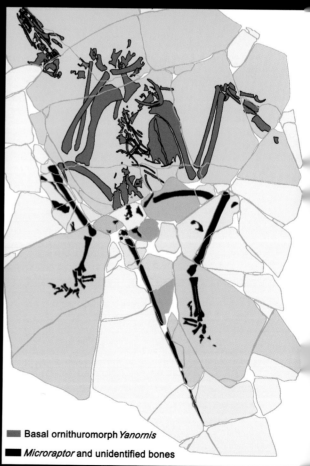

■ Basal ornithuromorph *Yanornis*
■ *Microraptor* and unidentified bones

because the size of a wing increases as the product of two dimensions, low values of wing loading can be more rapidly achieved by reducing the volume—the size, and hence the weight—of an animal, because its volume would decrease as the product of three dimensions.

An overall trend in size reduction is seen during the transition from nonavian maniraptorans to the most primitive birds. In fact, it is becoming apparent that a substantial reduction in size preceded the origin of birds. New fossils of basal members of several groups of maniraptorans have documented that most of these primitive members had sizes smaller than a meter—*Microraptor* is significantly smaller than its dromaeosaurid relatives, and new discoveries of primitive troodontids and oviraptorids also point at small sizes for these animals. José Luis Sanz and his team from the Universidad Autonoma in Madrid (Spain) calculated the wing loading of *Caudipteryx* to be twice the value of this parameter in *Archaeopteryx* and the wing loading of the latter to be substantially higher than that of more advanced Cretaceous birds. Sanz's estimations of weight and wing surface from the fossils were derived from statistical analyses of measurements taken from extant birds. Although using information from living birds to estimate the value of these parameters in very different and long-extinct animals is somewhat problematic, the general conclusion reached by the Spanish team seems well-fundamented: wing loading decreased during the nonavian theropod transition to birds. In fact, this transformation most likely occurred as the result of a substantial reduction in the size of the transitional forms, although the development of larger wings also contributed in adjusting the values of wing loading to those suitable for flight.

Even though the reduction in body size during the evolutionary transition from nonavian maniraptorans to birds is well-documented, the process underlying this transformation is not yet fully understood. Nonetheless, estimates of the rates of growth of nonavian dinosaurs and early birds are starting to shed some light on this question. A host of studies on living vertebrates has determined that the bone tissue of animals growing at rapid rates differs from the tissue formed under slow growth rates. The bone tissue formed during periods of rapid growth is rich in blood vessels, but the tissue formed when growth is slow is less vascular and is often interrupted by one or more concentric lines, similar to the growth rings of a tree. Various numbers of growth rings have been detected in the bones of a wide variety of nonavian dinosaurs and early birds, thus implying that the development of these animals alternated between fast and slow periods of growth. The factors determining the alternation of rapid and slow growth rates are not fully understood, but experimentation with animals in captivity and analyses of wild populations point at seasonal cycles—the reduction of daylight during winter and the concomitant decline in metabolic activity—as the most likely cause. As a consequence, growth rings are usually assumed to form annually. The characteristics of bone tissue—its density, cellular composition, and growth rings—are often exquisitely preserved in fossil bones. These microscopic features are usually studied by producing thin sections of the fossil bones, both across their shafts and along their main axes. All these factors have helped develop a field of research known as skeletochronology, which uses the number of growth rings laid down in the skeletal tissue to estimate the age of animals. These studies have been extremely helpful in estimating the ages at which fossil animals died; but limitations in dating individuals that have stopped growing—for example, in attempting to estimate the longevity of dinosaur species—persist. These shortcomings notwithstanding, numerous recent

investigations have produced a wealth of information about growth rates, ages of maturation, and other life history aspects of dinosaurs and other fossil vertebrates. Among these studies, some have provided a plausible explanation for the evolution of small size in the closest relatives of birds.

According to a hypothesis proposed by a team led by Kevin Padian, small adult size was acquired when the rapid rates of growth that characterized the majority of nonavian dinosaurs evolved developmental changes truncating the time it took an individual to reach adult size. This developmental transformation most likely played a key role in the trend towards miniaturization experienced during the transition from nonavian maniraptorans to birds. However, it is likely that developmental truncation was not the only process involved in this important transformation. Regardless,

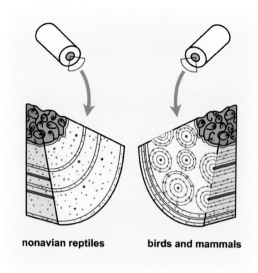

nonavian reptiles   **birds and mammals**

Two basic types of bone tissue. A tissue that is poor in blood vessels and interrupted by growth rings (left) indicates a lower rate of bone deposition and thus a slower growth rate than a tissue that is rich in blood vessels and devoid of growth rings (right). (Image: S. Abramowicz.)

we know now that sizes similar to those of the most primitive birds were attained well before the divergence of the group, a transformation that paved the way for the evolution of birds' most notable gift.

The Origin of Flight: Towards a Synthesis?

We have now seen how a diversity of studies consolidated the notion that birds evolved from bipedal, maniraptoran theropods, which are regarded as predominantly terrestrial animals. Like in most other major evolutionary transitions, from the invasion of land by the earliest vascular plants to the development of limbs and terrestrial locomotion in the earliest tetrapods, novel structures and functions had first to appear for flying abilities to be developed. Such a complex transformation is fortunately well-manifested in the fossils that so clearly document the transition from flightless nonavian theropods to flying birds. These fossils show how many of the structures and functions required for achieving flight originated before these animals became airborne and conquered their new ecological milieu. The development of these attributes suggests that these dinosaurs also evolved behaviors—from flapping wings while running and climbing inclines to controlling the complex movements of the wingstroke while balancing their lightweight bodies—essential for the origin of flight. In all, this evidence suggests that flight most likely evolved as a by-product of functions performed by a suite of aerodynamic structures originated among flightless animals. A plausible functional precursor of flight is wing-assisted running—a behavior many nonavian maniraptorans could have used to both increase their speed and ascend trees, boulders, or other heights.

We began our discussion of the origin of bird flight by highlighting the fact that all these explanations are conjectural by nature and that, as such, they are unlikely ever to be tested. Today, no theory framed within the maniraptoran ancestry of birds can be summarily excluded. The origin of flight from ground-dwelling maniraptorans that run while flapping their wings is plausible. Similar behaviors have been documented in a variety of living birds when ascending steep inclines. It is also possible that flight evolved from the "trees down" in some maniraptorans that had become tree-climbers and arboreal. In addition, it is feasible that flight evolved in contexts somewhere in between these opposing views. Powered flight need not have evolved on level ground, since the effects of running and leaping could have been enhanced if these jumps were performed from ridges, fallen trees, and other small heights; and flight could have also been assisted by specific behaviors such as running against the wind. All we can say with confidence is that the functional and aerodynamic considerations required for terrestrial animals to become airborne were present in the theropod forerunners of birds, and that gravity-assisted explanations are not necessarily the only approach to explaining the beginnings of birds' most notable function.

A classic rendering of the nonavian coelurosaur *Compsognathus* preying on *Archaeopteryx* by the renowned painter of prehistoric life, Charles R. Knight. Today we know that a variety of nonavian coelurosaurs were feathered and that they had evolved many of the structural prerequisites for achieving flight. (Image: American Museum of Natural History.)

CHAPTER THREE

The Bizarre *Mononykus* and its Kin: An Evolutionary Experiment

Paleontologists have long observed that the origin of a major group of organisms is accompanied by a wide range of evolutionary experimentation in which closely related lineages approach in greater or lesser degree the usual trademarks of the new group. Sometimes these lineages become genealogical wildcards, because the similarity they exhibit with respect to the novel group makes their placement in the family tree change from one study to the next. The origin of birds is no exception. In the transition from nonavian theropods to birds, we see several lineages of maniraptorans evolving features typically found among birds, and thus achieving varying degrees of birdness. Perhaps the most remarkable of these lineages includes the short-armed *Mononykus*—one of the few dinosaurs ever to make the cover of *Time* magazine—and its close relatives, a startling group of small to modest- sized maniraptorans recognized only in the last decade.

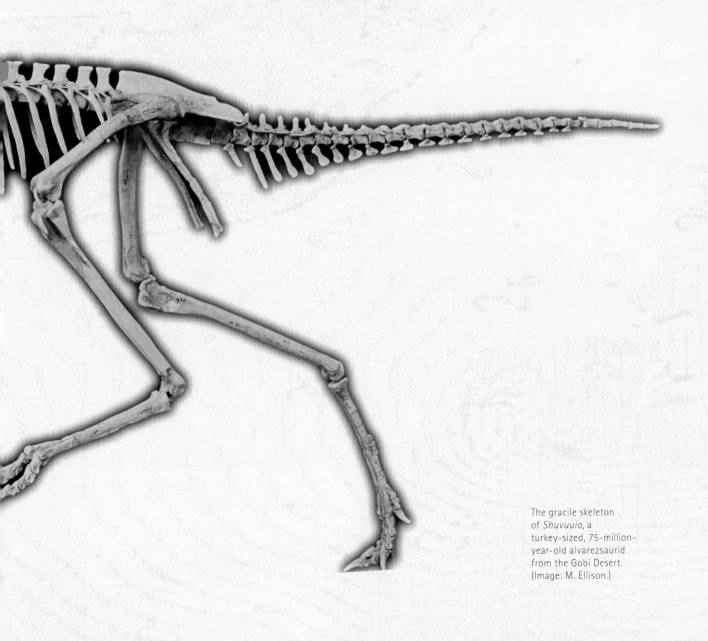

The gracile skeleton of *Shuvuuia*, a turkey-sized, 75-million-year-old alvarezsaurid from the Gobi Desert. (Image: M. Ellison.)

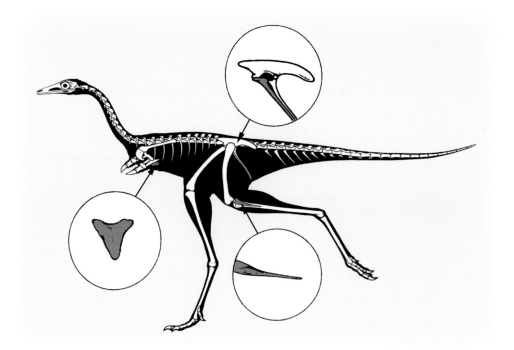

The skeleton of *Mononykus* shows a number of features typical of birds, including a keeled breastbone (left lower corner), a rear-facing pubis (top), and a splint-like fibula (center bottom). (Image after Perle et al., 1993.)

The First Hint: The Discovery of *Mononykus*

In the fall of 1992, Mongolian paleontologist Perle Altangarel was visiting New York City as a guest of the American Museum of Natural History. Perle was carrying an odd-looking fossil: a turkey-sized dinosaur he had discovered a few years earlier in the Late Cretaceous Nemegt Formation of Bugin Tsav, a 70-million-year-old fossil locality in the heart of the Mongolian Gobi Desert. Joint paleontological expeditions of the American Museum of Natural History and the Mongolian Academy of Sciences had been collecting fossils in the Gobi Desert since 1990. One such fossil, an immature specimen from the slightly older Djadokhta Formation, was remarkably similar to Perle's fossil, albeit much smaller.

The bones of these animals were lightly built, and in many respects they looked avian. Yet their stout, short forelimbs, which ended in a massively-clawed thumb, were drastically different from those of any bird. Undoubtedly, these fossils belonged to a highly specialized theropod lineage never before seen—but what was its precise place within the dinosaur family tree? Together with fellow Gobi expeditioners

Mark Norell and James Clark (George Washington University), Perle and I conducted a study of this lineage, and our conclusions surprised us as well as many of our colleagues. Our research suggested that these Mongolian fossils were evolutionarily closer to modern birds than was *Archaeopteryx*, thus indicating that these bizarre animals were members of a primitive lineage of flightless birds. In 1993, we used both these fossils to name the new species *Mononykus olecranus*, a name we coined in reference to the stout, and apparently single claw of its hand.

After our initial discovery of *Mononykus*, further expeditions to the Mongolian Gobi Desert collected over a dozen specimens of similar fossils (some with skulls carrying a large number of densely packed, tiny teeth) from various sites of the Djadokhta Formation and its equivalent Barun Goyot Formation. Soon, however, we realized that these fossils were somewhat different from the animal Perle had collected from the slightly younger Nemegt Formation. Differences in the design of the humerus, the neck vertebrae, and the astragalus (one of the anklebones adhered to the tibia) together with evidence indicating slower developmental rates led us to name a different species: *Shuvuuia deserti*, after the Mongolian word for bird (*shuvuu*) and

THE CENTRAL ASIATIC EXPEDITIONS:
IN SEARCH OF THE CRADLE OF MANKIND

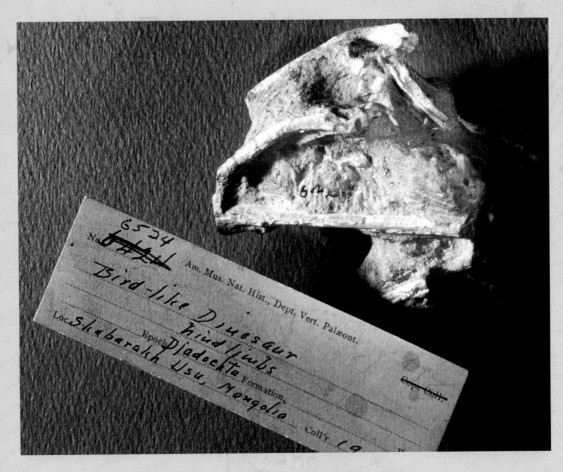

Mongolia is famous for being the home of Genghis Khan, who in the 13th century built the largest empire ever. This central Asian country, however, also encloses most of the Gobi Desert, which for paleontologists is famous for being one of the greatest dinosaur graveyards of all time.

Back in the late 1910s and early 1920s, this land overflowing with dinosaur fossils was hardly known to Western science. It was a time of intense interest in the search for human origins. The discovery of the "young" of Taungs was still to come, the African origin of humans was far from being accepted, and circumstantial evidence had led some researchers to believe that our evolutionary birth could lie in the heart of Asia. One of the supporters of the Asian origin of man was Henry Fairfield Osborn, a remarkable paleontologist, a well-connected aristocrat, and at the time,

the president of the American Museum of Natural History in New York City.

The Central Asiatic Expeditions of the American Museum of Natural History started with a reconnaissance trip in 1919, followed by a number of large-scale expeditions between 1922 and 1930. With his base in Beijing, and supported by a caravan of camels and motor vehicles, expedition leader Roy Chapman Andrews set off to collect specimens of the extinct and extant fauna of central Asia, to survey the geology of the Gobi Desert—and of course to hunt for the cradle of mankind.

The expeditions brought back an enormous wealth of information and specimens. The paleontological results were formidable, although far from significant for the origins of man. Several unknown species of dinosaurs and extinct mammals were found in

the Cretaceous and Tertiary beds of this previously uncharted territory. In 1923, the Central Asiatic Expeditions made their most famous discovery: the complete nest of a dinosaur from a site they called the Flaming Cliffs. Among other important finds were the first fossils of the now famous dinosaurs *Velociraptor* and *Oviraptor*. Less glamorous were the small and incomplete hindlimbs of an animal collected that same year from the Flaming Cliffs. Most likely because of its poor preservation, this fossil remained unprepared in a drawer of the dinosaur collection of the American Museum of Natural History. Almost 65 years later, this fossil was identified as an alvarezsaurid. Although its quality did not seduce the paleontological staff of the American Museum in the 1920s, its avian kinship did not pass unnoticed—the specimen label still reads "bird-like dinosaur." (Image: M. Norell.)

in reference to the semi-arid conditions that also prevailed at the time when it lived. Subsequent discoveries have made this 75-million-year-old alvarezsaurid the best-known member of this enigmatic lineage of theropods.

Alvarezsaurids Rediscovered: The True First Specimens Come to Light

After studying the central Asian collections made by the American Museum of Natural History in the first part of the 20th century, my colleagues and I realized that Perle's *Mononykus* was not the first alvarezsaurid ever discovered. In 1923, an expedition led by the legendary Roy Chapman Andrews (see *The Central Asiatic Expeditions: In Search of the Cradle of Mankind*) had discovered the incomplete fossil of an alvarezsaurid in the Djadokhta beds of the Flaming Cliffs, a renowned dinosaur locality in the Mongolian Gobi Desert. However, as the bones of this fossil were poorly preserved, they were never reported in a scientific forum, languishing forgotten in the museum's dinosaur collection until we brought attention to this lineage. Another fossil collected by the Central Asiatic Expeditions indicated that fossils of this lineage had also been discovered in apparently older, but still Late Cretaceous, deposits of Inner Mongolia, in northern China. Specifically, a single bone of the hindlimb—a fibula—of an alvarezsaurid had been found in 1922 at Iren Dabasu, near Erlian, the northernmost Chinese town of the Trans-Siberian railroad. This time, the specimen had been mistaken for a bone of the 3-meter-long ostrich-like dinosaur *Archaeornithomimus asiaticus*, an

The incomplete skeleton of *Alvarezsaurus*, a primitive alvarezsaurid that lived 80 million years ago in what is today Patagonia, Argentina. (Image: L. Chiappe.)

This well-preserved skeleton of *Shuvuuia* provides evidence of the anatomy of the hand of this peculiar animal. The inset shows the large claw of the thumb, sided by the tiny claws of the middle and outer fingers. (Image: R. Meier.)

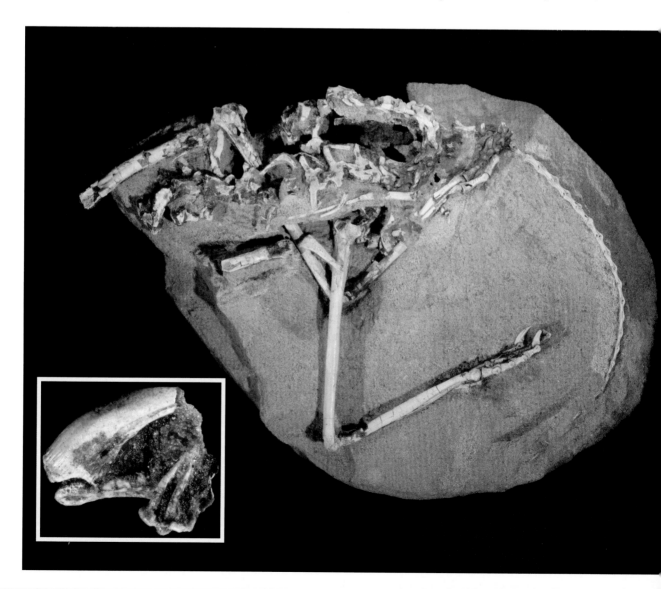

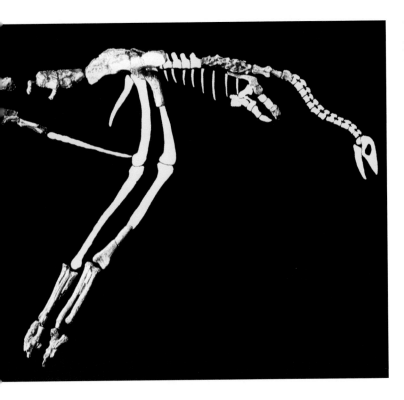

ornithomimid theropod. This mistaken identification not only confirmed that remains of these animals had been discovered many years earlier, but most importantly that these animals had reached sizes substantially larger than the turkey-sized *Mononykus* and *Shuvuuia*.

Southern Cousins: *Alvarezsaurus* and *Patagonykus*

While we were studying these singular fossils from the Gobi Desert, paleontologist Fernando Novas of the Museo Argentino de Ciencias Naturales in Buenos Aires was trying to understand the genealogical connections of two unusual theropods from Argentina. One of them was the 80-million-year-old *Alvarezsaurus calvoi*, an animal that had been unearthed a few years earlier from the Late Cretaceous exposures of the Bajo de la Carpa Formation of Neuquén, a city in northwestern Patagonia. The other had been recently discovered in the slightly older rock layers of a nearby site by an expedition led by Novas. As an expert on carnivorous dinosaurs,

he had identified his new, 85-million-year-old discovery as a maniraptoran theropod, but its precise genealogical relationships still eluded him. Our study of *Mononykus* gave him the clue he needed. Recognizing anatomical similarities, such as the stout and reduced forelimb, in both *Mononykus* and the maniraptoran theropod he had discovered earlier, Novas named his new animal *Patagonykus puertai*. Implicit in this name was his belief that although twice as big and from half a world away, the new Patagonian fossil was a close relative of *Mononykus*. The less-specialized anatomy of the older *Patagonykus*, however, indicated that this animal was more primitive than *Mononykus*—that it was genealogically closer to the common ancestor of the entire lineage. At the same time, Novas realized that the smaller, enigmatic *Alvarezsaurus* was also a primitive member of the same lineage as *Mononykus* and *Patagonykus*, and thus classified these three animals as alvarezsaurids.

More Findings in the Gobi Desert

The American Museum of Natural History was not the only foreign institution collecting fossils in the Mongolian Gobi Desert in the early 1990s. In 1992, a joint Russian–Mongolian expedition collected an incomplete, small dinosaur from the Barun Goyot sandstones of Khermeen Tsav, a site deep in the vast desert, not far from the Chinese border. This specimen remained unreported until four years later, when paleontologists Alexander Karhu and Alexander Rautian of the Paleontological Institute in Moscow recognized it as a close but much smaller relative of *Mononykus*, naming it *Parvicursor remotus*. Although from equivalent rock layers to those containing *Shuvuuia*, the much smaller *Parvicursor* shows anatomical differences that appear to support the conclusion of our Russian colleagues, that this fossil represents a small alvarezsaurid species living in the same arid environments and

at the same time as *Shuvuuia*. However, further studies are needed to be more conclusive, since the smaller size and apparent differences of *Parvicursor* may be age-related.

Other international teams have also collected alvarezsaurids from the Cretaceous deposits of the Gobi Desert. Most of these fossils, if not all of them, belong to *Shuvuuia*. These latest discoveries have afforded important anatomical information and provided a better picture of the appearance of this animal. For example, an exquisitely preserved, 75-million-year-old specimen recently collected by Japanese paleontologists from the Hayashibara Museum of Natural Sciences in Okayama has shown that the alvarezsaurid hand had the three typical fingers of maniraptorans, albeit with tiny middle and outermost digits in comparison to the stout thumb.

Discoveries in North America

Until 1997, alvarezsaurids were known only from the Late Cretaceous of central Asia and southern South America. That year, John Hutchinson, then a doctoral student at Berkeley's University of California, invited me to join him in the study of a small alvarezsaurid from the 65-million-year-old Hell Creek Formation of eastern Montana. The rich fossil-bearing deposits there document the collapse of the coastal ecosystems of the Pierre Seaway with the disappearance of the last known nonavian dinosaurs. Even though the Hell Creek alvarezsaurid was very incomplete, represented only by remains of the pelvis, the rod-like appearance of its pubis and ischium were unmistakable.

Like some of its Asian counterparts, this North American fossil had been found years before alvarezsaurids were identified as a distinct theropod lineage. Discovered in 1980, it remained unnoticed in the collection of the University of California's Museum

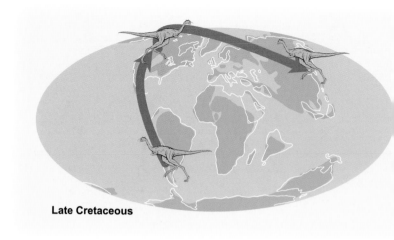

Late Cretaceous

of Paleontology. Since then, other fragmentary alvarezsaurids from the coastal environments of the American Late Cretaceous have been discovered, or identified in museum collections—these fossils are primarily represented by the stout thumb claws that are characteristic of the group. With these findings, the puzzle created by the separate distribution of alvarezsaurids in central Asia and Patagonia has come to an end. A wealth of data indicates that Asia and North America were undivided during the Late Cretaceous, and geological and paleontological evidence supports a geographical connection between North America and South America at this time. North American alvarezsaurids are more similar to their Asian relatives than to their older and more primitive Patagonian cousins, and fragmentary remains of possible alvarezsaurids have also been unearthed from Late Cretaceous rocks in Europe. Thus, the age, geographical distribution, and inferred genealogical relationships of these fossils support the view that Asian and North American, and perhaps European, alvarezsaurids diverged from a common ancestor not shared by their Patagonian relatives. The evidence suggests that this ancestor could have reached North America when some of the most primitive South American alvarezsaurids dispersed north through land connections that existed between these continents in the Late

The genealogical relationships, age, and geographic distribution of the different species of alvarezsaurids suggests that the group could have diverged in South America and migrated to the northern continents using a Late Cretaceous landbridge between the Americas. However, the distribution of alvarezsaurid fossils can also be explained as remnants of a group that was widely distributed over Jurassic Pangea. (Image: R. Urabe.)

Cretaceous. South America appears to have been isolated from the northern continents earlier in the Cretaceous, but because the origin of the alvarezsaurids is possibly much older, perhaps even Jurassic, alternative scenarios explaining the distribution of the Late Cretaceous alvarezsaurids as remnants of a pre-Cretaceous stock of Pangean distribution cannot be entirely dismissed.

Validation of hypotheses involving the geographical origin and dispersal of extinct organisms often require a rich and comprehensive fossil record. The geographical scenario highlighted above is congruent with what is known from the fossil record and genealogical interrelationships of alvarezsaurids, as well as the geographical connections of the continents during the Late Cretaceous;

but undoubtedly, only a fraction of the evolutionary history of these animals is represented by the fossils so far unearthed, and only further discoveries will allow this hypothesis to be fully tested.

Avian or Nonavian? A Controversial Classification

Despite the anatomical resemblance between alvarezsaurids and birds, our initial proposal of avian relationships for *Mononykus*—here referred to as the "avian hypothesis"—was received with much skepticism. Critics claimed that *Mononykus* and its relatives were not birds but nonavian coelurosaurs—the group of dinosaurs that includes the ostrich-like ornithomimids and all maniraptorans. For those endorsing the theropod origin of birds and understanding birds as avian theropods (and coelurosaurs), the difference between the avian hypothesis and the genealogical relationships proposed by these critics was not so dramatic. After all, our initial study argued that alvarezsaurids were the most primitive birds after *Archaeopteryx*. Placing alvarezsaurids among nonavian coelurosaurs required a small switch in their position within the family tree of theropods from being closer to modern birds than *Archaeopteryx* to being further from modern birds than *Archaeopteryx*. However, for those rejecting a dinosaurian origin of birds, the difference between being avian and nonavian was enormous.

From the beginning, the specialized skeleton of alvarezsaurids complicated understanding of their genealogical relationships. Perhaps because the features they uniquely share with birds are expressed in details, most initial objections to the avian hypothesis highlighted the odd appearance of these animals. In other words, critics dismissed the avian hypothesis simply because alvarezsaurids were generally "un-birdlike." Another line of criticism focused on the burrowing lifestyles some had inferred from the mole-like

Remains of North American alvarezsaurids (bottom) have been found in 66-million-year-old rocks of the Hell Creek Formation of Montana. Although very fragmentary, these pelvic remains (pubis and ischium) show great similarity to the pelvic bones of Asian alvarezsaurids such as *Shuvuuia* (center and top). (Image: R. Urabe.)

golia

Montana

UKHAA TOLGOD:
AN EXTRAORDINARY CRETACEOUS ECOSYSTEM

In mid-July of 1993, the joint expedition of the American Museum of Natural History and the Mongolian Academy of Sciences made yet another forced stop: one of the trucks was stuck in the soft terrain of the Nemegt Valley, in the Gobi Desert. Tired after days of cross-country driving, the expedition set up camp on a plateau overlooking the valley. Not far from camp, you could see a small set of brownish badlands, not impressive by Mongolian standards. Even so, the next morning a party went down to prospect these outcrops. Within minutes, they realized that the place was littered with the remains of 75-million-year-old creatures: dozens of skulls of tiny lizards and mammals as well as many skeletons of dinosaurs were found in that single foray! I will always regret missing that exciting first strike. With a badly injured foot, I had stayed back in camp, but the following day found me (foot bandaged and with a shovel as cane) crawling the ravines of Ukhaa Tolgod—"brown hills"—as the team named this locality.

Ukhaa Tolgod is truly an exceptional fossil site. Hundreds of skulls and skeletons of mammals, lizards, and turtles—as well as the complete remains of ankylosaurian, protoceratopsian, and theropod dinosaurs—have been collected from an area of less than 5 square kilometers. Several exquisite skeletons of alvarezsaurids have also been found there. Theropod dinosaurs, often rare in other dinosaur localities, are extremely abundant and diverse. They include adults, juveniles, and embryos of the birds' forerunners—dromaeosaurids, troodontids, oviraptorids, alvarezsaurids, and ornithomimids—as well as archaic birds. This site is also extraordinary for the preservation of its fossils, which in some instances give literal "snapshots" of ancient behaviors. Such optimal preservation suggests rapid burial and minimal disturbance. Sandstorms of catastrophic dimensions are commonly cited as the physical agent responsible for the exceptional preservation in other Mongolian fossil localities, but geologists David Loope (University of Nebraska) and Lowell Dingus (American Museum of Natural History) have argued that the fossils of Ukhaa Tolgod were buried by water-saturated avalanches from enormous dunes, flows fast enough to bury this magnificent menagerie Pompei-style. (Images: L. Chiappe/M. Ellison.)

appearance of the forelimbs of these animals. This argument ruled out the possibility of the avian relationship of these animals simply because of the inferred lifestyle. Certain penguins and burrowing owls notwithstanding—birds that spend a fair amount of their lives underground—this argument overlooked the fact that structures that perform different functions can still share a common evolutionary origin. In other words, arguing that alvarezsaurids could not be birds because they were ground diggers (an assumption that as we will see is poorly supported) is as logical as arguing that our middle ear ossicles cannot have a common evolutionary history with the bones forming the jaw articulation of fish because we use them for hearing and fish use them for feeding purposes.

The Avian Hypothesis Accumulates Support

Discoveries following the initial study of *Mononykus* brought the avian hypothesis additional support. The "un-birdlikeness" of alvarezsaurids was considered trivial by the researchers who, by studying the new fossil discoveries, were becoming familiar

Although the genealogical relationships of the different species of alvarezsaurids are relatively well-understood, the position of the group in the theropod family tree remains unsettled. Alvarezsaurids were initially regarded as primitive long-tailed birds (avian hypothesis; top cladogram) but today, most researchers interpret them as a bizarre, bird-like group of nonavian maniraptorans (bottom cladogram). (Image: S. Orell/R. Urabe.)

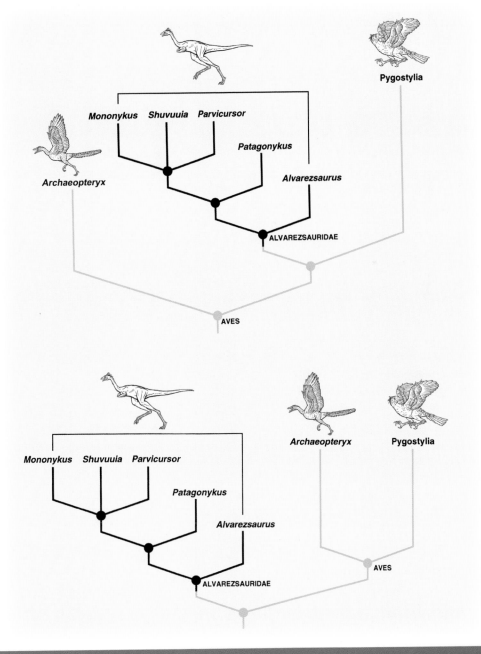

with the anatomy of these animals. In his study of Patagonian alvarezsaurids, Fernando Novas recognized that some of the attributes thought to be shared by *Mononykus* and birds were likely the result of evolutionary convergences, because these features were absent in the more primitive Patagonian fossils. Even so, the result of his extensive comparative analysis was providing the same answer: the simplest evolutionary explanation for the similarities shared between alvarezsaurids and birds was to nest the former within the latter. Just as snakes are considered odd-looking tetrapods, whales unusual mammals, and sea-horses bizarre ray-finned fish, anatomical evidence suggested that alvarezsaurids had to be considered birds regardless of their discrepancies with our stereotypical concept of birds.

For a while, the avian hypothesis continued to survive the scrutiny of other independent studies, but as our knowledge of the primitive members of other maniraptoran lineages increased, this view started to be considered in need of revision. We have seen how cladistic analyses in which characteristics are matched against each other depend on the inclusion of new lineages—statements about the homology of physical characteristics depend on how congruent these statements are with similar interpretations about other characteristics. In the late 1990s, expanded genealogical studies of theropods that included a host of newly discovered fossils—many of them from China and Mongolia—started to undermine the support the avian hypotheses had enjoyed for a number of years. Fernando Novas was the first to relinquish his earlier views of alvarezsaurids as birds, proposing a more basal coelurosaur position for these singular animals. Along these lines, although working independently, Paul Sereno, a paleontologist at the University of Chicago, argued for a close

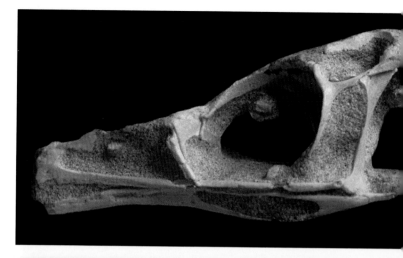

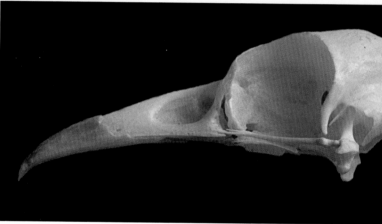

relationship between alvarezsaurids and ornithomimids. In New York, Mark Norell and his associates began interpreting these animals as an early divergence of maniraptorans; and my own investigations placed them as the most immediate theropod split predating the origin of birds.

There is not yet conclusive evidence for either of these views (avian or nonavian), but cladograms placing alvarezsaurids outside birds are now more parsimonious than the avian hypothesis. Yet the alvarezsaurids' position within theropod evolution is clearly not resolved. Crucial information unveiling the kinship of alvarezsaurids is likely hidden in the primitive members of this highly specialized lineage, because very early alvarezsaurids—those near the ancestry of the entire lineage—have not yet been found.

The skull of *Shuvuuia* compared to that of a modern bird. (Image: M. Ellison.)

Burrower or Ant Eater?

The lifestyle of alvarezsaurids, with their short and stout forelimbs, has puzzled paleontologists since the discovery of *Mononykus*. Even if these animals were to be seen as birds, their peculiar forelimbs were obviously hopeless for flying. Would they have been used for defense, mating, predation, or some other action? The small size of these arms would make them unsuitable weapons against the much larger predators that inhabited the same environments. Alternatively, although one could envisage these forelimbs being used by males to grab females during copulation, the absence of sexual dimorphism—such as male specimens sporting larger forelimbs than female specimens—renders this scenario unlikely. The superficial resemblance between these forelimbs and those of ant eaters, moles, and other digging mammals has led some to suggest that these bizarre theropods were burrowers. However, other parts of the skeletons of these animals are quite different from those of digging mammals. Burrowing mammals, for example, have short necks, compact heads with orbits that face sideways, and short hindlimbs that assist the forelimbs in digging. None of these features are present in alvarezsaurids, whose highly movable skulls with enormous orbits, long S-shaped necks, and elongate hindlimbs suggest these animals were agile runners that relied on their vision to find food.

The lifestyles of extinct creatures are always difficult to ascertain. Who would imagine, looking at the skeleton of a Magellan penguin, that this bird digs holes in the ground large enough to house a pair of adults and their hatchlings, and that they spend several months living in these holes? What *is* clear about alvarezsaurids is that, except for the forelimbs, these animals do not show any of the features found in digging animals. In contrast, their elongated hindlimbs with short toes suggest they were swift runners. The large muscles that originated in the strong pelvis and long tail provided the power for fast locomotion. This is particularly evident in the Asian alvarezsaurids, in which the foot was specialized for high speed. Like extant running birds, such as ostriches and rheas, these alvarezsaurids have very long metatarsals and short toes. By a combination of elongated feet and reduced toes, these animals were able to perform longer strides and minimize friction with the ground.

Alvarezsaurids may have used their robust forelimbs in more ways than we can imagine, but it is clear that some of their functions required precise and powerful movements. It is feasible that they used their forelimbs for digging in the context of predatory activities such as breaking off nests of termites or tearing apart the bark or surface of

Forelimb of *Shuvuuia* compared to that of an ant eater. (Image: M. Schwengle/ S. Abramowicz.)

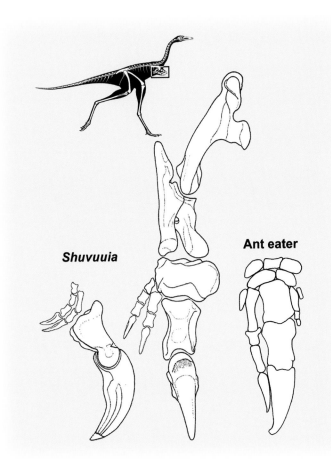

Shuvuuia

Ant eater

As for most extinct vertebrates, little is known of the external appearance of *Mononykus* and its kin. Unlike the fine-grained shales of northeastern China, where feathers and soft tissues are commonly preserved, the coarser deposits of the Gobi Desert and other localities with alvarezsaurids are not ideal for the preservation of such structures. Nevertheless, an investigation led by Mary Schweitzer of the State University of North Carolina has shed important light on this issue. Schweitzer's microscopic and immunological studies of fibrous structures overlying a skeleton of *Shuvuuia* converged to indicate that these tiny fibers are remnants of feathers. Their microscopic appearance was shown to be comparable to rachises and barbs, and the immunological analysis identified them as beta-keratin structures, as are feathers. Schweitzer's studies of this and other fossil dinosaurs have demonstrated that the molecular and structural characteristics of proteins can survive for tens of millions of years and that ancient proteins can thus be detected using the same immunological procedures as modern proteins. The *Shuvuuia* fibers were exposed to antisera specific for different kinds of keratin. The binding of the beta-keratin antiserum to the constituents of the fibers—recognized as a fluorescent signal in a test tube—indicated the nature of these soft structures. Schweitzer's innovative studies documented the presence of feathers in alvarezsaurids and highlighted once again the widespread distribution of this integumentary covering among nonavian maniraptorans. (Image: E. Heck.)

plants to expose insects or their larvae. A recent study of forelimb motion in *Mononykus* has supported this predatory role, equating the function of the stout arms of this animal to the scratch-digging of ant eaters and pangolins. Furthermore, the long and narrow snout and large size of the tongue bones (hyoids) of these animals also agrees with this interpretation, as does the presence of numerous teeth. Creatures with small, multiple teeth are typically associated with an insectivorous diet, and studies on living animals have shown that those species with predominantly insectivorous diets have a greater number of teeth than those that find their nutrition elsewhere.

Birdness in Parallel

The degree of similarity between the lightly-built alvarezsaurids and birds is in some respects more pronounced than that evolved by any other lineage of nonavian theropods. The skulls of these animals have enormous orbits and a long, narrow snout that is loosely connected to the braincase. The shape, joints, and connections of the bones around the orbit suggest that, as in most modern birds, the alvarezsaurid snout moved independently from the braincase, a type of movement called prokinesis. Although the biological role of this functional property is not entirely clear, it is possible that such a sophisticated degree of skull kinesis enhanced the jaw-grasping ability of these animals. Regardless, the snout mobility of alvarezsaurids represents a clear example of evolutionary experimentation towards the avian trademark—which, given their genealogical gap with respect to modern avians, must be interpreted as evolving in parallel.

Other examples of evolutionary convergence towards the appearance of modern birds are evident throughout the skeletons of these animals. In the hindlimbs of the Asian and North American species, the fibula lost its contact with the ankle and evolved the characteristic pin-like shape of present-day birds. In the pelvis of these animals, the ischium approached the length of the pubis—a proportion contrasting with the longer pubis of nonavian maniraptorans and early birds—and the contact between the ends of the rear-facing pubes became greatly reduced. The peculiar sternum of alvarezsaurids also resembles in some ways the breastbones of birds—in *Mononykus* and *Shuvuuia* this bone has a keel-like, downward projection that must have augmented the surface for the origin of the breast muscles that powered the arms. And even in the stout and abbreviated forelimbs of these animals we find details that evoke birdness—one of these is the fusion of wristbones (the semilunate carpal of maniraptorans) and metacarpals forming a compound bone, the carpometacarpus.

The discovery of the alvarezsaurids remains an important event in our understanding of the origin of birds. The history of this process illustrates the scientific basis of cladistics, in which previous hypotheses are constantly tested in light of cumulative new evidence—in the form of additional lineages and physical characteristics that are incorporated into analyses. Beyond methodology, the outstanding evolutionary convergences between these animals and birds are a vivid testimony of the amount of experimentation that characterized the transition from nonavian theropods to birds; and in particular, the evolution of this peculiar lineage of short-armed maniraptorans.

CHAPTER FOUR

Archaeopteryx: The Earliest Bird

Archaeopteryx lithographica constitutes the opening page in the ancient saga of bird evolution. With its first skeleton unearthed soon after Darwin expounded his theory of evolution by natural selection, this archaic, Late Jurassic bird has played a key role in all discussions about the descent of the group and the origin of their flight. Despite being an icon —a paleontological Mona Lisa— this historically and biologically priceless animal is only known by a handful of largely two-dimensional fossils that in many ways are best compared to ancient roadkill. Indeed, our entire knowledge of this popular animal is based on just ten fossil skeletons and an isolated feather, all collected during the last 150 years of quarrying the Solnhofen Lithographic Limestones of Bavaria, in southern Germany.

The London specimen of *Archaeopteryx*. (Image: American Museum of Natural History.)

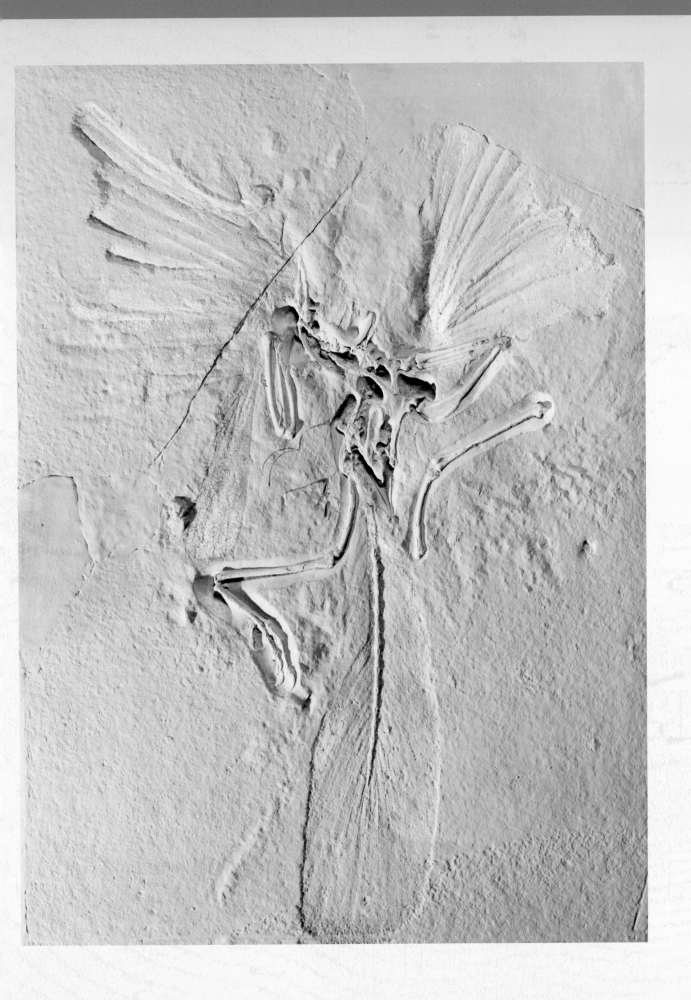

For centuries, the human inhabitants of this region have quarried the flat-bedded limestones for a variety of purposes. Murals and sketches were created on these rocks as early as the Stone Age and they were used for building everything from Roman baths to medieval cathedrals. The limestones are still widely used for construction, but their heyday came in the 1700s with the discovery that some of these compact, fine-grained rocks were ideal for a printing technique known as lithography. This technique involves the differential carving of the rock by acid, creating a bas-relief that, when inked and pressed on paper, leaves a print of the carved motif. Illustrating once again the interconnectedness of history, science, and technology, the development of lithography significantly increased the quarrying of the lime-rich Solnhofen slates, leading to the discovery of a large array of precious fossils and opening one of the most enlightening windows on a Mesozoic ecosystem.

Today we revere the Solnhofen Lithographic Limestones as one of the most complete images of a remote moment in Earth's history. These finely-grained rocks have yielded thousands of fossils, from microorganisms to dinosaurs, illustrating the evolution of nearly every major group of animals and preserving many of their soft structures in exquisite detail. However, despite the multitude of fossils discovered in these ancient rocks, remains of vertebrate animals are uncommon. The quarrying is done entirely by hand, and although some quarries are richer than others, quarrymen often work an entire week before finding a single fossil fish. The discovery of an *Archaeopteryx* is extremely rare.

A Handful of Rare Discoveries

The known history of birds begins near the end of the Jurassic Period, some 150 million years ago. The scientific community first learned of the existence of such ancient birds in the summer of 1860, when a stonemason split in two a slab of lime-rich slate near the Bavarian village of Solnhofen to reveal a nearly complete, blackish feather. More evidence came a year later, when physician Karl Häberlein of Pappenheim, a village 30 kilometers northwest of Solnhofen, received a unique fossil in exchange for his medical services. It was the nearly complete skeleton of a crow-sized bird also unearthed near Solnhofen. Clear impressions of feathers fanned out from its forelimbs, and multiple pairs of equally long feathers branched from its rod-like, bony tail. Häberlein brought the fossil to the attention of Hermann von Meyer, a leading German paleontologist, who in the same year applied the name *Archaeopteryx lithographica* to feather and skeleton, exalting both their ancient age (*Archaeo*) and winged nature (*pteryx*). In 1862, Häberlein sold the skeleton to the British Museum (now the Natural History Museum) in London—allegedly to raise a dowry for the wedding of his daughter. The *Archaeopteryx* skeleton was the centerpiece of a handsome collection of Solnhofen fossils that Häberlein had amassed over years of medical practice, and the British Museum paid a hefty but worthy sum for his collection. This skeleton is today known as the London specimen.

Yet history has shown that these momentous fossils were not the first remains of birds to be discovered in the Solnhofen Lithographic Limestones. As early as 1820, a German article commented on the existence of "a few bird remains and feather impressions in the Sohlenhofer and Pappenheimer calcareous slates." These fossils have not survived to the present but one that did survive was a fragmentary skeleton found in 1855 in a quarry near the town of Riedenburg, some 90 kilometers east of Solnhofen. In 1857, this fossil was classified as the flying reptile *Pterodactylus crassipes* by von Meyer, and three years later it was purchased by the Teylers Museum of Haarlem in The Netherlands. Surprisingly, von Meyer never associated the physical attributes

of this fossil with those of the one he reported only four years later—the London *Archaeopteryx*—and its true identity remained masked until 1970, when John Ostrom discovered the mistake and added to the pride of Haarlem the possession of the only *Archaeopteryx* outside Germany in continental Europe.

After the London specimen, the next *Archaeopteryx* was found in 1877, at a quarry near the town of Eichstätt, which is also not far from Solnhofen. Soon after its discovery, the fossil was acquired by Karl Häberlein's son Ernest, who regarded it as a pterosaur until he chiseled away the limestone—together with a good portion of the fossil itself!—and uncovered the delicate traces of feathers. Further preparation revealed a complete skeleton, with its forelimbs spread wide and most of the plumage of its elliptical wings and tail imprinted on the limestone. In 1881, following years of negotiations with a number of interested buyers, including Yale University's Peabody Museum paleontologist O. C. Marsh, Ernest Häberlein sold the spectacular fossil to the Humboldt University's Naturkunde Museum in Berlin, a transaction financed by Ernst Werner von Siemens, the founder of the renowned German electrical and communications

A Solnhofen Limestone quarry around the end of the 19th century. Today, these limestones are still quarried by hand. (Image: © Bayerische Staatssammlung für Palaeontologie und Geologie, Munich.)

company. The Berlin specimen, as it is known, still resides in this museum and is undoubtedly one of the most famous of all fossils.

Modern Discoveries

More than 80 years passed until the next *Archaeopteryx* was recognized. In 1959, Florian Heller from the Bavarian Erlangen University reported on an incomplete fossil with poor feather impressions that had been found a few years earlier around Solnhofen. For 15 years this specimen resided at the Maxberg Museum near Solnhofen, but it was then returned to its owner, Eduard Opitsch of Pappenheim. Opitsch did not

< Unearthed in 1860 and here photographed under ultraviolet light, this *Archaeopteryx* feather provided the first evidence of the existence of Jurassic birds. Comparisons suggest this feather to be a secondary feather, one which attached to the outer portion of the forearm. (Image: H. Tischlinger.)

∨ The Berlin specimen of *Archaeopteryx* (Image: L. Chiappe.)

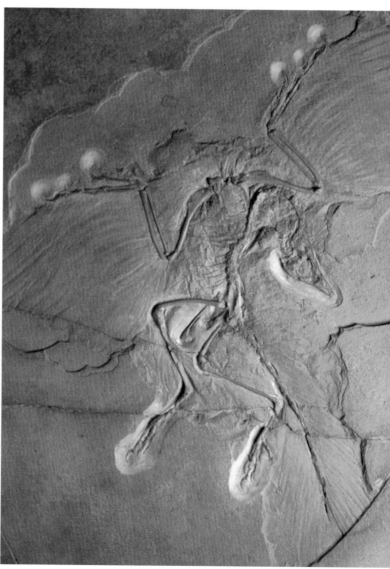

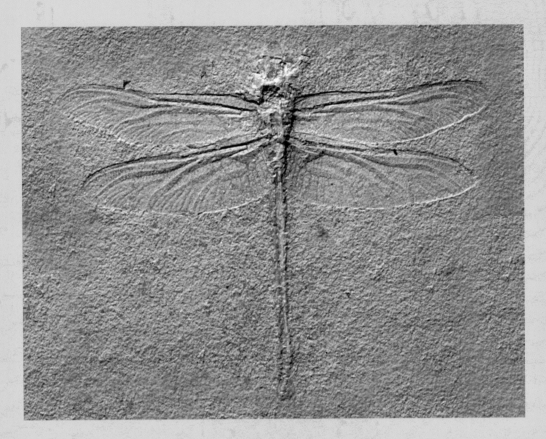

THE SOLNHOFEN LIMESTONES: UNDERSTANDING THEIR EXCEPTIONAL FOSSIL PRESERVATION

At Solnhofen, creatures that died 150 million years ago are preserved in exquisite detail. From the hair-like appendices of tiny shrimps to the soft mantles of jellyfish and the film-like wings of pterosaurs, a host of delicate structures have remained intact in these fine-grained deposits. This extraordinary preservation is intimately linked to unusual environmental conditions that developed in this lagoonal environment. With limited water exchange with the open sea, the bottom of the Solnhofen lagoons was regularly stagnant. The surrounding warm climate led to heavy evaporation, thus increasing water salinity, particularly near the bottom. During periods of high salinity and intense stagnation, the oxygen concentration at the bottom dropped to nearly zero. The salinity and oxygen concentration of the surface was subject to drastic changes through the action of storms mixing surface and bottom waters. Thus, although in normal conditions some fish and other organisms were able to tolerate the less saline surface waters, rapid changes in salt and oxygen concentration led to events of mass mortality. Large clusters of fish lying close together on the same fossil bed, and others with prey intact in their stomachs or even in their mouths, testify to these sudden episodes of death.

Several models for the sedimentary origin of the Solnhofen Lithographic Limestones have been proposed, but a model originally proposed by the late K. Werner Barthel from Berlin's Technical University and later expanded by Günter Viohl of the Jura Museum (Eichstätt) is today the most accepted. This model interprets the sediments of Solnhofen as lime-rich mud formed and accumulated in and around coral reefs. Wave action during occasional storms would have stirred this ooze and washed it into even the most distant parts of the lagoons. Along with these particles came the diverse biota of the coral reefs, and occasionally, animals washed in from the open ocean. From time to time, these powerful storms brought down animals that flew over the lagoons—*Archaeopteryx* and more than a dozen species of pterosaurs are some of the fliers that likely drowned in these stormy waters. The elevated salinity and low oxygen concentration of the water poisoned most of the reef animals that were washed into the Solnhofen lagoons. Some of their corpses sank to the bottom and were rapidly buried by the lime-rich mud. Others remained floating on the surface waters, leading to their complete disarticulation. A few bottom dwellers that survived the poisonous surface waters died of asphyxia in the oxygen-depleted vicinity of the lagoon floors. On the bottom, the same conditions that killed these animals kept the microbial activity very low, thus limiting decay and favoring the preservation of soft tissues.

make it available for scientific study and upon his death, in 1992, it was declared lost. Another *Archaeopteryx* was discovered in 1951 at a quarry near Workerszell, north of Eichstätt, but this complete skeleton was misinterpreted as a juvenile of the nonavian theropod *Compsognathus*. It was not until 1970 that this fossil was properly identified as another *Archaeopteryx*, when the weak impressions of its plumage were revealed by light shone at a low angle across the fossil surface. Smaller than all other specimens, and exhibiting age-related features, the Eichstätt *Archaeopteryx* constitutes the centerpiece of the outstanding collection of Solnhofen fossils displayed at this town's Jura Museum. A much larger *Archaeopteryx*—twice the size of the Eichstätt specimen—has also been found near Workerszell. Being stouter than other specimens and lacking most of the skull, neck, and trunk, the bones of this fossil were also misinterpreted as those of the theropod *Compsognathus* until correctly identified by Günter Viohl of the Jura Museum. Since it resides at the Burgermeister-Müller Museum

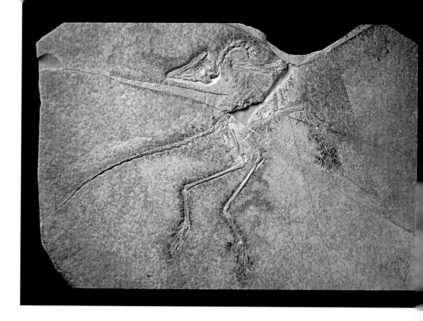

in Solnhofen, this interesting fossil is usually known as the Solnhofen *Archaeopteryx*.

A seventh *Archaeopteryx* skeleton was found in 1992 only a few hundred meters from the discovery sites of the London and Maxberg specimens. Nearly complete and with clear impressions of feathers on its wings and tail, this exquisite fossil has clarified important issues about the earliest bird. For a number of years, it was named after the quarry's owner (the Solnhofer Aktien-Verein Company) but it became known as the Munich specimen in 1999

∧ Slab and counterslab of the Eichstätt specimen of *Archaeopteryx*. The smallest of all known specimens, this fossil is considered to be of a juvenile. (Image: L. Chiappe.)

∨ One of the two slabs of the Munich specimen of *Archaeopteryx*. Some researchers have interpreted this fossil as a species (*Archaeopteryx bavarica*) different from *Archaeopteryx lithographica*. (Image: G. Janssen; © Bayerische Staatssammlung für Palaeontologie und Geologie, Munich.)

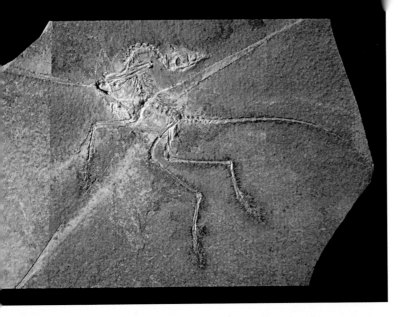

∧ This partially articulated wing is one of the most recent discoveries of *Archaeopteryx.* Although similar in size to the Berlin specimen, close examination of the bone surface reveals a pattern of diminutive pores characteristic of juveniles, thus suggesting that the animal was still growing at the time of its death. (Image: L. Chiappe.)

the forelimbs, this eighth specimen is particularly important because of its age. It is believed that all other known specimens of *Archaeopteryx* lived within a time interval representing less than 20 percent of the 250,000–500,000 years estimated for the deposition of the Solnhofen Lithographic Limestones—but the Daiting specimen is from the somewhat younger Mörnsheim Formation, which extends the known existence of *Archaeopteryx* by hundreds of thousands of years. To date, most scientists have seen only a cast of this specimen, and whether the owner will allow it to be studied remains unknown. However, a more promising future awaits the latest discovery. In 2004 the isolated right wing of an *Archaeopteryx* was recovered from a quarry near Solnhofen. Although still privately owned, this specimen is currently on display at the Burgermeister-Müller Museum. Intermediate in size between the London and Berlin specimens, the wingbones of this ninth *Archaeopteryx* are three-dimensionally preserved, with feathers still attached. Plans are in place for this fossil to become part of the stunning collection of the Burgermeister-Müller Museum, but the practice of privately owning *Archaeopteryx* specimens has sadly become fashionable.

Recently, a specimen long in the hands of a Swiss collector was acquired by a German collector associated with the Wyoming Dinosaur Center in Thermopolis. This tenth specimen consists of a nearly complete skeleton and provides critical information on the anatomy of the skull and feet. Although the specimen is expected to become available for research and to be exhibited at the Wyoming Dinosaur Center, it is still privately owned, so whether it will continue to be available for scientific scrutiny in the future remains uncertain.

when it was purchased by the Bavarian State Collection of Paleontology and Historical Geology in Munich.

Other recent discoveries include an incomplete fossil found in 1997 at a quarry near Daiting, some 12 kilometers southwest of Solnhofen. Although poorly preserved and represented only by a flattened skull and portions of

The Most Reptilian of All Birds

The handful of *Archaeopteryx* fossils constitute the most enlightening testimony of the earliest phases of avian evolution and the only evidence of Jurassic birds. However, the quality of preservation of these treasured fossils varies greatly, and many anatomical details remain obscure. Some *Archaeopteryx* specimens are complete, with portions of their plumage and many bones still articulated—others are much more deficient, with no traces of their integument. As the Solnhofen fossils are found when a stonemason cracks a limestone block in half, they are often split into two slabs and locked in the encapsulating limestone. Their two-dimensional perspective complicates the appreciation of physical details, especially when compared to fossils that, because they have been completely extricated from the surrounding rock, are accessible for viewing from every angle. Despite these shortcomings, the numerous studies have left us with an accurate picture of the anatomy of *Archaeopteryx*, confirming both the archaic nature of this bird and its common heritage with maniraptorans such as dromaeosaurids and troodontids.

Many features of the skull reinforce *Archaeopteryx's* primitiveness. Not only are its jaws toothed, but the configuration of its snout is utterly primitive. Equally notable is the design of the rear portion

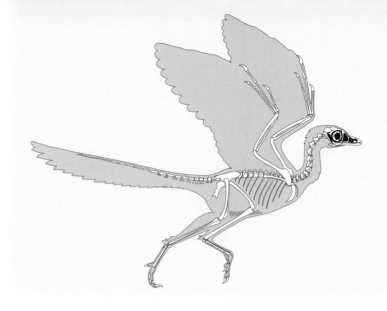

Skeletal reconstruction of *Archaeopteryx*. (Image after Rowe et al., 1998.)

The skull of the Eichstätt specimen of *Archaeopteryx* under ultraviolet light. The inset shows details of its snout and teeth. (Image: H. Tischlinger.)

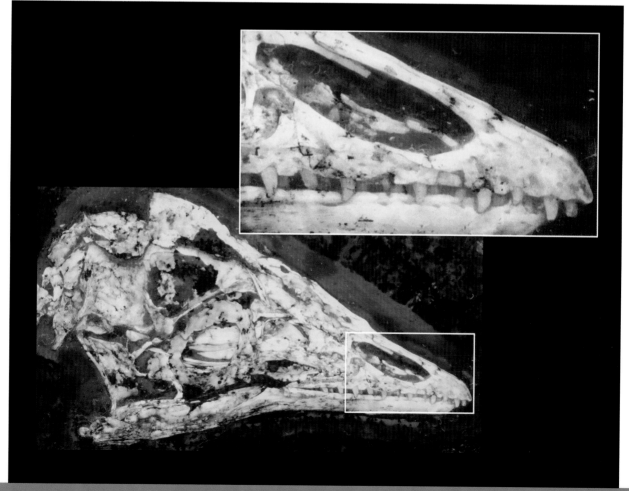

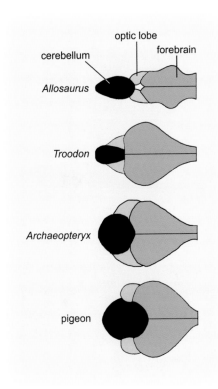

Basic structure of the brain in nonavian theropods and birds. The size of the forebrain, which controls most sensorial perceptions, approximately doubled with the divergence of coelurosaurs (compare *Allosaurus* with the coelurosaur *Troodon*). *Archaeopteryx* shows that even the earliest birds had a cerebellum of substantially increased relative size. The cerebellum controls muscular activity and balance, and in the Late Jurassic bird it already approached the relative size typical of modern birds (pigeon). (Image after Chatterjee, 1997.)

of the skull, which essentially retains the temporal openings—and presumably the associated musculature—of early diapsid reptiles. However, most interesting for our central story of bird evolution is the presence of many features that *Archaeopteryx* evidently inherited from its maniraptoran ancestors. Indeed, a cursory observation of the skulls of these animals shows that the Solnhofen bird inherited the small head, large orbits, and

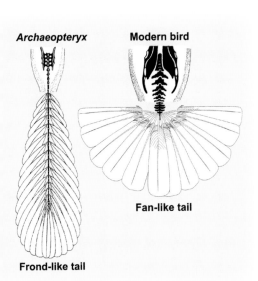

Tail design in *Archaeopteryx* and a modern bird. The frond-like arrangement of the tail feathers of *Archaeopteryx* produced only a fraction of the lift generated by the fan-like tail design of its modern counterparts. (Image after Gatesy and Dial, 1996.)

vaulted braincase of its most immediate relatives. However, it is in the details that the astonishing resemblance between these animals becomes most evident. These features include the configuration of the snout, which is perforated by identical antorbital openings; the presence of small triangular bones called interdental plates that line the tooth spaces inside the lower jaws; and also the smooth-crowned teeth of *Archaeopteryx*, which seem to have had a similar pattern of replacement to the teeth of some troodontids. Despite these holdovers of its theropod ancestry, a number of features of the skull of *Archaeopteryx* highlight its avian nature. The number of teeth decreased substantially and these evolved a waist separating the crown from a barrel-shaped root. Other important innovations evolved in the brain, which became larger and more modern in design. Estimates of the volume of the brain of *Archaeopteryx* indicate that although it was smaller than those of modern birds of comparable size, it was proportionally much larger than those of nonavian reptiles. Furthermore, new computerized tomography studies of the braincase and inner ear of the London specimen have shown that *Archaeopteryx* had an acute sense of vision and that the regions linked to hearing and three-dimensional perception approached those of its living relatives.

The slight departure from its closest maniraptoran relatives is also expressed in the anatomy of the vertebral column. The spine of the Solnhofen bird lacks the modifications typical of modern birds and its neck is comparable in relative length and vertebral number to its most immediate predecessors. The saddle-shaped joins controlling the accurate neck movements of modern birds had not yet evolved—the surfaces of the vertebrae of *Archaeopteryx* are essentially flat. The rest of the vertebral column is also much more primitive than that of its living relatives. The long trunk and short synsacrum provided less strength to the

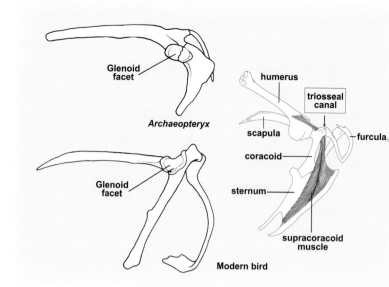

ribcage and pelvis, and its long tail still played an important role in terrestrial locomotion. Yet the tail of *Archaeopteryx* was shorter than that of its forerunners, even if the size of the airfoil was increased by the addition of feathers. With the exception of the first few vertebrae, each tail vertebra anchored a pair of long and symmetrically vaned feathers arranged in a frond-like pattern. Despite the resulting larger airfoil, this design would have produced only a fraction of the lift generated by the tails of modern birds of similar size. The design afforded only a limited ability to spread the feathers and control the overall surface area of the tail. Because modern birds use quick adjustments to their tail area to maneuver, turn, and brake, it is possible that the restricted adjustability of *Archaeopteryx's* tail surface circumscribed its maneuvering ability during flight.

Archaeopteryx's shoulder also remained virtually unchanged from that of its maniraptoran precursors. The coracoid is not much longer than those of these dinosaurs and it is still firmly connected to the scapula. Like in its predecessors, these bones were angled at 90 degrees to each other and constituted an arm socket that faced sideways, both important aerodynamic features that allowed high amplitude wingstrokes—but these bones failed to form the pulley-like (triosseal) canal that in modern birds guides the tendon of the supracoracoid muscle from its origin in the sternum (far below the shoulder) to its attachment near the head of the humerus. In living birds, the pulley-like action of the canal allows this muscle to elevate the wing during the upstroke, an action achieved by the rapid rotation of the humerus upon muscular contraction. The absence of a triosseal canal in *Archaeopteryx* suggests that other muscles must have been responsible for the upstroke of its wings, a conclusion that agrees with the apparent absence of a bony sternum and its relatively large wing loading, since

birds with high wing loading values tend to have weaker muscles for elevating the wings during upstroke. The shape of the furcula (wishbone) also highlights the primitive nature of the Solnhofen bird. Not only is this stout and boomerang-shaped bone similar to those of nonavian maniraptorans, but functionally it is poorly designed for performing the coordinated movements that in modern birds provide additional energy to the

The laterally facing glenoid of *Archaeopteryx* would have resulted in wingstrokes of less amplitude than those allowed by the dorsolaterally facing glenoid of its modern counterparts. The shorter coracoid and the absence of a triosseal canal suggests that the supracoracoid muscle of the Solnhofen bird would have had limitations for acting as the primary elevator of the wing. (Image after Jenkins, 1993.)

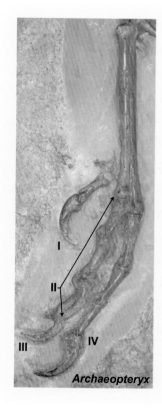

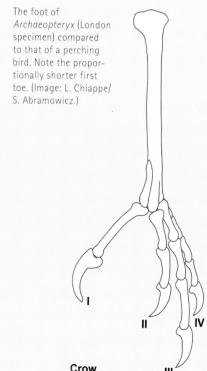

The foot of *Archaeopteryx* (London specimen) compared to that of a perching bird. Note the proportionally shorter first toe. (Image: L. Chiappe/ S. Abramowicz.)

wingbeat cycle.

The clawed forelimb of *Archaeopteryx* is also very similar to that of its maniraptoran relatives, albeit proportionally longer. As with these avian predecessors, the humerus is longer than either the ulna or the radius, and shorter than the hand. Other primitive features are present in the wrist. For example, the carpal bones are not interlocked with the metacarpals, unlike the completely fused carpometacarpus of modern birds. In contrast to the skeletal framework of the wing, more significant changes are seen in its plumage. Not

descendants. For example, more primaries of *Archaeopteryx* attach to the second metacarpal than in extant birds, and the wing lacks an alula (bastard wing). Furthermore, despite the fact that the bases of the primaries and secondaries are covered by rows of contour feathers, the sharp claws of the three fingers project outside the wing plumage. Such a design has convinced many that the claws of the Solnhofen bird could have been partially functional, with grasping, climbing, and preening among some of the possible uses.

More primitive features are evident in the pelvis and hindlimbs of *Archaeopteryx*, parts of the body that also show similarities with those of nonavian maniraptorans. The ilium and ischium are proportionally much shorter than in extant birds and the vertically oriented pubic bones are firmly fused to each other at their ends—the tips of these bones in *Archaeopteryx* also retained the boot-like expansion of its predecessors. In the hindlimbs, the slender fibula still articulates with the anklebones, and details of the femur, tibia, and metatarsals also contrast those of living birds. These primitive features are probably related to a wide range of functions, yet this design suggests that the landing gear of the Solnhofen bird was still rudimentary and that its way of walking was not much different from that of its dinosaurian predecessors. Yet a significant departure from its maniraptoran relatives is visible in the foot, where the short first toe has partially rotated backwards. It has been consistently interpreted that the backwards rotation of this toe was similar in degree to that seen in many living birds, in which it points in the opposite direction to the three main, front-facing toes. However, the foot of the most recently studied *Archaeopteryx*—the so-called Thermopolis specimen—suggests that the first toe was only half-way rotated. Although this innovation heralds the evolution

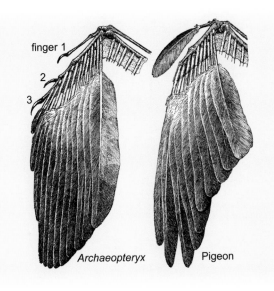

finger 1
2
3

Archaeopteryx   Pigeon

The hand of *Archaeopteryx* (including primary feathers) compared to the hand of a living bird. (Image after Heilmann, 1926.)

only is the design of these feathers fully modern, with a central shaft and asymmetrical vanes formed by hundreds of parallel barbs, but their arrangement is also akin to that of extant birds. Indeed, the number of flight feathers attached to the wing approaches the modern range: about 11–12 primaries and 12 or more secondaries. Although the number of primaries is comparable to those of the primitive dromaeosaurid *Microraptor*, the number of secondaries became much smaller during the transition from similar nonavian maniraptorans to the earliest birds. However, the plumage of the wing of *Archaeopteryx* still retains some primitive features when compared to those of its living

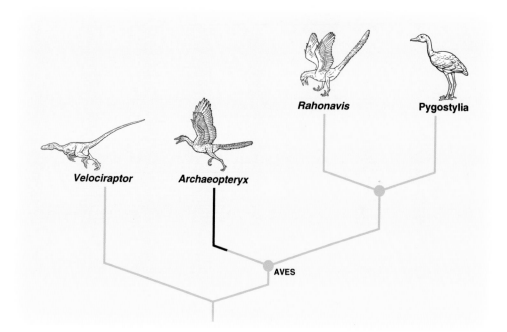

Velociraptor   Archaeopteryx   Rahonavis   Pygostylia

AVES

Archaeopteryx is today
regarded as the most
primitive known bird.
In cladistic terms, it
is the sister species
(not the ancestor)
of all other birds.
(Image: E. Heck/S. Orell.)

of a perching foot, the apparent partial rotation of the "hind" toe, together with its small size and elevated position, indicate that the foot of the Solnhofen bird was not yet specialized for perching.

Immediate Stardom: Center Stage in the Evolutionary Debate

It did not take long before the Solnhofen bird captured the fascination of both the scientific community and the public. After all, the discovery of *Archaeopteryx* in the 19th century documented the existence of birds much more ancient than those previously known, and with its toothed skull, clawed forelimbs, long bony tail, and saurian-like hip, the Solnhofen bird provided undeniable evidence of the reptilian origin of the group. Yet not everybody immediately accepted either its avian nature or its evolutionary significance. In 1862, Andreas Wagner, a prominent German paleontologist, regarded its feathers as "peculiar adornments," christening the London specimen *Griphosaurus* (meaning "enigmatic saurian"). Wagner was a staunch anti-evolutionist, and in the same article he tried unsuccessfully to persuade evolutionists to use *Archaeopteryx* in defense of their views.

Not surprisingly, *Archaeopteryx* soon became embroiled in the 19th century evolutionary debate, a controversy that would persist for a good portion of that century. What is perhaps more surprising is that although many scientists of the time quickly embraced the body plan of *Archaeopteryx* as transitional between "reptiles" and birds, heralding it as unquestionable evidence in support of evolution, two main figures in the evolutionary debate, Charles Darwin and Thomas Henry Huxley, remained cautious and to some extent unimpressed by the significance of the Solnhofen bird. Darwin made only a brief mention of *Archaeopteryx* in the sixth edition of the *Origin of Species*. Although he may have thought that *Archaeopteryx* provided evidence in support of evolution, it did not provide evidence for his theory that natural selection is the driving evolutionary force. Darwin knew that his fundamental concept applies to and acts across an entire population of organisms—thus, the individual organism represented by a fossil *Archaeopteryx* does not provide any evidence for or against the existence of natural selection. Huxley's dismissal of *Archaeopteryx* as a crucial argument for his evolutionary crusade can also be understood when his views on the origin of birds are examined more

THE PILTDOWN CHICKEN:
IS *ARCHAEOPTERYX* A HOAX?

n 1985, British astronomer and Nobel Prize laureate Sir Fred Hoyle and his associates made an alarming claim: that the London and Berlin specimens of *Archaeopteryx* were the result of a forgery involving not only Karl and Ernest Häberlein but also Richard Owen, superintendent of the Natural History Museum and a strong adversary of Darwin's theory of evolution by natural selection. In Hoyle's view, this conspiracy was designed as a trap for Darwin, who was expected to herald *Archaeopteryx* as a confirmation of his evolutionary views—the predicted intermediate form between "reptiles" and birds. Darwin would then have been publicly humiliated by the exposure of "his" unethical tactics. According to Hoyle, the alleged forgeries were perpetrated by skilled craftsmen from Pappenheim, men with a good knowledge of lithographic techniques, who applied thin layers of grained Solnhofen Limestone mixed up with glue to slabs containing skeletons of the Solnhofen nonavian theropod *Compsognathus*, and then imprinted modern feathers on these layers to create the "feathered *Archaeopteryx*." The wishbone of the London specimen was also thought to be crafted with the idea of emphasizing its avian status.

The paleontological community quickly retaliated. A team from the Natural History Museum in London demonstrated that the rock surfaces of the slab and counterslab of the London specimen fit together perfectly. They argued that if a layer of glue had been applied to the surfaces of these slabs, they would not fit together with such precision. They also showed that this perfect match applied as well to several deep cracks on both slabs: the cracks are in an identical position, which would have been impossible to forge by simply adding a layer of cement. The British team went on to show that even the pattern of branching deposits of dark inorganic minerals called dendrites was the same in both slabs—and that some of these dendrites were formed within the feather impressions. Other scientists pointed at inconsistencies within

Hoyle's rationale. As emphasized by the late Steven Jay Gould, among others, Owen was not an anti-evolutionist who rejected the transformation of one species into another by means of natural forces; he simply disagreed with the driving force proposed by Darwin (natural selection). As pointed out by Gould, "although a fierce anti-Darwinian—[Owen] was also an evolutionist."

Perhaps the last nail in the coffin of Hoyle's heretical allegation came from history itself. As pointed out by John de Vos from the Netherlands' National Museum of Natural History in Leiden, the Haarlem specimen (photo) was found four years prior to the publication of Darwin's *Origin of Species* and the onset of the controversy that, in Hoyle's opinion, led to the falsification of *Archaeopteryx*. The Haarlem specimen was purchased by the Teylers Museum in 1860, a year before anything was published about the London specimen.

Since then, this specimen has remained at the Teylers Museum, and it was not until 1970 that its identity was revealed by John Ostrom. The chronology of the Haarlem specimen—the dates of its discovery, purchase, and proper identification—precludes its implication in the alleged fraud. Yet the Haarlem *Archaeopteryx* shows clear traces of feathers. (Image: L. Chiappe.)

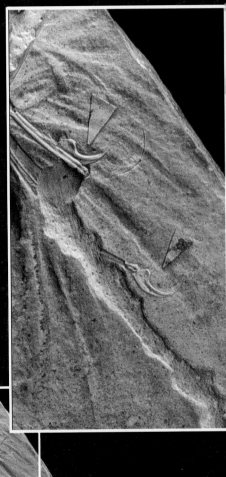

carefully. Although Huxley was deeply convinced of the dinosaurian ancestry of birds, he did not regard *Archaeopteryx* as a transitional ancestor of modern birds but rather as a true bird. In his view, *Archaeopteryx* was further from the transitional line between dinosaurs and birds than some modern ratites, the flightless birds that include the ostrich and the rhea. For Huxley, modern ratites—and not *Archaeopteryx*—were the true relics of the earliest stages of avian history. He believed that ratites had descended directly from dinosaurs and that similar birds had subsequently evolved into their flying counterparts. In Huxley's mind, the transition from dinosaurs to ratites had occurred many millions of years before the existence of *Archaeopteryx*, in the Paleozoic Era. Thus, although he regarded *Compsognathus* as the most bird-like of all dinosaurs and *Archaeopteryx* as the most reptile-like of all birds, he perceived these fossils as too young to have taken part in the transition. Evidently, the commonly stated view that Huxley heavily relied on *Archaeopteryx* in his struggles against Richard Owen and other anti-Darwinians is an oversimplification—for Huxley, *Archaeopteryx* was just the shadow of the "true" transitional forms.

Once the evolutionary paradigm was established, *Archaeopteryx* became the paramount focus of discussion on bird origins—either from dinosaurs, pterosaurs, or primitive archosaurs or archosauromorphs—and its avian nature was almost universally accepted. John Ostrom epitomized this view in his 1975 statement that "the question of bird origins and the origin of *Archaeopteryx* are one and the same problem." Our current view is a different one, however. The origin of *Archaeopteryx* is no longer equated to the origin of birds. Although regarded as the most primitive known bird, *Archaeopteryx* is not the common ancestor of all birds. Its starring role in the search for bird origins is indisputable, but its shine has started to fade in light

of a large number of discoveries filling out more vividly the transition from nonavian maniraptorans to modern birds.

An Unsettled Classification

Few other fossils in the history of paleontology have a classification as twisted and controversial as the handful of specimens of *Archaeopteryx*. The London specimen alone has been baptized *Archaeopteryx*, *Griphosaurus*, and *Griphornis*, and misspellings of these names and a myriad of different specific denominations have also appeared in the scientific literature. Names other than *Archaeopteryx* have also been created for the Berlin, Eichstätt, and Solnhofen specimens. For example, in his initial description of 1897, German paleontologist Wilhelm Barnim Dames honored industrialist Werner von Siemens by applying the name *Archaeornis siemensii* to the Berlin specimen; and in 1985, British paleontologist Michael Howgate adopted the name *Jurapteryx recurva* for the Eichstätt specimen. These taxonomic discriminations have relied on minor and sometimes subjective differences, and until recently there was a clear consensus that all known specimens of *Archaeopteryx* were members of the same species, *Archaeopteryx lithographica*.

In the last few years, however, *Archaeopteryx* specialists have put forward two other taxonomic proposals. In 1993, Peter Wellnhofer of the Bavarian State Collection of Paleontology and Historical Geology recognized the Munich specimen as *Archaeopteryx bavarica*; and more recently, Polish paleontologist Andrzej Elzanowski of Wroclaw University placed the Solnhofen specimen within the new species *Wellnhoferia grandis*. On the one hand, Wellnhofer highlighted a bone interpreted as the sternum of *Archaeopteryx bavarica*, an element thought to be unossified (and hence, missing) in other specimens, and the somewhat longer hindlimbs of the Munich specimen as compared to those

of the remaining specimens. On the other hand, Elzanowski based his claim on the stoutness and larger size, and the presence of four rather than five phalanges in the fourth toe of the Solnhofen specimen. Undoubtedly, these physical discrepancies among specimens are intriguing, but my opinion is that most can be explained as life history differences between individuals. My experience studying specimens of ancient birds has made me wary of using the presence of one bone or another as a criterion to nominate new species. Among dozens of superbly preserved and articulated fossils of the primitive Chinese bird *Confuciusornis sanctus*, certain bones are present in some but absent in others. Clearly, even when dealing with articulated skeletons, such as those of some *Archaeopteryx*, the absence of a particular bone may still be a preservational artifact or the result of incomplete preparation. In fact, recent preparation of the Munich specimen has revealed that the bone previously interpreted as a sternum is instead a portion of the coracoid. Thus, the question of whether *Archaeopteryx* ever had a bony sternum remains unanswered. Only the Munich, Solnhofen, Berlin, Thermopolis, and Eichstätt *Archaeopteryx* are articulated enough to provide a reliable assessment of the existence of a bony sternum, and the immature nature of the latter specimen would in itself be sufficient to explain why it appears to be missing. Even though new studies of the Berlin specimen suggest the existence of a very poorly preserved sternum, perhaps represented by calcified cartilage, additional well-preserved skeletons are needed to solve this conundrum. Likewise, the difference in hindlimb proportions between the Munich specimen and the other specimens, as well as the larger and stouter nature of the Solnhofen *Archaeopteryx* and the presence in this specimen of four rather than five phalanges in the fourth toe may well be related to differences in the developmental stages of the specimens,

physical variations among sexes, individual differences, or a combination of some of these possibilities. Ostrom has interpreted the different number of toebones of the Solnhofen specimen as a sex-related feature. This and some of the other minor dissimilarities may be sex-related, but since we do not have any reliable evidence as to the specific sexes of the specimens, these claims are difficult to evaluate. Nonetheless, the available sample of fossils of *Archaeopteryx* is so small that the possibility of these differences being abnormalities or individual variations is also difficult to ascertain.

Size as a Classificatory Tool

Differences in limb proportions could well be the result of life history variation among individuals of *Archaeopteryx*. The existence of distinct developmental stages among specimens is hinted at by their remarkable disparity in size. The Eichstätt *Archaeopteryx*, the smallest so far discovered, is about half the size of the largest Solnhofen specimen. Most ornithologists working on living birds would argue that size differences of this magnitude indicate species differentiation. Nevertheless, comparable size variations are not unprecedented among other species of primitive birds—in the Chinese *Confuciusornis sanctus* and *Jeholornis prima*, some of the specimens are 60 and

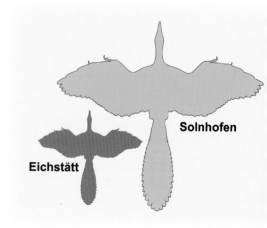

Relative sizes of the Eichstätt and Solnhofen specimens of *Archaeopteryx*. Statistical studies using a wide range of measurements from the bones of these and other specimens of *Archaeopteryx* suggest that, differences in size notwithstanding, these specimens are part of the growth series of a single species. (Image: L. Chiappe.)

Solnhofen

Eichstätt

75 percent, respectively, the size of others.

Clearly, the taxonomic weight given to these size differences should depend on our interpretation of the pattern of growth of *Archaeopteryx* and other primitive birds. Extant birds typically reach full body size a few months after they hatch. If *Archaeopteryx* had a growth pattern similar to its modern counterparts and all its fossils were to be interpreted as belonging to a single species, then the ten known skeletons would represent a sample of only the first few months in the lifespan of this ancient bird, because their sizes fall along a continuum between the Eichstätt and the Solnhofen specimens. Given the rarity of *Archaeopteryx* fossils, the odds that these ten specimens actually range in age from hatchling to a few months old are incredibly low. *Archaeopteryx*, however, may have had a pattern of growth different from that of its modern counterparts.

Our knowledge of the growth rates of Mesozoic birds is quite limited. Some preliminary insights into this question were provided by a study I conducted a decade ago in association with Anusuya Chinsamy from the South African Museum and Peter Dodson from the University of Pennsylvania. By sectioning the femur of a few Mesozoic birds, we observed a series of growth rings preserved in the bone tissue of these fossils, a pattern that today has been documented for several other primitive birds. Bone growth rings are much like those of trees—they are concentric lines that mark the cross-section of a bone. These lines are often assumed to be deposited annually, and their presence indicates a decrease in the growth rate or even a complete pause in bone formation. Growth rings are common in the bones of extant nonavian reptiles and they have also been detected in cross-sections of the bones of many nonavian theropods. Extant nonavian reptiles grow throughout their lives, although their rate of growth decreases when they are old. The presence of growth rings in primitive birds thus suggests that, unlike their living counterparts, these animals took more than one year to achieve their adult size. Even though we do not know the specifics of the bone tissue of *Archaeopteryx*, it is likely that this ancient bird also took more time than its living relatives to reach its full-grown size.

Another observation that points to substantial differences between the pattern of growth of *Archaeopteryx* and that of its living analogs is the fact that the ends of the limb bones of the smallest Eichstätt specimen are as well-formed as those of the other specimens. The limb bones of extant birds increase in length via the elongation of their ends. In contrast, in nonavian reptiles, the elongation of the limb bones is produced by bone growth within their shafts. The presence of well-formed ends in the Eichstätt specimen does not necessarily indicate that this specimen is of a different species. On the contrary, it suggests that the growth pattern of the limb bones of all the specimens of *Archaeopteryx* was more like that of extant nonavian reptiles, although this ancient bird certainly grew much faster.

Support for the idea that several of the subtle differences observed among different specimens of *Archaeopteryx* are the result of growth differences came from a study led by Marilyn Houck from the University of Arizona. In 1990, Houck and her associates took a large number of measurements of the bones of all known *Archaeopteryx* specimens of the time (the London, Berlin, Maxberg, Haarlem, Eichstätt, and Solnhofen specimens). They first compared several skeletal dimensions of the skull, wing, hindlimb, and tail to the length of the femur and determined that these measurements differed only in absolute size. They estimated that 98 percent of the differences among specimens were the result of variations in general size, thus concluding that those six specimens of *Archaeopteryx*—which included the Solnhofen specimen, later used by Elzanowski to erect *Wellnhoferia grandis*—were part of the size series

of a single species, *Archaeopteryx lithographica*. A similar metric study was recently done by Phil Senter and James Robins from the Northern Illinois University. Just as Houck's study a decade earlier, this statistical analysis concluded that all previously used differences in the proportions of the bones could be explained as variation within the growth series of a single species. Included in the analyzed sample was the Munich specimen, which at the time of Houck's study had not been discovered, the specimen used by Wellnhofer to name *Archaeopteryx bavarica*.

The Ancient Environment of Solnhofen

One hundred and fifty million years ago, Europe was an archipelago of large islands surrounded by a shallow tropical sea. Submerged under these shoal waters, Bavaria was farther south than its present position, wedged between two large islands to the north and the vast Tethys Ocean to the south. As the sea level declined, large mounds of sponges and algae developed, in time forming a strong underwater topography of valleys and ridges—remnants of these ancient mounds are still visible as towering exposures of light-gray limestone spotting the forested valleys of east-central Bavaria. With further shallowing of the sea at the end of the Jurassic, these mounds were gradually replaced by coral reefs, which hosted a remarkable diversity of fish and marine invertebrates. This system of near-shore, submarine ridges also encircled a network of relatively shallow basins (20–60 meters deep) that exchanged a limited amount of water with the deeper Tethys Ocean. The Solnhofen Limestones and their astonishing diversity of exquisite fossils accumulated in the bottom of these basins (see *The Solenhofen Limestones: Understanding their Exceptional Fossil Preservation*).

Although the Solnhofen lagoons developed in a warm climate at 25–30 degrees latitude north—a tropical range comparable to Florida today—the Bavarian

Geographic location (green box inside inset map) and environmental reconstruction of the area containing the Solnhofen Limestones. Six different environments can be recognized: (1) emerged areas, (2) well-aerated surface waters, (3) shallow, soft-bottom areas, (4) shallow, hard-bottom areas, (5) hard-bottom areas in deeper water, and (6) coral reefs. The fossils of the Solnhofen Limestones became buried in the hostile bottom (gray), whose deep, stagnant waters were hypersaline and devoid of oxygen. (Image: G. Viohl.)

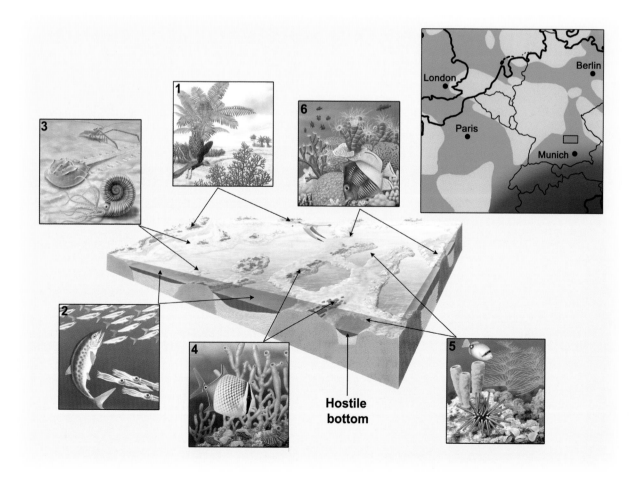

Hostile bottom

climate of the Late Jurassic was much less humid than modern Florida. Evidence of this is found in the absence of terrestrial sediments, which suggests a rather flat landscape devoid of large rivers, and consequently, with limited rainfall; and in the presence of plants with structures characteristic of those from arid climates—thick cuticles, scale-like leaves, and stems adapted for water storage. Yet ephemeral streams and seasonal ponds must have been present, because of the abundance and diversity of fossil insects with a larval cycle dependent on fresh water. Furthermore, although a handful of plant remains and the existence of insects with wood-boring larvae support the presence of trees, particularly conifers, this meager record suggests that forests were scarce at best.

Considered together, the land-dwelling organisms fossilized in the Solnhofen lagoons and the characteristics of the sediments entombing them suggest that the region was dominated by a dry climate with a short, monsoonal season of limited rainfall, and a sparsely vegetated landscape with shrubs of seed-ferns forming the undergrowth and small groves of bush-like conifers and extinct relatives of the living cycads. The island realm of *Archaeopteryx* also supported a diverse array of other vertebrates, including the nonavian theropod *Compsognathus*, small lizards, and a great variety of fish-eating and filter-feeding pterosaurs.

Archaeopteryx's Lifestyle

In recent years, much of the debate about *Archaeopteryx* has focused on whether it was a terrestrial animal or an arboreal, tree-climbing one, and whether it was able to fly and if so, how well. A wide range of opinions about these interdependent issues has been advanced in dozens of scientific articles and books. At one end of the spectrum, *Archaeopteryx* is regarded as an essentially flightless animal, with little or no capacity for flight and a terrestrial existence similar to the one commonly inferred for nonavian maniraptorans such as *Velociraptor* and *Deinonychus*. At the other end of the spectrum, the Solnhofen bird is envisaged as a percher, with a design for flight approaching the condition found in its extant counterparts, and an ability to climb tree trunks using its clawed wings and feet. Between these two extremes lie a myriad of different opinions and scenarios as to the mode of life of *Archaeopteryx*.

Detailed studies of the design of the pelvis, the shape of the claws of both its hands and feet, and the proportions of its toebones have been at the center of this discussion. Arguing in favor of an arboreal existence, John Ruben and collaborators proposed a correlation between rear-facing pubes and tree-dwelling habits, although their conclusions were based on an unrealistic design of the pelvis of *Archaeopteryx*. Because of differential preservation among individual fossils, an old controversy has surrounded the precise orientation of the pubis of this ancient bird. However, in recent years it has become clear that skeletal reconstructions where the pubes are almost parallel to the ischium and ilium, thus adopting a near-modern orientation, were based on preservational artifacts of the Berlin and London specimens. Detailed studies by John Ostrom and Peter Wellnhofer have independently shown that the pubes of *Archaeopteryx* were nearly vertically oriented, at an angle of about 100–110 degrees relative to the long axis of the ilium. These conclusions alone undermine the basis of Ruben's case for reconstructing *Archaeopteryx* as an arboreal bird, but the essence of his argument—the correlation between rear rotation of the pubes and arboreality—is moot at best. From the stout-armed *Mononykus* to the hesperornithiform divers of the Cretaceous seas, a variety of animals have pubes that point farther backwards than those of *Archaeopteryx*, but nothing about their anatomy indicates

arboreal specializations.

Perhaps the strongest arguments in support of an arboreal habit for *Archaeopteryx* and its purported tree-climbing abilities have been presented by Derek Yalden and Alan Feduccia. We have already discussed how these researchers compared the general shape and geometry of the claws of *Archaeopteryx* to those of modern birds. Yalden concluded that the strong curvature, extreme compression, and thickening of the top edge of the hand claws of *Archaeopteryx* best compare to those of the foot of extant tree-climbing birds such as woodpeckers. Feduccia pointed out that while the claw geometry of the foot of *Archaeopteryx* was comparable to that of perching birds, the claw geometry of its hand falls within the range found in the foot of tree-climbing birds. Larry Martin took this tree-climbing scenario to an extreme, portraying *Archaeopteryx* in an erect, squirrel-like posture, one never adopted by any modern bird nor substantiated by the skeletal design of their extinct relatives. In reality, even the feature that Martin considered to be the cornerstone of his interpretation—the absence of a facet in the back of the hip socket called the antitrochanter—is actually present in *Archaeopteryx*, as can be seen in the Eichstätt specimen.

While these tree-climbing scenarios depend on contrasting the claws of the hands of *Archaeopteryx* with those of the feet of modern birds, we have already argued that a more obvious comparison would be between the hand claws of *Archaeopteryx* and the hand claws of nonavian maniraptorans. The claws of the hands of these dinosaurs are also strongly curved, remarkably compressed, and bear a thickened top edge that extends to their tips—their overall geometry approaches that of *Archaeopteryx*. Equally important is the fact that the correlation between claw geometry and lifestyles remains unclear. In an extensive study of claw shape, Stefan Peters and Ernst

Arguments in favor of an arboreal lifestyle for the earliest birds led to speculative reconstructions of *Archaeopteryx* (left) and the Early Cretaceous bird *Confuciusornis sanctus* (right) in squirrel-like poses. (Image after Martin et al., 1998.)

Görgner (of the Senckenberg Museum in Frankfurt) concluded that the hand claws of *Archaeopteryx* are comparable not only to those of the feet of modern tree-climbing birds but also to those of a variety of birds with very different ecological preferences. Peters and Görgner also highlighted the fact that the claws of the Solnhofen bird (on both hands and feet) do not show any signs of wear—something that would be expected in a bird that actively uses its claws to climb trees.

In addition, the design of the foot of *Archaeopteryx* also disagrees with the specialized arboreal habits proposed by some paleontologists. I have previously pointed out how comparisons between the anatomy of the feet of *Archaeopteryx* and that of living perching birds suggest that the Solnhofen bird was not a specialized percher. The relative lengths of the first and second toes of *Archaeopteryx* are dramatically different from those of extant perchers. While the first toe of the Solnhofen bird is roughly half the length of the second, the first toe of specialized perchers is proportionally much longer—in perching songbirds, for example, this

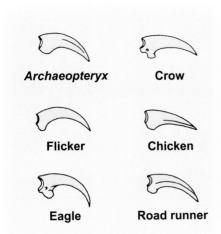

Comparisons of claws of the third toe in *Archaeopteryx*, a perching bird (crow), a tree-climber (flicker), a ground bird (chicken), an aerial predator (eagle), and a ground predator (road runner). Inferences of lifestyles on the basis of claw curvature remain largely inconclusive. (Image after Ostrom, 1976.)

Studies of the relative phalanges lengths in the third toes of living birds show *Archaeopteryx* and *Confuciusornis* to be consistent with birds that are neither fully arboreal nor fully terrestrial. (Image after Hopson, 2001.)

toe is usually longer than the second toe. In addition, the well-preserved feet of the Thermopolis specimen suggest that the first toe of *Archaeopteryx* was not fully opposed to the other toes. Furthermore, the relative lengths of the phalanges (excluding the claw) of its third toe also contrast with those of extant birds with grasping feet. In species with grasping feet, the third toebone of the third toe (the penultimate phalanx of this toe) is longer than either the first or the second phalanges, but in *Archaeopteryx* the length of these three toebones decreases toward the tip of the toe.

We mentioned earlier how in a comprehensive study of the toe proportions of living birds, paleontologist James Hopson documented a close correlation between the relative lengths of the phalanges of the third toe and the habitat preferences of a wide range of birds. Hopson analyzed groups of extant birds containing both specialized arboreal and terrestrial species. For example, among largely arboreal birds such as parrots, while the toebones of the arboreal species showed the relative lengths typical of grasping feet, those of New Zealand's ground-dwelling kakapo parrot showed proportions characteristic of terrestrial organisms, with phalanges decreasing in length toward the end of the toe. Hopson showed that the relative lengths of the non-clawed phalanges of the third toe of *Archaeopteryx* were intermediate between those of arboreal and terrestrial birds. The values he obtained for *Archaeopteryx* were comparable to those for pigeons—birds that spend time both in trees and on the ground. The penultimate phalanges of the other main toes of *Archaeopteryx* are also shorter than the first phalanges of those toes, another difference from extant perchers, in which the proportion between these phalanges is the opposite. So the anatomical evidence suggests that *Archaeopteryx* may have occasionally perched on trees, but its relatively short and elevated first toe, and the relative lengths of the phalanges of the third toe, suggest limited grasping capabilities and cast doubt on it having been a predominantly arboreal animal.

The idea that *Archaeopteryx* was a specialized arboreal bird and a tree-climber has also relied on the paleoenvironmental misconception that there were trees of significant height surrounding the Solnhofen lagoons. Countless illustrations of the Solnhofen bird show it perching in the high canopy, disregarding the lack of evidence indicating forests of any significant size

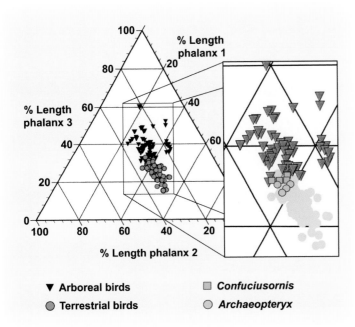

▼ Arboreal birds   ▢ *Confuciusornis*

⬤ Terrestrial birds   ◯ *Archaeopteryx*

and height in the vicinity of the lagoons. The stem of the largest fossil plants quarried in the Solnhofen Lithographic Limestones shows a special design—a wooden central rod encircled by a spongy cortex—which, albeit efficient for arid environments, substantially weakens the stems. Indeed, it has been estimated that due to these constraints, these plants would not have grown much higher than 2–3 meters. In light of this evidence, the often-used rationale that *Archaeopteryx* needed to climb trees in order to either fly or avoid predators is nonsensical, because with trees of that height, the Solnhofen bird would have gained neither significant aerodynamic advantage nor protection from potential predators such as large nonavian theropods. In the absence of evidence reconcilable with the purported arboreal lifestyle, some supporters of this view have argued that *Archaeopteryx* may not have lived in the vicinity of the Solnhofen lagoons, but rather in areas farther inland, where trees are supposed to have been more abundant and taller. This is possible, but since our knowledge of *Archaeopteryx* is entirely based on a handful of fossils quarried from a relatively small area in Bavaria, no evidence is available to support such a geographic distribution.

The Flight of *Archaeopteryx*

Although the anatomy of *Archaeopteryx* does not support tree-climbing specializations and its feet were not well-designed for perching, this does not imply that the Solnhofen bird was exclusively a ground-dwelling animal. Most specialists concur that *Archaeopteryx* was capable of some sort of flight, although there is less agreement about how it took off into the air. Certain features appear to suggest that *Archaeopteryx* was a good flier. Its wings were large—the wingspan of the Berlin specimen is 50 centimeters—and fully formed by asymmetric feather vanes, a design that has consistently been regarded as of aerodynamic significance.

In the shoulder, its glenoid faces laterally, indicating that its wings were capable of wavering about this joint with substantial amplitude. The presence of airholes in its vertebrae suggests a respiratory system in some respects similar to that of living birds, where the linkage of the lungs to air sacs that penetrate the vertebral column facilitates the large volume of oxygen required during flight. Computerized tomography of the braincase and inner ear of the London specimen also shows that the brain of *Archaeopteryx* evolved a greater sense of vision, audition, and three-dimensional perception, all features typical of flying animals. Other features, however, suggest that *Archaeopteryx* still had aerodynamic limitations. For example, the skeletal support of its wings retained primitive proportions that probably resulted in less lift than that generated by the wings of its descendants, and the frond-like appearance of its feathered tail was equally ill-suited for lift production. Likewise, its wrist (not fused into a solid carpometacarpus) and the overall design of its hand are significantly less fitted for supporting the air pressures of flying than the rigid, interlocked hand of its living equivalents, and its apparent cartilaginous sternum lacks the strength of those that in modern birds anchor powerful flight muscles.

Although *Archaeopteryx* was probably not a proficient flier when compared to any of its flying relatives, no evidence in its skeletal design suggests that it was incapable of take-off from the ground. In 1991, John Ruben envisioned a cold-blooded *Archaeopteryx* able to expend short bursts of energy powerful enough to launch it into the air from a standstill position. Such a claim is hard to evaluate, because interpreting the energetic output of extinct organisms is problematic, if not impossible. Still, the combination of aerodynamic and anatomical data of extant birds and *Archaeopteryx* suggests that the Solnhofen bird was indeed able to take off from the ground. We have

already discussed how the aerodynamic calculations involved in the hypothetical take-off of *Archaeopteryx* support the idea that this bird was able to lift itself from the ground after a short propulsive run. Although these results are ultimately approximations, since they are based on physical parameters (weight, wingspan, and others) that cannot be calculated with precision for extinct organisms, they confirm that *Archaeopteryx* had the aerodynamic equipment for taking off from the ground.

Additional support for *Archaeopteryx*'s flying capabilities comes from the fact that its fossils became entombed in the bottom of the Solnhofen lagoons. Although some have theorized that these animals died on the shore, became desiccated on the beach, and were later washed into the lagoon, a more plausible explanation suggests that they were drowned while flying over the lagoons, brought down by monsoons that regularly affected the region. Arguments in favor of the first idea highlight the distinct arrangement of the skeletons of the Berlin, Eichstätt, Solnhofen, and, to some extent, Munich specimens. The preservation of these fossils resembles that of the dried-out carcasses of shorebirds: the shrinkage of the skin and the contraction of the breast muscles as they dry out under the sun pushes the wings away from the body, the head over the back, the tail upwards, and draws the legs towards the belly. Yet this scenario fails to explain how a desiccated carcass can sink while retaining its characteristic bone arrangement. On the other hand, the bodies of these birds would have easily sunk if they were caught in a storm while flying, drowning with their lungs full of water and their plumage soaked. These floating carcasses could have also adopted the poses typical of desiccating corpses, with their wings and neck hanging down from the body.

Direct evidence of the functional and behavioral aspects of extinct animals is rarely preserved in the fossil record. Our only tools for understanding the flight capabilities of *Archaeopteryx* are its anatomy, whatever aerodynamic inferences we can make from it, and our interpretation of how these birds died.

Anatomy alone does not provide clear answers to functional questions such as flight performance, and extrapolating our knowledge of the aerodynamics of modern birds to creatures that lived millions of years ago is risky. In the end, any conclusion about the functional properties of extinct organisms carries a dose of speculation. Conjecture notwithstanding, our understanding of the lifestyle of *Archaeopteryx*, its anatomy, and the physics of flying suggest that this ancient bird retained the predominantly ground-dwelling habits of its forerunners but developed novel aerodynamic capabilities enabling it to occasionally fly to trees and across the shallow lagoons.

Challenging the Oldest Bird: the Triassic *Protoavis*

In 1985, both the scientific community and the public were astounded by Sankar Chatterjee's announcement that two partial skeletons of a bird more advanced than *Archaeopteryx* had been found in sediments of the Late Triassic Dockum Formation of Post Quarry (Texas), rocks formed some 75 million years earlier than the Solnhofen Lithographic Limestones. Chatterjee was not just pushing back the history of birds by many millions of years but was also implying that the group had a beginning as deep as the origin of dinosaurs.

Not surprisingly, the discovery of *Protoavis texensis*—as these fossils were eventually named—fueled concerns about the theropod origin of birds. Fossils can only supply limited information about the temporal occurrence of an evolutionary event— the 150-million-year-old *Archaeopteryx* tells us that the origin of birds is at least this ancient, but the fossils of this archaic bird cannot pinpoint precisely when the group actually originated. However, since the earliest dinosaurs are Late Triassic and *Archaeopteryx* lived during the Late Jurassic, the

divergence of birds is often placed within the Jurassic Period. If the fossils of *Protoavis* were truly avian (and more advanced than *Archaeopteryx*), the established chronology for the origin of birds had to be drastically modified. Birds would have had their origin around the time of, or even earlier than, the first known dinosaurs, something viewed as a caveat to the theropod ancestry of birds.

Acceptance of novel scientific ideas usually requires the discovery of a substantial body of new information. In this case, Chatterjee's study of *Protoavis* has failed to provide the evidence necessary to support his claim. Many of the illustrations used in his publications (including the one reproduced here) are

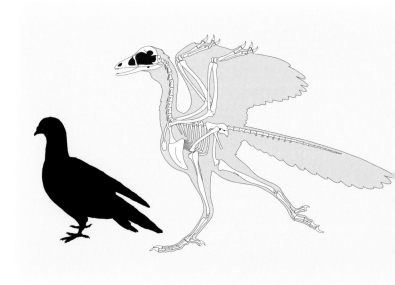

Reconstruction of *Protoavis* scaled to a pigeon (black silhouette). (Image after Chatterjee, 1999.)

misleading. In fact, several of the fossils from Post Quarry are so badly preserved that recognizing specific bones remains a challenge even for the specialist. Additional remains from a slightly older site (Kirkland Quarry) located kilometers away from Post Quarry cannot be confidently assigned to *Protoavis*. Of particular interest is a bone identified as a large, keeled "sternum." This bone—if indeed it is a breastbone—does not have a comparable element within the original bones of *Protoavis* from Post Quarry (in

fact, two flat bony plates were interpreted as the sternum of *Protoavis* in Chatterjee's original publication). In light of this, one wonders how Chatterjee arrived at the conclusion that this bone, found isolated several kilometers away, was in fact the keeled sternum of *Protoavis*.

The situation gets even more complicated when we are reminded that there is neither evidence indicating that all the "*Protoavis*" bones from Post Quarry belong to only two individuals, nor that these fossils belong to a single species. Several paleontologists have expressed their belief that *Protoavis* is an assemblage of various small Triassic creatures, making *Protoavis texensis* a composite species. John Ostrom argued that the initial two specimens were an ensemble of bones from lizards, pterosaurs, and crocodiles, among other nonavian reptiles. Recent studies have compared the skull and neck of *Protoavis* to those of the Late Triassic *Megalancosaurus* from Italy, an animal that lies outside the origin of archosaurs, and other studies have argued that some of the limb bones belong to the primitive theropod *Coelophysis*, also from the Late Triassic. It is true that the issue at stake is not whether the two specimens from Post Quarry belong to a single species but whether any of those bones can be identified as avian. Perhaps the only bird-like bones from this assemblage are a few neck vertebrae, whose joints are slightly saddle-shaped, but similar vertebrae are present in the neck of *Megalancosaurus*, so the issue remains unresolved.

In the end, the avian nature of *Protoavis* remains far from clear. The fossils have become a paleontological Rorschach test of one's training, methodological bias, and predisposition. *Archaeopteryx* still stands firm as the oldest known bird.

THE NORTH KOREAN "*ARCHAEOPTERYX*"

In 1993, while searching for fossils at a construction site in the northwestern corner of North Korea, some high school students made an exceptional discovery. Flattened against a dark slab were the remains of a feathered animal, with its skull, neck, and wings spread out. The discovery was soon reported by the Japanese magazine *Korean Pictorial*, which published a murky photo of the feathered fossil and a brief description of it. Alleging it to be of Late Jurassic age, the magazine dubbed it the North Korean "*Archaeopteryx*"—challenging *Archaeopteryx*'s long-held title of the oldest bird.

According to *Korean Pictorial*, and to the extent it can be seen from the printed photograph, the new fossil is relatively well-preserved. Yet the scarce information provided in the caption does not allow the fossil to be properly interpreted. Its long wings and traces of feathers suggest it is a maniraptoran theropod but whether it is also a bird remains uncertain. Its alleged kinship with *Archaeopteryx* is even more tentative. In fact, the proportions between the forelimb elements suggest the North Korean fossil is not *Archaeopteryx* but a different animal. Like the renowned Early Cretaceous fossil sites of Liaoning, barely 320 kilometers from the North Korean border, the rock layers entombing the North Korean "*Archaeopteryx*" have furnished the remains of plants, insects, and fish. In fact, it would not be surprising if future stratigraphic studies were to correlate these Korean beds with those across the border in China, although this research will have to wait for change in the political climate of North Korea.

Not surprisingly, North Korea treated this ancient fossil with great reverence. Soon after its discovery, President Kim Il Sung ordered the fossil to be named "*Proornis coreae*"—the founding bird of Korea! Species names, however, are not decreed but formally erected in well-distributed publications, along with adequate descriptions and illustrations to justify the need for recognizing a new species. Kim Il Sung's wishes notwithstanding, the name "*Proornis coreae*" means nothing outside his own country. This fossil, however, is unquestionably an important piece of evidence for understanding the origin and early evolution of birds, even if its full meaning will not be revealed until the international scientific community is able to examine it.

CHAPTER FIVE

Rahonavis and Other Long-tailed Birds

For more than 130 years, the Late Jurassic *Archaeopteryx* stood alone as the only known long-tailed bird. In the last decade, however, this situation has changed with the discovery of several other Mesozoic birds with extensive bony tails. These discoveries have filled an important gap in our knowledge of the earliest phases of the evolutionary history of birds. Not only do these new fossils document several lineages of primitive long-tailed birds that survived beyond the Jurassic, they also show them thriving in the Cretaceous, overlapping in time and sharing the same environments as many short-tailed birds. One of these long-tailed birds is the sickle-toed *Rahonavis ostromi* from the Late Cretaceous of Madagascar, perhaps the most primitive known bird other than *Archaeopteryx*, and one discovered amid a fantastic menagerie of island-dwellers.

The 75-million-year-old *Rahonavis* was unearthed from this quarry in the north-western corner of Madagascar. (Image: S. Sampson.)

A Sickle-toed Bird

In the summer of 1995, a joint expedition of the State University of New York (SUNY) and the University of Antananarivo in Madagascar made a formidable discovery in a small, 75-million-year-old quarry near the Malagasy village of Berivotra, in the northwestern corner of this southern African island. Most dinosaur-bearing rocks that crop out in this area are hidden under thick grass. The fire set to clear the vegetation covering a small hillside revealed the jumbled bones of a titanosaur dinosaur, a member of a far-flung lineage of long-necked sauropods. Between the large bones of this animal, the expedition unearthed the delicate remains of a primitive, long-tailed bird—like the Phoenix, *Rahonavis* had been brought back to life.

The only known specimen of *Rahonavis* is composed primarily of the rear half of the animal, including the trunk, pelvis, hindlimbs, and numerous tail vertebrae, many of them found in articulation. Portions of the forelimb and shoulder lay adjacent to these bones—in fact, all that is known of *Rahonavis* was found in an area smaller than a letter-sized page. Although they were not joined to the rear portion of the skeleton, it is believed that the bones of the forelimb and shoulder belong to the same fossil individual, since they match the size of the hindquarters and exhibit the same type of preservation.

Details of these bones exhibit a striking combination of characteristics that are typically avian with others that are usually not. SUNY paleontologists Catherine Forster and David Krause, Scott Sampson from the University of Utah (then at the New York College of Osteopathic Medicine in Old Westbury), and I christened this bird *Rahonavis ostromi*, a name drawn from the Malagasy term "Rahona," meaning "menace from the clouds," and in honor of the late John Ostrom, the relentless student of *Archaeopteryx* and the origin of birds.

Rahonavis' Skeleton: A Step Further on the Avian Path

One-fifth larger than the Solnhofen specimen of *Archaeopteryx*, the skeletal design of *Rahonavis* is strikingly similar to that of its 75-million-year-older relative. In many respects, the hindquarters of this raven-sized bird remained essentially unchanged, and so did some features of its forelimbs and spine. The pelvis retained the same design and general proportions as in *Archaeopteryx*—the vertically oriented pubis is much longer than the ischium, and has a foot-like expansion at its end—and its bony tail, although incompletely known, appears to be as long as that of its much more ancient cousin. The long and slender bones of the hindlimbs have similar proportions to those of *Archaeopteryx* but the fibula has become splint-like, approaching its modern design. Although the foot shows the same degree of grasping capacity as *Archaeopteryx*, the robust second toe carries a much larger sickle-shaped claw, reminiscent of the predatory weapon of dromaeosaurids

The pelvis and feet of *Rahonavis* compared to those of *Archaeopteryx*. Note the large, sickle-shaped claw of the second toe of *Rahonavis*. (Image after Forster et al., 1998; Wellnhofer, 1974.)

> Reconstruction of the raven-sized *Rahonavis* next to a sketch of how its bones were found. (Image after Forster et al., 1998.)

> The sickle-shaped claw (bottom) of the second toe of the dromaeosaurid *Velociraptor* next to its skull. *Rahonavis* may have used a similarly lethal toe to dispatch its prey. (Image: L. Chiappe.)

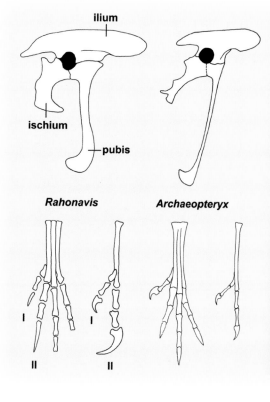

ilium

ischium

pubis

Rahonavis **Archaeopteryx**

I I

II II

and troodontids. The shape of the joints indicates this sickled toe was capable of long-range, powerful movements, which could have been used to dispatch prey in a fashion similar to these maniraptorans.

The common characteristics between the hindquarters of *Rahonavis*, *Archaeopteryx*, and their maniraptoran predecessors suggest that the basic movements involved in walking and running remained the same across these species. In the last decade, Steve Gatesy of Brown University has supplied important insights into the hindlimb motion of theropods. The studies he conducted document notable differences in the movement of the hindlimbs of birds and crocodiles, suggesting that nonavian theropods walked more like crocodiles than modern birds. The

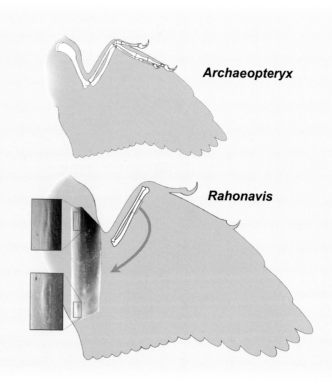

reasonable to assume that the femur of long-tailed birds was held more vertically and that the walking style of these animals was more similar to that of their close nonavian relatives than to their modern counterparts. Gatesy interpreted the abbreviation of the tail and associated musculature as the primary factor involved in the shift between these locomotory mechanisms. The final developments leading to the modern type of terrestrial locomotion appear to have evolved rather late in avian history, and to some extent in concert with the shortening of the tail.

More significant modifications took place in the forequarters of *Rahonavis*. Although it is similar in shape to the corresponding bone in *Archaeopteryx*, the dimensions of the ulna of the Malagasy bird indicate that the relative size of its wing was much greater—the ulna of *Rahonavis* is 50 percent longer than that of *Archaeopteryx* when compared to the femur. Another significant difference of the forelimb is the presence of quill knobs. Although highly variable among

< In crocodiles and primitive archosauro-morphs like *Euparkeria* (top), the hindlimb stride is principally powered by tail muscles that pull the vertically held femur backwards. Non-avian theropods and primitive birds such as *Rahonavis* and *Archaeopteryx* (center) probably retained this primitive style of gait. In modern birds (bottom), the muscles that powered their predecessors' stride became substantially reduced with the shortening of the bony tail. In these birds the femur is held more horizontally and the swinging of the tibiotarsus carries on the stride. (Image after Gatesy, 1990.)

Archaeopteryx

Rahonavis

functional differences between these animals are principally related to the size of the tail and its musculature. The long tail of crocodiles is much involved in their locomotion. During the stride, powerful muscles of the tail of these animals pull the vertically held femur backwards. In contrast, the short tail of modern birds is minimally involved in their terrestrial progression—its musculature is reduced and mostly limited to controlling the position of the tail feathers. While walking or running, the femur of modern birds is held nearly horizontally and it is the swinging of the tibiotarsus that carries on the stride. Based on these observations, it is

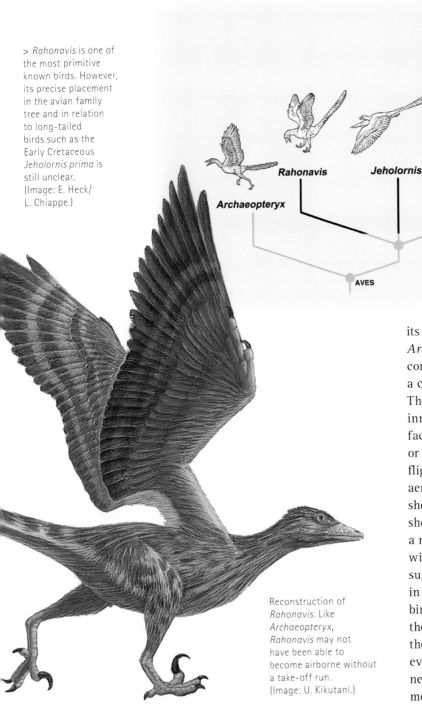

> *Rahonavis* is one of the most primitive known birds. However, its precise placement in the avian family tree and in relation to long-tailed birds such as the Early Cretaceous *Jeholornis prima* is still unclear. (Image: E. Heck/ L. Chiappe.)

Archaeopteryx

Rahonavis

Jeholornis

Confuciusornithidae

Ornithothoraces

PYGOSTYLIA

AVES

Reconstruction of *Rahonavis*. Like *Archaeopteryx*, *Rahonavis* may not have been able to become airborne without a take-off run. (Image: U. Kikutani.)

< The wing of *Rahonavis* compared to that of *Archaeopteryx*. The Malagasy bird had larger and longer wings, and its flight feathers were firmly attached to quill knobs on the wingbones. (Image after Forster et al., 1998/S. Abramowicz.)

its upper end shows that unlike *Archaeopteryx*, this bone joined the coracoid through ligaments that allowed a certain degree of independent mobility. The functional significance of this innovation is not entirely clear, but the fact that these two bones are either fused or tightly connected to each other in most flightless birds highlights the apparent aerodynamic significance of a mobile shoulder joint. Interestingly, although the shoulder joint of *Rahonavis* approaches a modern configuration, comparisons with primitive short-tailed birds suggest this design most likely evolved in convergence with more advanced birds. These early transformations of the avian flight system drive home the complexity of this remarkable evolutionary feat and how much still needs to be discovered. Regardless, the modern design of *Rahonavis'* shoulder and the size of its feathered forelimb indicate that this ancient Malagasy bird was able to fly, although it probably did so in a more clumsy way than its modern counterparts.

Bird, Dinosaur, or Both?

Even if the discovery of feathered nonavian maniraptorans put to rest the belief that feathers are unique avian attributes, *Rahonavis* has several skeletal features that place it within the realm of

extant birds and extremely rare among Mesozoic birds, these attachments indicate that the wing of *Rahonavis* carried robustly attached flight feathers—the number of secondaries attached to its ulna even falls within the range of feathers attached to the ulna of modern birds. Yet the most notable difference between *Rahonavis* and *Archaeopteryx* lies in the shoulder. Even though we know only the scapula of *Rahonavis*,

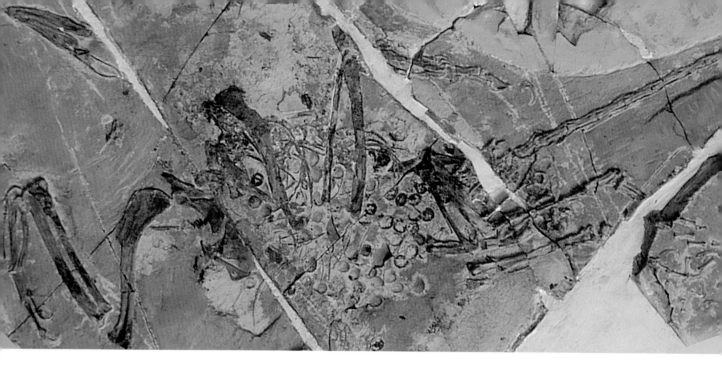

birds. The rear direction of its first toe and the development of quill knobs are unknown among nonavian dinosaurs; and the splint-like condition of its fibula is typical of birds. Its wings are also longer than those of *Archaeopteryx* or any nonavian theropod, and the degree of air sac invasion into its backbone is greater (manifesting a further step towards the respiratory system of living birds). At any rate, the anatomy of *Rahonavis* indicates a degree of evolution comparable to that seen in *Archaeopteryx*, if not slightly more advanced.

When we announced the discovery of *Rahonavis*, our genealogical interpretation supported a close relationship with *Archaeopteryx*. Yet our study went further, suggesting that the anatomical support for this proposal might also substantiate placing *Rahonavis* as the next branching lineage of the cladogram—evolutionarily closer to modern birds than to *Archaeopteryx*. This view was later supported by my own research on the evolutionary relationships of primitive birds. Detailed studies of *Rahonavis* are still pending and interpretations of its genealogical placement may vary significantly as we augment our knowledge of the anatomy of this animal. Paleontologist Peter Makovicky of Chicago's Field Museum

has recently argued that *Rahonavis* belongs to a group of specialized South American dromaeosaurids, thus lying outside the early diversity of birds. This view needs to be carefully considered, but in my opinion, future genealogical interpretations are likely to show that despite its young geologic age, the Malagasy fossil ranks among the most primitively known birds. Perhaps the early geographic isolation of Madagascar created an ecological refuge for what were, even in the Late Cretaceous, archaic species; birds that had long vanished from other parts of the world.

Soon after our announcement of *Rahonavis*, skeptics of birds' maniraptoran origins rushed to herald it as a fossil "chimera." Paleontologist Larry Martin considered the hindquarters of *Rahonavis* as those of a nonavian dinosaur and regarded the few bones from its wing and shoulder as those of a flying bird. Although the hindquarters of *Rahonavis* were not found connected to its shoulder and forelimb, they were discovered in great proximity. Martin's assertion was based on the apparent mismatch of the more modern appearance of the shoulder and wing with the more primitive characteristics of the hip, legs, and tail. However, close examination of the shoulder and wing

The 120-million-year-old *Jeholornis prima* is one of several long-tailed birds recently unearthed in China. (Image: L. Chiappe.)

> Breakdown of the Jurassic supercontinent of Gondwana during the Cretaceous. Most researchers believe that Madagascar (in red) became detached from other landmasses as early as 90 million years ago, but some argue that a landbridge connecting this island to Antarctica may have been in place until the end of the Mesozoic. (Image: R. Urabe.)

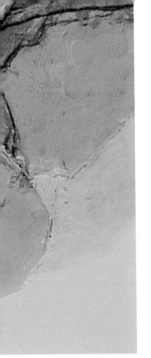

also reveals their primitive constitution. Anatomical innovations notwithstanding, these bones are clearly not from any advanced lineage of Cretaceous birds. Most importantly, our genealogical study showed that the placement of *Rahonavis* within birds was equally supported even if considering only its hindquarters. Thus, the avian nature of *Rahonavis* was supported by all parts of its skeleton. The discovery of *Rahonavis*, with its lethal sickle-like foot, highlighted once again the maniraptoran ancestry of birds. In the end, Martin was partly right—*Rahonavis* was both a bird and a dinosaur.

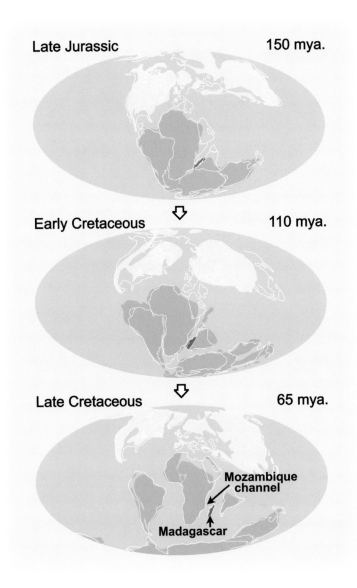

Late Jurassic   150 mya.

Early Cretaceous   110 mya.

Late Cretaceous   65 mya.

Mozambique channel

Madagascar

The Cretaceous Fauna of the Great Red Island

Fourth in size among the world's islands and covered by reddish, iron-rich soils, Madagascar has provided one of the clearest windows on a Cretaceous island ecosystem. In the Late Jurassic, some 150 million years ago, the Great Red Island was part of the Mesozoic supercontinent of Gondwana—a giant land mass that encompassed all emerged territories of the southern hemisphere. Within the next 30 million years, in the Early Cretaceous, Africa began to split apart from Madagascar and India, thus creating a deep submarine rift. Over tens of millions of years, the tectonic forces that had set this process in motion continued to generate new sea floor at this ancient rift, pushing the joined territories of Madagascar and India toward the southeast. By around 90 million years ago, Madagascar had already reached its current position and soon after India began its long voyage toward the northeast—millions of years later, its collision against Asia would lift the Himalayas. Separating Madagascar from Africa was the Mozambique Channel, a formidable oceanic barrier that in some places was more than 1,500 meters deep. The 90-million-year-old geographic isolation of Madagascar is reflected in the uniqueness of its present fauna and flora: 85 percent of the animals and plants of the Great Red Island exist nowhere else in the world.

The 75-million-year-old sediments that entombed *Rahonavis* accumulated in the sandy bed of a slowly running river that dissected the Mahajanga Basin, a depression that existed in the northwestern corner of Madagascar. The Mahajanga Basin was then more or less where it is today. The exceptional preservation and partial articulation of the only known specimen of *Rahonavis* indicate that the river's current minimally transported its skeleton—the bird died and was buried within a few

kilometers from the ocean shore. We know that this seaside environment harbored a diverse fauna. The 30-square-meter quarry where *Rahonavis* was recovered has yielded a large array of other exquisitely preserved fossils. In addition to the titanosaur skeleton that surrounded *Rahonavis* bones, there were remains of the small abelisaur theropod *Masiakasaurus* and teeth of its larger cousin, *Majungatholus*. Fossil bones and teeth of mammals, crocodiles, turtles, snakes, lizards, frogs, and fish were also excavated from this and other neighboring quarries. A variety of other, less complete, fossil birds were also excavated from these sites, including at least three species of enantiornithines and the peculiar *Vorona berivotrensis*. This last bird appears to be among the most primitive members of the Ornithuromorpha – the lineage that includes the familiar *Hesperornis regalis* and *Ichthyornis dispar* as well as all extant birds. This unique primitive fauna gives us a glimpse of Madagascar 75 million years ago, and

the mammals, frogs, and birds found in the Late Cretaceous deposits of the Great Red Island. These distinct genealogical links among fossils today separated by vast oceans provide a clear testimony to continental drift and its role in shaping Earth's fauna and flora.

Skeletal reconstruction of *Jeholornis*. (Image: S. Abramowicz.)

many of these ancient, isolated animals have evolutionary ties with other faunas that populated the southern continents during the Cretaceous. The predatory abelisaurs are also known from South America and India. One of Madagascar's bizarre ancient crocodiles, the plant-eating, pug-nosed *Simosuchus clarki*, also has close relatives in Africa and South America, and so do some of

More Long-tailed Birds

In the last few years, the discovery of several spectacular fossils in the Early Cretaceous shales of Liaoning has augmented our knowledge of long-tailed birds, as well as the roster of other feathered animals from this corner of China. The first two of these birds were unearthed from the 120-million-year-old Jiufotang Formation and reported almost simultaneously by two rival teams of Chinese paleontologists. One of them was named *Jeholornis prima*—the primitive bird from Jehol (in reference to the renowned Jehol fossil fauna)—the other is *Shenzhouraptor sinensis*, meaning "China's ancient raptor." Size differences notwithstanding, for the largest specimen of *Jeholornis* is about a third larger than *Shenzhouraptor*, these archaic birds are anatomically very similar to each other. Indeed, it is possible that they are the same type of bird and that their size variation simply reflects the more juvenile age of *Shenzhouraptor*, a situation comparable to that discussed earlier for *Archaeopteryx*.

Jeholornis and *Shenzhouraptor* are relatively large birds. Although smaller than *Rahonavis*, the largest known specimen of *Jeholornis* is somewhat bigger than the largest *Archaeopteryx*—the

Solnhofen specimen—and with wings that are proportionally longer. Their skulls have an overall primitive design. For example, a large antorbital opening perforates the side of the snout and a robust bony strut separates the orbit from the infratemporal opening, a configuration that, as we have seen, must have limited the independent elevation of the snout that characterizes the skull of modern birds. Yet the head of these birds is specialized in its abbreviated dentition, limited to a few small teeth at the tip of stout lower jaws. In many aspects, the skeleton of these long-tailed birds is similar to that of *Archaeopteryx* and *Rahonavis*. The neck is short and the trunk is long; the furcula is strong and boomerang-shaped; the wide and short sternum is followed by a large corset of zigzag-patterned gastralia; the ischium is much shorter than the pubis, and the latter bone is oriented nearly vertically, ending at a boot-like expansion. In the hindlimb and tail, these birds also resemble *Archaeopteryx* and *Rahonavis*, although their feet did not have the notable predatorial specializations of the latter. Interestingly, the stiff tail of these Chinese birds is proportionally longer (including also a greater number of vertebrae) than that of *Archaeopteryx*, and feather impressions of a *Jeholornis* fossil show that its plumage projected as a distal fan,

A tuft of blackish feathers can be seen at the end of the long tail of this juvenile specimen of *Jeholornis*. (Image: L. Chiappe.)

similar to those attached to the end of the tails of *Caudipteryx*, *Microraptor*, and other nonavian maniraptorans. Such a resemblance in length and plumage pattern suggests that the ancestor of all birds could have had a tail more similar to its maniraptoran predecessors than to the frond-like feathered tail of *Archaeopteryx*, although additional fossils of long-tailed birds may need to be found before we can place more confidence in this idea. More intriguing is the notion that such a tufted tail probably generated less lift than the wider airfoil formed by the shorter feathered tail of the Solnhofen bird. The general appearance of the long tail of *Jeholornis* and *Shenzhouraptor* highlights the fact that the evolution of aerodynamic features was far more complicated than previously expected—while the long wings of these birds became more specialized for flying, their long tails retained the primary locomotory role played by their terrestrial maniraptoran forerunners.

Although these birds echo *Archaeopteryx* in many skeletal features, the shoulder and forelimb of *Jeholornis* and *Shenzhouraptor* point to a substantial improvement in flight performance. The shoulder is mobile like in *Rahonavis* and its glenoid extends further upwards than in *Archaeopteryx*; the coracoid has a more strut-like appearance; and the hand is shorter and more compact, with the bones of the wrist completely interlocked with each other. The elongation of the coracoid and the strengthening of the wrist suggest a powerful wingstroke. Furthermore, even though these Chinese bony-tailed birds still bore three clawed fingers—which probably projected from the outline of their asymmetrically feathered wing—the relative length of their forelimbs was about 25 percent greater than in *Archaeopteryx*, with proportions closer to those of their short-tailed counterparts. Based on the relative lengths of the ulna and femur (both bones preserved in *Rahonavis*), the wing of *Jeholornis* and *Shenzhouraptor* was only slightly shorter than that of the Malagasy

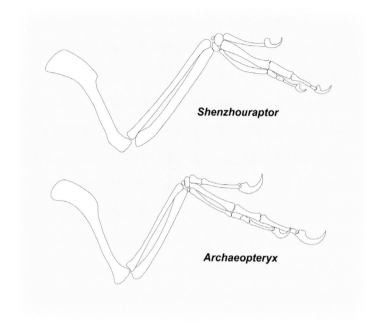

The wingbones of *Shenzhouraptor* compared to those of *Archaeopteryx*. Comparable in size to the largest individual of *Archaeopteryx*, the Solnhofen specimen, the wing proportions of *Shenzhouraptor* are generally more modern— the forearm (ulna) is longer than the upper arm (humerus) and the hand is relatively shorter. The unfused nature of *Shenzhouraptor*'s wrist may reflect the estimated young age of the holotype. (Image: L. Chiappe.)

bird. Yet given the current anatomical information, it is difficult to ascertain whether these Chinese birds share a closer kinship with living birds than *Rahonavis*—more fossils and additional studies are necessary to answer this question. Nonetheless, evident in the design of the forelimb and shoulder of all these long-tailed birds is the beginning of a pattern that characterizes the early evolution of birds. Namely, very early in their evolutionary history, birds began enhancing their remarkable flying abilities—and this was long before they developed their characteristic mode of terrestrial progression.

The unusual preservation of stomach contents in a specimen of *Jeholornis* has also given us clues about the diet of these primitive birds. The presence of dozens of seed-like structures, barely a centimeter across, inside the belly of *Jeholornis* adds to our scant knowledge of the food preferences and feeding behaviors of early birds. Not only does the large number of these undigested structures point at a seed-based diet, but the fact that they are contained in such a great number inside the digestive tract of *Jeholornis* suggests the existence of a large crop for food storage.

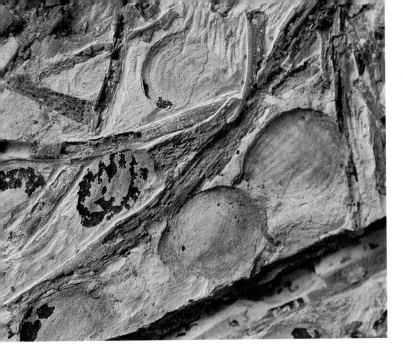

Seeds inside the belly of the holotype of *Jeholornis prima* are one of the very few instances of direct evidence about the dietary habits of ancient birds. (Image: L.Chiappe.)

The half-meter-long skeleton of the holotype of the Chinese Early Cretaceous *Jinfengopteryx elegans*. Further studies are necessary to determine whether this animal is a short-winged, primitive bird or a member of a lineage of nonavian maniraptorans. (Image after Ji et al., 2005.)

The Latest Discoveries

The fossil-rich shales of northeastern China continue to yield more evidence of the newly detected diversity of long-tailed birds. Recent discoveries have unearthed fossils of long-tailed birds older than *Jeholornis* and *Shenzhouraptor*—and thus closer in time to *Archaeopteryx*. One of these is the controversial *Jinfengopteryx elegans*, a half-meter-long animal from the Qiaotou Formation, fossiliferous beds interpreted to be somewhat older than the renowned, 125-million-year-old Yixian Formation. *Jinfengopteryx* is known from an articulated, feathered specimen, whose skeleton and preserved plumage has been described as remarkably similar

to that of *Archaeopteryx*. This fossil shows features that could well make it next of kin to the Solnhofen bird, but the abbreviated nature of its forelimbs and its large number of teeth—characteristics absent in the most primitive birds—have sent a wave of skepticism among researchers, who believe this fossil lies outside birds. Indeed, in these and other respects, *Jinfengopteryx* resembles the troodontids, a group of nonavian maniraptorans very closely related to the origin of birds. A good amount of experimentation characterizes the early evolution of birds, and it could well be that *Jinfengopteryx* reverted very early to a wing and dental design that is more typically found among some nonavian maniraptorans. We may have to wait until further studies on this most recent discovery put these views to the test.

Another recent discovery, in this case unquestionably a long-tailed bird, is *Jixiangornis orientalis*, a short-snouted, beaked animal, whose forelimbs are 30 percent longer than its hindlimbs. Although older than *Jeholornis* and *Shenzhouraptor*, this 125-million-year-old fossil shows features suggesting a greater flying ability. For example, the hand, albeit still with three clawed fingers, is shorter and more compact, and the sternum is larger, with a short keel that expands the surface available for the origin of flight muscles. While the developments of the shoulder and wing of *Jeholornis* and *Shenzhouraptor* suggest these birds were genealogically closer to modern birds than *Archaeopteryx*, the transformations in the skeleton of *Jixiangornis* place it even closer to its living counterparts. Thus, despite its old geologic age, *Jixiangornis* is the most advanced member in the family tree of long-tailed birds, with a skeletal design not much different from the most primitive *short*-tailed birds.

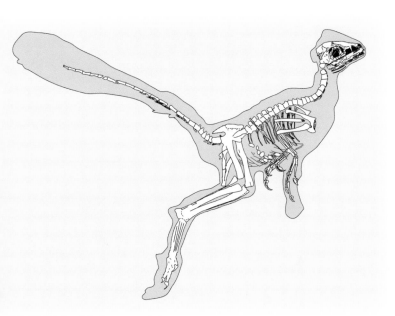

CHAPTER SIX

The First Short-tailed Birds

Virtually nothing is known about the evolutionary abbreviation of the avian tail. The fossil record of long-tailed birds shows no trend indicating a reduction in the size of the bony tail, nor does it suggest the beginning of vertebral fusion to form a pygostyle—the rump-bone at the end of the spine of short-tailed birds. The appearance of these birds in the fossil record is abrupt—125 million years ago, they make their debut with a fully formed pygostyle. The secret to this important transformation is likely buried in the 25 million years that separate the first short-tailed birds from *Archaeopteryx*, but our knowledge of this critical segment of the history of birds is almost nil. Nowhere in the world have informative skeletons of birds from the earliest stages of the Cretaceous been found.

With hundreds of specimens unearthed in the last decade, *Confuciusornis* is the most abundant Mesozoic bird. (Image: M. Ellison/L. Chiappe.)

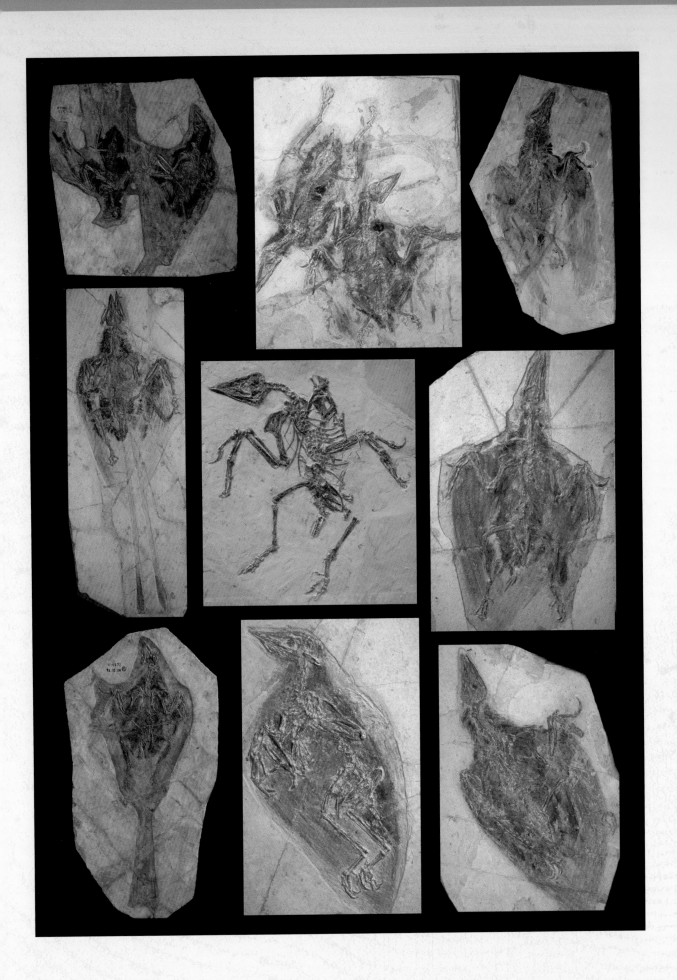

THE EVOLUTION OF TOOTHLESSNESS:
A COMMON THEME IN AVIAN HISTORY

Early discoveries of Mesozoic birds assumed that even the oldest representatives of the group had the beaked nature that characterizes all their living relatives. Nineteenth century scientists realized that *Archaeopteryx* was indeed toothed only years after the discovery of the first specimen, but even decades later, the toothed nature of some other early avian lineages was still questioned. Since then, many fossil discoveries have proved that the skulls of most Mesozoic birds carried teeth and that the diversity of their dental designs was substantial. These fossils have also revealed that the evolutionary loss of teeth was a recurrent phenomenon in the evolution of both birds and their theropod predecessors. While the lack of teeth in all modern birds is interpreted as the result of a single event, the evolutionary transformation by which toothed jaws were replaced by a sharp beak occurred several other times independently. The available fossils tell us that among pre-modern birds, teeth were lost in at least two and perhaps three lineages that straddle much of the avian family tree—in confuciusornithids, in one lineage of enantiornithines (*Gobipteryx minuta*), and perhaps in early ornithuromorphs such as *Hongshanornis longicresta* (the absence of teeth in this bird remains controversial). Molecular studies have also documented that in living birds, the developmental program that leads to tooth formation has remained dormant—this program presumably also remained dormant in the extinct lineages of toothless birds. Although the frequency of this transformation suggests that beaks must have endowed birds with a valuable way of handling food items and nesting material, understanding the functional advantage of an evolutionary innovation is often complex. Regardless, future discoveries are likely to reveal an even greater number of Mesozoic lineages that convergently evolved the beak that characterizes their living relatives. (Image after Chiappe et al., 1999.)

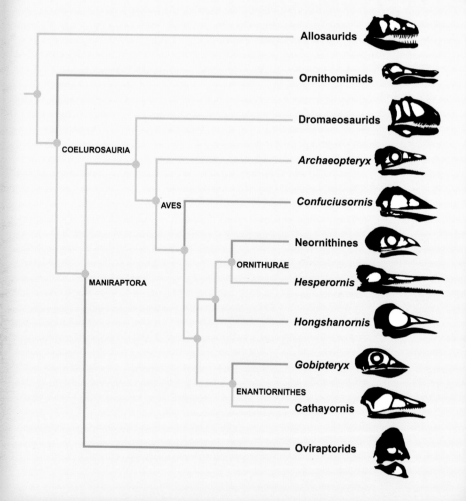

A SOPHISTICATED BREATHER

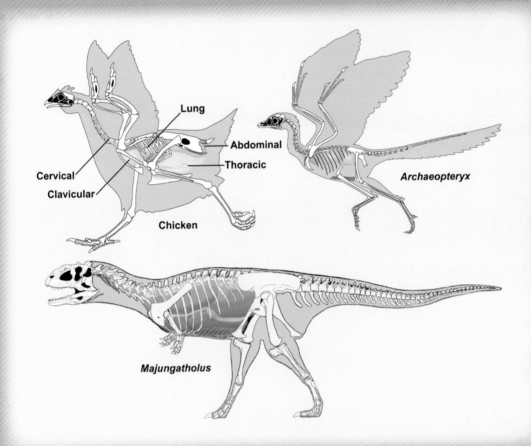

The reptilian lung is characterized by internal subdivisions formed by clusters of small air passages. In birds, this ancestral reptilian lung became modified by developing an opening on each end and connections with a series of air-filled sacs—cervical, clavicular, abdominal, and thoracic—that envelop the viscera like the shells of walnuts. Tube-like evaginations of these air sacs reabsorb the bony tissue of neighboring bones, including the vertebral column, humerus, wishbone, and portions of the pelvis and hindlimbs. Even if the lightening of bones through this process of pneumatization is an obvious advantage for airborne animals, lung ventilation is likely the primary function of this system of air-filled sacs. Changes in the volume of the air contained in the sacs generate the airflow that ventilates the lungs. As new, oxygen-rich air enters the lungs from one end, used, oxygen-low air is pushed into the system of air sacs. Thus, during ventilation, the air passes through the lungs in only one direction. This unidirectional airflow is coupled with a system of highly efficient cross-current gas exchange at the level of the smallest subdivisions that delivers a great amount of oxygen to the blood stream. This sophisticated respiratory system has not only allowed birds to meet the high energy demands of flight but also to be able to breathe at altitudes where the percentage of oxygen is significantly lower than at sea level.

Evidence of this respiratory system in the fossil record is unusual. Inferences about the presence of pneumatization in fossil bones may be made by detecting the openings that served as an entrance for the tube-like evaginations of air sacs. Until recently, it was thought that *Archaeopteryx* lacked the sophisticated respiratory mechanism of its modern counterparts, but the discovery of tiny airholes in the neck and trunk vertebrae of the London, Eichstätt, and Berlin specimens has suggested otherwise. Evidence of bone pneumatization appears to be present in the pelvic region of the London specimen as well. These observations suggest that the sophisticated respiratory mechanism of extant birds was at least partially developed in the earliest member of the group—present evidence indicates that the lungs of *Archaeopteryx* were connected to cervical, clavicular, and abdominal air sacs. In fact, evidence of vertebral pneumatization in several nonavian theropods (for example, *Majungatholus*) also suggests that the evolution of avian pulmonary air sacs predated the origin of birds and that birds inherited aspects of their unique breathing system from their dinosaurian forerunners. (Image after Rowe et al., 1998; O'Connor and Claessens, 2005.)

AVIAN FOOTPRINTS
IN PRE-JURASSIC ROCKS

Isolated footprints that resemble to some degree those made by birds on a mudflat have been found at a handful of localities that predate the Late Jurassic. Avian footprints are often small, and wider than they are long, with slender toemarks and a wide angle between the outer and inner toes. However, the appearance and hence the identification of fossil footprints is affected by a large number of factors, including the type and quantity of minerals in the ground, the substrate's water salinity and humidity, and the climatic conditions that prevail in the area. Identifying particular groups of animals from their fossilized footprints

is much harder and more tentative than recognizing them from their fossilized bones. Interpretations of fossilized footprints are often controversial.

By far the most startling set of avian-like footprints predating the Late Jurassic comes from the Santo Domingo Formation of western Argentina (photo), rocks claimed to have formed approximately 55 million years before the time of *Archaeopteryx*. Approaching in appearance those of quails and other landfowls, these small footprints show the high-density pattern (over 500 footprints per square meter) and frequent overprinting

seen in many birds that live alongside ponds or bodies of water. Perhaps most remarkable is the fact that their distribution and lack of a preferred direction suggest these animals were able to take off from a standstill. Although few would doubt these footprints were avian if they had been found in later rocks, the purported Late Triassic age of these fossils still needs to be demonstrated with greater confidence. The significance of these remarkable footprints is not only tinted by the vagaries of fossilized footprints but by the lack of solid evidence documenting their Late Triassic age. (Image: R. Melchor.)

There is little doubt, however, that the shortening of the tail is one of the greatest steps in the history of birds, a successful transformation that had profound aerodynamic implications and paved the way for the group's first major evolutionary radiation. Future discoveries may fill the pages of this early chapter of birds' ancient saga, but for now we can only speculate about the complexity of this fundamental transformation. This disappointing gap in our knowledge has been somewhat mitigated by abundant discoveries from China, which have provided considerable information on the early diversity and appearance of the ancientmost short-tailed birds. Among these fossils are the confuciusornithids—one of the best known and most spectacular groups of archaic birds.

A New Ancient Lineage

Towards the end of 1994, three fragmentary fossils from Liaoning's 125-million-year-old Yixian shales were brought to the attention of Zhou Zhonghe, and his senior colleague Hou Lianhai, of Beijing's Institute of Vertebrate Paleontology and Paleoanthropology. The fossils had been excavated by local farmers near Sihetun and purchased by a friend of Zhou at a local flea market.

The toothless skull of the holotype of *Confuciusornis sanctus*. (Image: L. Chiappe.)

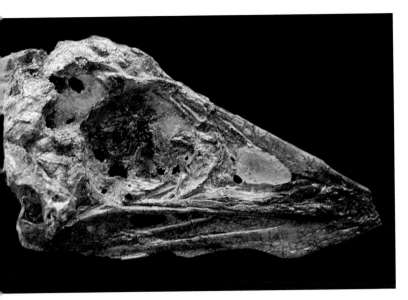

One of them consisted of a toothless skull connected to a nearly complete wing; the other two fossils included portions of the pelvis and hindlimb as well as contour feathers of a very primitive bird. Although Yixian fossils had been known for years and these new finds were not the first Mesozoic birds from Liaoning, their singular characteristics did not pass unnoticed. The combination of stout and beaked jaws with large and powerfully clawed hands was sufficient evidence that the fossils belonged to a previously unknown bird. The following year, Hou baptized these fossils *Confuciusornis sanctus*—meaning "the sacred bird of Confucius"—in reference to the celebrated 5th century BC scholar.

That first purchase of a *Confuciusornis* fossil at a Liaoning swapmeet was an omen of what was to come. In the ensuing years, beautiful crow-sized fossils of this bird—their exquisite skeletons surrounded by a dark halo of plumage—started to flood the international fossil market, which, fueled by the sale of the *Tyrannosaurus rex* "Sue" for more than $8,000,000, was growing exponentially. It is estimated that local farmers have unearthed nearly a thousand specimens of *Confuciusornis* from multiple localities of the Yixian Formation and that, despite China's severe penalties for those caught dealing commercially in fossils, only a fraction of this enormous sample has made it into public institutions. This alarming loss has been to some extent mitigated by the creation of a Fossil Administration Office within the Bureau of Land and Resources of Liaoning, which protects a large area around Sihetun, but it is likely that around 80 percent of all discovered *Confuciusornis*—along with plenty of other fossils—have been smuggled into the hands of fossil dealers and private collectors. Still, the large number of these excellent fossils that have found repository in public museums offer the best opportunity yet to study the range of physical variation and life history of a very ancient bird. We can

only hope that further control from the Chinese government, along with goodwill from private fossil collectors, will continue to augment the number of specimens kept at public institutions and available for future studies.

The Great Wall Bird: *Confuciusornis* Gets a Sibling

In the summers of 1997 and 1998, I spent many days examining the spectacular collection of fossil birds from Liaoning at the National Geological Museum of China in Beijing. My daily fascination with these fossils was enhanced by the sound of the morning Buddhist prayers at the Guangji Temple, across the street from the museum. One of those mornings, paleontologist Ji Shu'an showed me a small *Confuciusornis* specimen the museum had recently purchased from farmers in Jianshangou, a village only a few kilometers from Sihetun. With its skeleton lined by dark-gray feathers, this jay-sized fossil was split in two slabs with many of its bones still encased in the rock. Although few anatomical details were visible, there was something different about this "*Confuciusornis*." Ji and I decided to add it to the list of fossils that he was taking to the American Museum of Natural History in New York City (where I had my office) to be prepared as part of our collaborative research program. Once carefully prepared in New York, it became evident that this was a specimen of a yet unknown species, one we ended up naming *Changchengornis hengdaoziensis*—the "Great Wall bird."

Not surprisingly, *Changchengornis* is genealogically more closely related to *Confuciusornis* than to any other known bird—they are the only convincingly identifiable confuciusornithids. However, despite the anatomical likeness between these animals, considerable differences make the slightly smaller *Changchengornis* stand out. For example, the structure of its head is clearly less

robust than that of its abundant cousin; its beak is shorter and prominently curved; its upper jaw is much longer than the lower jaw; and in the foot, *Changchengornis* exhibits a first toe that is noticeably longer and thus, better designed for grasping, than *Confuciusornis*. To date, all we know about this ancient short-tailed bird is based on just one fossil, that of the National Geological Museum of China. The contrast in the abundance of the only known confuciusornithids—hundreds of fossils of *Confuciusornis* to a single example of *Changchengornis*—is truly staggering, and while it may reflect a sparse population of the latter bird, future fossils of *Changchengornis* are likely to be distinguished as more and more fossils of *Confuciusornis* are studied in detail.

The Confuciusornithid Body Plan

Although the powerful toothless jaws of these birds were sheathed by a horny beak, the overall architecture of the skull and the rest of the skeleton of *Confuciusornis* and *Changchengornis* retain many ancestral features found in the more primitive long-tailed birds and their dinosaurian relatives. For example, these Chinese beaked birds share with *Archaeopteryx* and its precursors a well-developed postorbital bone and a squamosal bone that does not take part in the construction of

A close relative of *Confuciusornis*, the smaller *Changchengornis hengdaoziensis*, is thus far known only by this specimen split in two slabs. (Image: L. Meeker.)

Confuciusornis (foreground) compared to *Changchengornis* (background). Note the shorter and curved beak of the latter. (Image: E. Heck/S. Abramowicz.)

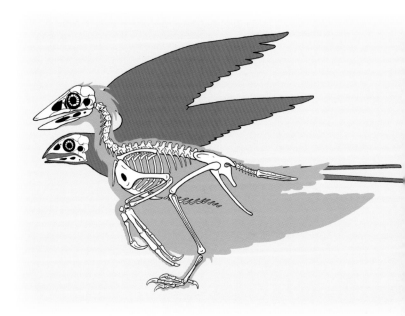

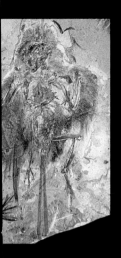

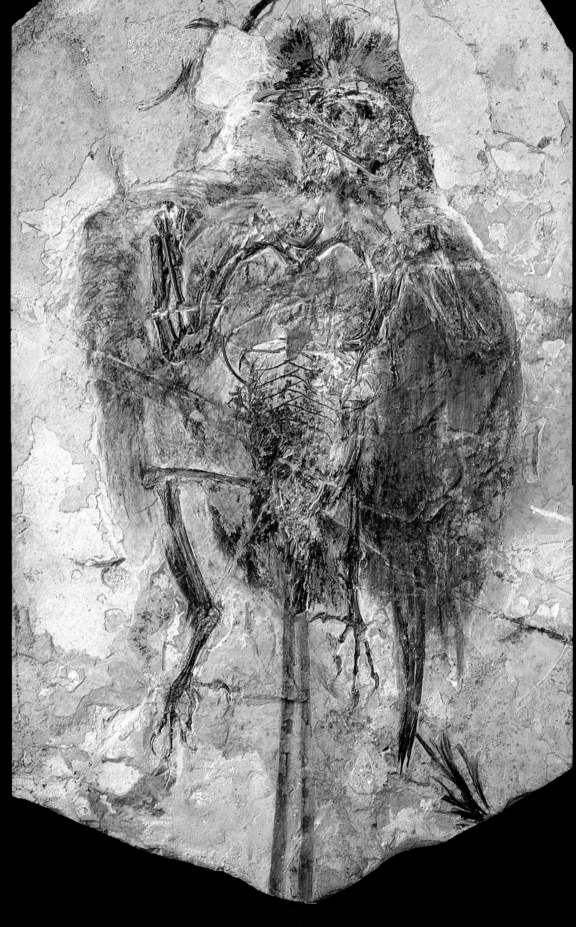

the braincase—which suggests that the jaws were powered by much larger muscles than those in modern birds. Yet perhaps the most striking aspect of the confuciusornithid skull is the contribution of the postorbital to a stout bony bar that completely encloses the infratemporal opening behind the orbit, leading to a configuration reminiscent of the primitive diapsid design of nonavian reptiles. The discovery of such an unusual design had a tremendous anatomical significance, because the highly modified diapsid skull of living birds had always been regarded as an important avian trademark, one common to all members of the group. Interestingly, when the genealogy of *Confuciusornis* and *Changchengornis* is taken into consideration, the complete separation between orbit and infratemporal opening needs to be interpreted as a specialization, because the placement of these fossils within the avian family tree indicates that such a structure evolved from birds whose skulls had already acquired a less "diapsid" appearance.

Archaeopteryx is a good example of a bird more primitive than the confuciusornithids, whose orbit is only partially separated from

the infratemporal opening. The complete separation found among the confuciusornithids also has important mechanical consequences. As discussed earlier, the skull of modern birds is characterized by a system of bony struts that grants the snout some degree of independent motion relative to the braincase, a functional attribute known as skull kinesis. Movements of the snout are controlled by the forward and backward rotation of the quadrate, the bone that connects the lower jaw to the rest of the skull, and the action of a series of bony struts in the palate. Yet the stout bony bar that in the beaked confuciusornithids separates the orbit from the infratemporal opening likely blocked any forward movement of the quadrate bone, thus limiting, if not impeding, any kind of independent motion the snout could have had.

No substantial differences are seen in the proportions of the neck and trunk of *Confuciusornis* and *Changchengornis* with respect to *Archaeopteryx* or other long-tailed birds. Yet the shape of the vertebral joints in the front portion of the neck had begun to more closely resemble the saddle-like articulations of living birds. Together with evidence from other Mesozoic fossils, this suggests that the evolution of the unique saddle-shaped vertebrae of birds progressed backwards from the beginning of the neck.

A more modern design is also present in the sacral region, where the synsacrum incorporated additional vertebrae and became longer than that of the long-tailed birds. Such an enlarged synsacrum provided further strength to the pelvis, a transformation that became particularly important during landing. However, the most notable change in the spinal column of these birds involved the abbreviation of the tail. Indeed, most of the tail vertebrae of *Confuciusornis* and *Changchengornis* became fused into a rump bone called the pygostyle, that portion of the skeleton which supports the fatty parson's nose of fowl and other

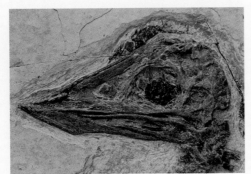

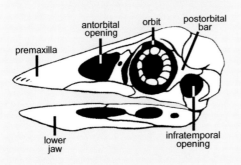

modern birds. The long pygostyle of *Confuciusornis* and *Changchengornis* anchors a feathered tail that greatly differs from the frond-like appendage of *Archaeopteryx*, holding only a short tuft of thin feathers that gave the fleshed-out tail a diamond-like shape. Such a compact tail could hardly have generated any significant lift—the powerful wings of these birds must have been responsible for producing most of it—but we should not assume that the abbreviation of the tail had no aerodynamic significance. On the contrary, there is little doubt that this evolutionary innovation must have had profound aerodynamic implications. With the substantial abbreviation of the tail, not only were the hindlimbs and tail functionally decoupled, but the latter was also able to enhance steering during flight.

These primitive beaked birds also retained a number of features of the shoulder and ribcage of their long-tailed predecessors. Nonetheless, several important transformations with respect to the skeleton of *Archaeopteryx* are clearly noticeable. Their stout coracoids, for example, approached the strut-like appearance of living birds, and the sternum became significantly larger, even though it had not yet developed any significant crest for the origin of enlarged flight muscles. Another change was a reduction in the number of rods that form the gastralia—a corset of interlocked bony rods that lines the abdominal wall of many nonavian reptiles and early birds and assists them to ventilate their lungs during breathing. While the existence of this bony corset in *Confuciusornis* and other early birds suggests that aspects of a typical reptilian breathing system may have still been at play during the early phases of avian evolution, the reduced number of rods from *Archaeopteryx* onwards suggests the gradual evolution of a modern type of breathing mechanism that relies mostly on the movements of a large sternum to aerate the lungs.

The forelimb skeleton, albeit supporting a wing of modern feathers, also retained the basic configuration of birds' theropod forerunners. Although the relative length of the whole forelimb was greater than in theropods, the ratios between confuciusornithid wingbones, and the structure of their clawed hands, were comparable to those of long-tailed birds and nonavian theropods. A notable exception is the partial fusion of the wristbones, a feature probably acquired in the context of enhanced flying capabilities. Like in *Archaeopteryx* and other long-tailed birds, the sharp claws

pygostyle

Confuciusornis is one of the most primitive known short-tailed birds. The long pygostyle of this bird anchored a feathered tail drastically different from both the frond-like tails of its predecessors and the fan-like tails of many of its living counterparts. The short pintail of this bird probably generated a limited amount of lift. (Image: L. Chiappe.)

of *Confuciusornis* and *Changchengornis* probably projected from the outline of their wing, perhaps retaining some of the grasping capabilities of their forerunners.

In most aspects, the basic framework of the pelvis of these beaked birds was also similar to that of *Archaeopteryx* and other long-tailed birds. The acetabulum is large, the ischium is much shorter than the pubis, and the counterparts of the latter bone converge to contact one another over their lower fourth. More significant differences evolved in the hindlimbs. For example, the tibia of *Confuciusornis* and *Changchengornis* is completely fused to the upper anklebones,

thus forming a true tibiotarsus—a compound bone characteristic of birds. Likewise, the short metatarsals of the foot became fused together in their uppermost portions and to the lower anklebones, thus forming the avian tarsometatarsus. The development of these two compound bones—tibiotarsus and tarsometatarsus—must have had significance in the evolution of the modern pattern of stance and gait, as well as in strengthening birds' landing gear. Like that of most birds, the foot of confuciusornithids had three main toes facing forward and an opposable hind toe or hallux. The fact that the latter was short and lay in an elevated position suggests that these beaked birds did not have the sophisticated grasping ability and perching capacity of many of their modern counterparts.

In sum, when confuciusornithids are compared to more primitive long-tailed birds such as *Archaeopteryx* and *Jeholornis*, we witness key transformations in the flight apparatus, together with changes strengthening the sacrum and hindlimbs. These modifications likely played a role in the aerodynamic performance of these birds while also improving their landing ability. However, the most significant change is manifested in the abbreviation of the tail and its functional migration, from its use in terrestrial locomotion to its new role during flight.

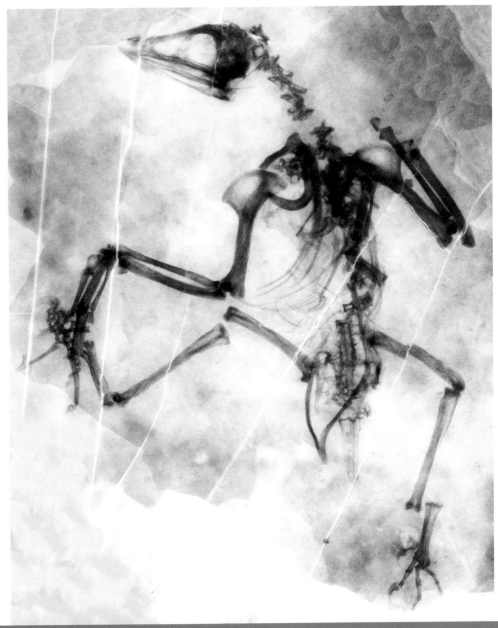

X-ray of a *Confuciusornis* specimen. X-rays and computerized tomography imaging often allow anatomical details otherwise obscured by the encasing rock to be seen. These techniques are also helpful in guaranteeing the integrity of many Chinese fossils. (Image: A. Goernemann.)

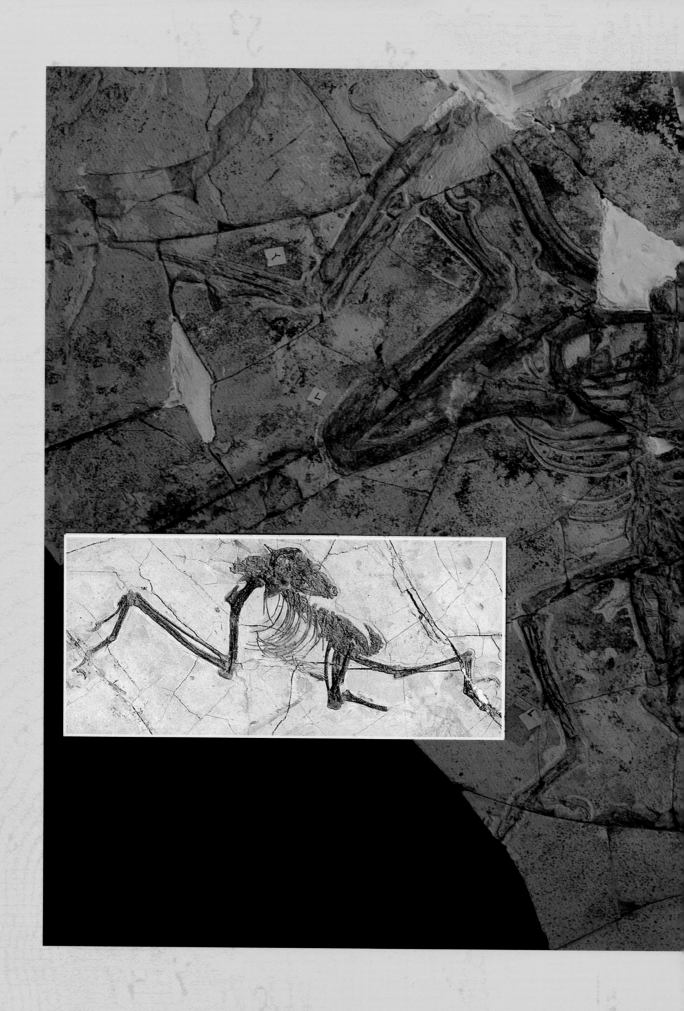

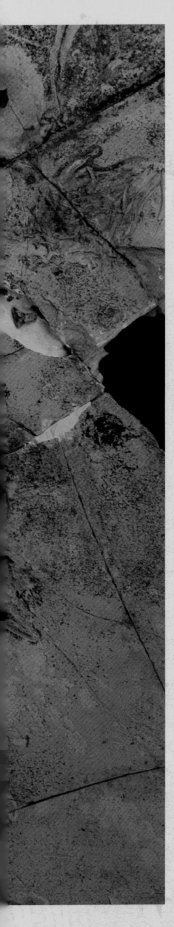

SAPEORNIS:
A LARGE SHORT-TAILED BIRD

Fossils of a primitive short-tailed bird much larger than *Confuciusornis* have been recently unearthed in western Liaoning. With a wingspan of nearly a meter, *Sapeornis chaoyangensis* is larger than any other Early Cretaceous bird. Only five million years younger than *Confuciusornis*, this bird represents a similarly early stage of avian evolution. In fact, the evolutionary development of its skull, vertebral column, pelvis, and hindlimbs is quite like that of confuciusornithids, although *Sapeornis* has large, protruding teeth on its upper jaw, a somewhat longer neck, and a shorter pygostyle. Other significant differences exist on its shoulder, ribcage, and wing, where *Sapeornis* evolved a mosaic of features (some more primitive and some more derived) with respect to the confuciusornithids. Indeed, while its short and wide coracoid more closely resembles that of *Archaeopteryx* than the strut-like coracoid of the confuciusornithids, its long wings and wishbone were built in a more modern fashion.

Half a dozen exquisite specimens have told us a great deal about the appearance of *Sapeornis*, but this peculiar bird remains in many ways a challenge. The remarkable length of its wings and the functional significance of its abbreviated tail are clearly indicative of

aerodynamic proficiency, as is the shortening of its outermost and innermost fingers—the beginning of a characteristic of more skillful fliers. However, these aerodynamically important features contrast with the primitive construction of the shoulder region, which in the absence of a breastbone of any significant size, likely powered the wings with muscles originating on the expanded coracoids and the stout wishbone. The functional consequences of such a design are not fully understood, but the flight style of such a bird can hardly be compared with those of its modern counterparts.

Sapeornis' lifestyle is not much clearer. Its feet are better designed for grasping, and thus perching, than those of *Confuciusornis*, but the remarkable length of the wings is better fitted for open environments. The stout design of its teeth and numerous stomach stones tell us that *Sapeornis* probably fed on hard structures. Stomach stones are typical of seed-eaters and other herbivorous animals, where their grinding action assists gastric acids in digesting tough items, but the food range of this archaic bird remains elusive.

Even its placement within the avian family tree is not yet settled. Paleontologist Zhou Zhonghe believes *Sapeornis* diverged before the confuciusornithids, making it the most primitively known short-tailed bird; but a combination of features exclusive to these lineages (such as a large foramen piercing the upper portion of the humerus) and others clearly more advanced (the shape and proportions of the wing, the smaller pygostyle, and the design of its furcula and feet) cast some doubt on this hypothesis.

Undoubtedly, more fossils and additional studies are needed to understand the flying ability, ecological milieu, and kinship of this remarkable bird—but the discovery of *Sapeornis* has consolidated the notion that as early as 120 million years ago, some of the most archaic short-tailed birds had already achieved sufficient flight power to overcome the aerodynamic problems associated with large body sizes. (Image: L. Chiappe/S. Abramowicz.)

The feet of *Confuciusornis*. The short and elevated position of its hind toe—opposed to the others—must have limited the perching capabilities of this bird. (Image: L. Chiappe.)

The evolution of locomotor modules from primitive tetrapods to birds. The large locomotor module of early tetrapods became reduced in the bipedal predecessors of birds. Further modifications to this module occurred during the early evolution of birds, and the origin and fine-tuning of flight led to the development of two new modules (tail and wing). (Image after Gatesy and Dial, 1996.)

Short Tails and Locomotion

When examining the evolution of short-tailed birds and the significance of this notorious transformation in the fine-tuning of flight, it is useful to introduce the notion of "locomotor modules," a concept elaborated by functional morphologists Steve Gatesy and Kenneth Dial. Locomotor modules are anatomical regions of the combined skeletal and muscular systems that act as integrated units during locomotion. The earliest tetrapods had a single locomotor module that controlled the lateral undulations of their body and the limbs responsible for their locomotion. The bipedal ancestors of dinosaurs evolved a smaller locomotor module that comprised their hindlimbs and tail—their forelimbs became disengaged from locomotion. In time, the nonavian theropods close to the origin of birds developed a forelimb locomotor

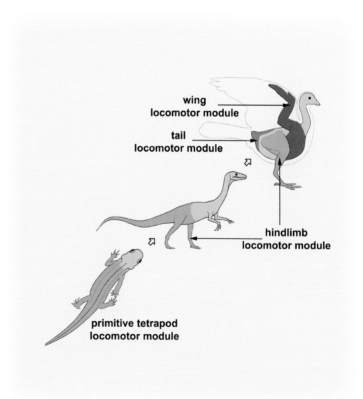

wing locomotor module

tail locomotor module

hindlimb locomotor module

primitive tetrapod locomotor module

module that possibly allowed them to increase their running speed and to ascend a variety of natural inclines. The evolution of this wing locomotor module paved the way for the origin of avian flight. However, further modifications to the primitive locomotor module—the one composed of the tail and hindlimbs—were also paramount in the development of flight. As Gatesy and Dial have shown, this module became subdivided during the transition from nonavian theropods to birds.

Living archosaurs—crocodiles and birds—move their legs in profoundly different ways. In crocodiles, most of

the hindlimb swing during the gait cycle occurs at the level of the femur, which pivots around the hip socket. In birds, however, there is minimal movement at this level and most of the gait is generated by the tibiotarsus, which swings around the knee joint. Studies led by Gatesy have documented that the most important factor in these two different kinds of gait is the reduction of the tail and its concomitant musculature. Changes in musculature as inferred from a variety of muscle attachments on the hindlimbs of nonavian theropods indicate that the transformation from a crocodile-like gait to the avian type of

Confuciusornis had very long and tapered wings, formed by fully asymmetric feathers (inset). (Image: M. Ellison.)

stride began to happen way before the tail became the abbreviated appendage of short-tailed birds. The reason behind this transformation is not entirely clear but what *is* evident is that it eventually led to the subdivision of the ancestral locomotor module of early dinosaurs into two different modules: while the hindlimb continued to operate within the terrestrial realm, the tail portion became functionally decoupled from its hindlimb component. Such an important innovation allowed the tail to function in concert with the forelimbs, a novel association that facilitated the remodeling of the tail—including its abbreviation—and the modern integration of locomotor modules.

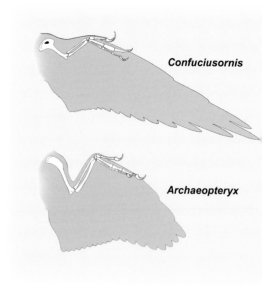

Confuciusornis evolved wings that were much longer and larger than those of *Archaeopteryx*. The relative size of the wing of this Chinese bird could have mitigated the minimal lift generated by its short tail. (Image: S. Abramowicz.)

An Ornate Plumage

Much of the plumage of *Confuciusornis* and *Changchengornis* is preserved as a dark halo surrounding their fossilized skeletons, and like in *Archaeopteryx*, it looks remarkably modern. The long flight feathers of the wing have asymmetrical vanes and together form a sharp airfoil of high aspect ratio that calls to mind living terns. The short and tightly packed feathers of the tail surround the pygostyle in a design that resembles the pintails of some extant birds. However, the most remarkable feature of the plumage of these birds is the presence of a pair of very long and stiff, streamer-like feathers adorning the tail, which lack a distinct shaft and barbs through most of their length and compare best to the ribbon-like feathers of the tail of the birds of paradise of New Guinea.

Interestingly, the long feathers of the tail of confuciusornithids do not ornament every individual—they are missing in certain fossils and this absence appears not to be related to preservation. Indeed, the completeness of many "tailless" specimens of *Confuciusornis*, often preserved with intact, articulated skeletons and surrounded by a full complement of flight and contour feathers, rules out the possibility that the tail feathers

of these fossils became detached after death. In all probability, these "tailless" individuals did not bear these feathers, at least at the time of their death. This distinctive difference between specimens of the same species has generally been interpreted as a sexual characteristic, a feature expressed differently between genders. We will return to this issue, but first, let us examine some other types of individual variation and their significance in understanding the evolution of confuciusornithids.

Is There More Than One Species of *Confuciusornis*?

Species recognition is at the heart of the paleontological enterprise. However, deciding whether two particular fossils belong to the same species is not an easy task. Not only do the expected individual differences among specimens of a single species need to be accounted for, but nearly all fossils show some degree of deformation as a consequence of being at the mercy of a wide range of physical agents for millions of years. The case of *Confuciusornis sanctus* is not an exception—its fossils are typically flattened and distorted because of their compression within a slab. In such cases, the line between what is an individual difference between two specimens of the same species and what is a difference between two individuals of different species becomes a thin one.

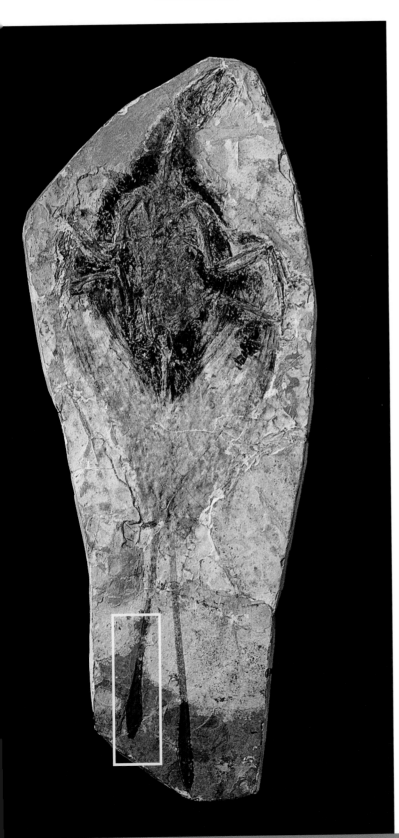

A number of specimens of Confuciusornis—as well as the only known specimen of Changchengornis—carried a pair of very long feathers that projected from the tail. (Image: M. Ellison.)

Since the discovery of the first *Confuciusornis sanctus*, a handful of specimens of virtually identical appearance have been used to erect either additional species of *Confuciusornis* or species alleged to be near relatives of these birds. In some instances, the new species were distinguished on the basis of perceived differences in the thickness of limb bones, the relative lengths of the tarsometatarsus, or the degree of curvature of the hand claws with respect to those exhibited by fossils typically regarded as *Confuciusornis sanctus*. However, examination of these fossils suggests that the alleged differences should be treated with caution. Indeed, when these fossils are examined in the context of a larger sample of specimens, the discriminatory features fall within a continuum—in other words, within the range of variation expected for *Confuciusornis sanctus*. In other cases, the alleged features used to erect new species were simply based on anatomical misinterpretations. The latter problem, for example, applies to *Confuciusornis chuonzhous*, a species based on a fossil thought to have had three phalanges in its hind toe, a design unknown for any bird. Yet detailed examination of this fragmentary fossil reveals that the "first" phalanx of its hind toe is in fact this digit's metatarsal—the fossil used to erect *Confuciusornis chuonzhous* actually has the same two phalanges as all other fossils of Confuciusornis sanctus, and similar hind toe proportions.

In reality, once the deformation of all published specimens is taken into account, the only obvious difference within the large sample of *Confuciusornis* fossils appears to be in their size. While the average specimen has a wingspan of about 80 centimeters and a size comparable to that of a magpie, the lengths of the long bones of the smallest individuals are 60 percent the size of the largest ones, even though none of them exhibit features characteristic of young individual ages. Most ornithologists working on living birds would argue that size differences of this magnitude indicate species differentiation, but comparable size variations are not unprecedented among other species of primitive birds. Earlier, I noted that despite the Eichstätt *Archaeopteryx* being almost half the

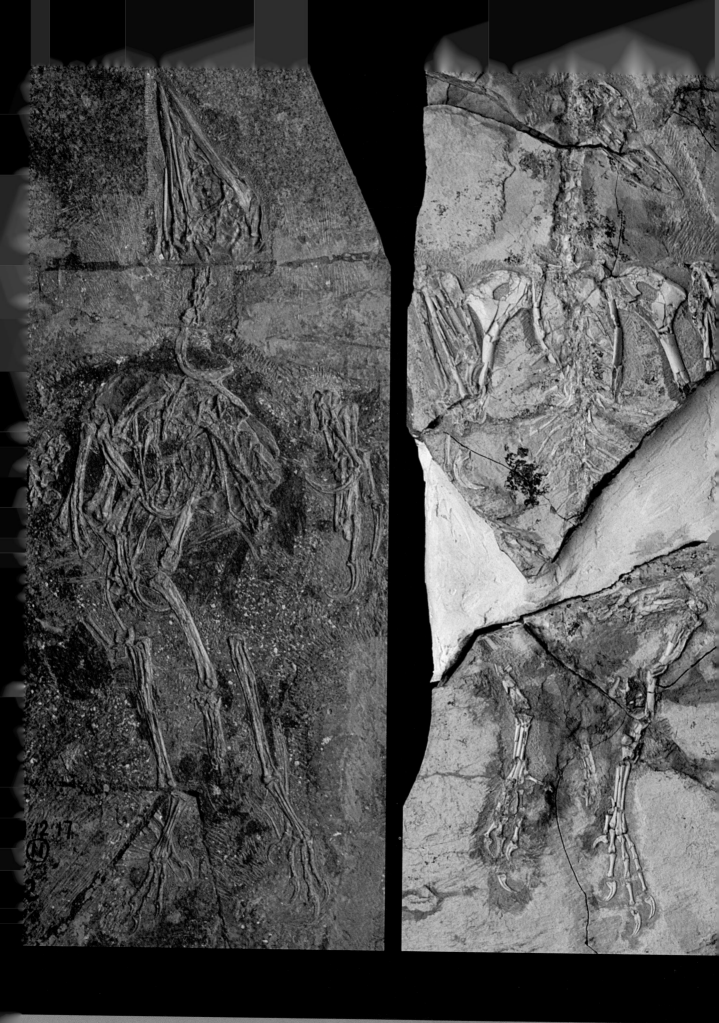

Some of the previously named close relatives of *Confuciusornis*—*Jinzhouornis zhangjiyingia* (left) and *Jinzhouornis yixianensis* (right)—are most likely specimens of the same species, *Confuciusornis sanctus*. The alleged differences between these species are likely individual variations within a single population. (Image: L. Chiappe.)

size of the Solnhofen specimen, these fossils are usually interpreted as the same species. I have also highlighted how two statistical studies of the relative dimensions of the bones of several *Archaeopteryx* specimens have supported the notion that all these fossils are part of a single growth series—in other words, that they all belong to the same species. If the significant size variation of *Archaeopteryx* can be explained as the result of sampling a growth series, size differences in *Confuciusornis* could also be explained in the same fashion. Given that no convincing anatomical argument for the existence of more than one species of *Confuciusornis* has ever been proposed, it is more reasonable to regard the many fossils of this bird as belonging to a single species, *Confuciusornis sanctus*.

Courtship Attire

The discovery of multiple fossils of *Confuciusornis* with their complete feather attire offers the first opportunity to look into the plumage variation of very ancient birds. As mentioned earlier, a conspicuous difference among well-preserved specimens involves the presence or absence of a pair of long, ribbon-like feathers projecting from the tail. This difference has been heralded both as a sexual characteristic and as evidence for ancient avian courtship, because males and females of living birds often differ in their plumage. Linking this obvious difference to sex is undoubtedly a reasonable alternative, but let us consider some other possible explanations for the plumage patterns observed in *Confuciusornis*.

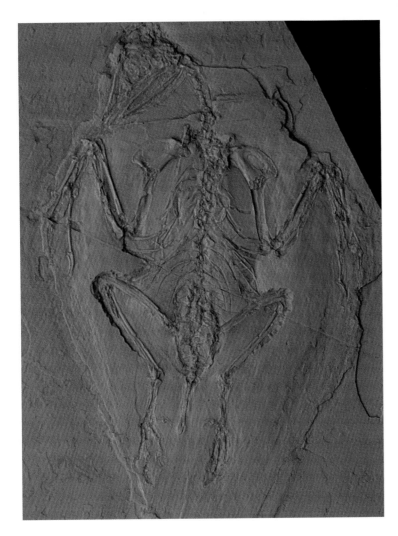

The anatomical evidence used in favor of the existence of more than one species of *Confuciusornis* is largely unwarranted. Perhaps the only exception is *Confuciusornis dui*, whose upcurved beak stands out from the straight upper bill of *Confuciusornis sanctus*. Unfortunately the holotype of *Confuciusornis dui* is unavailable, being kept in a private collection; a cast of this specimen is shown here. (Image: L. Chiappe.)

As in other primitive birds and their dinosaurian predecessors, anatomically indistinguishable specimens of *Confuciusornis* exhibit a wide range of sizes. The two sizes shown here are at the ends of a spectrum interpreted as the growth series of this bird. (Image: L. Chiappe.)

One may involve molting, the periodic replacement of feathers. This typical avian phenomenon was written about during Roman times, when the shedding of the peacock's tail in winter and its rebirth in spring linked these birds to immortality. In fact, most extant birds molt their entire plumage, or portions of it, at least once a year. As a result of their exposure to sunlight and other physical agents, feathers wear out and eventually break. During molting, the deteriorated feathers are replaced by new ones that grow within the same skin follicles, pushing out the damaged ones. All molt stages can be found within a population during the molting season. An extreme example is found in birds of paradise, in which males displaying exuberant plumage coexist with males that have just begun to grow their flamboyant attire. Definitive evidence of molting has never been documented for any Mesozoic bird, but the fact that the plumage of *Confuciusornis* and other early birds shows no evidence of intense wear suggests that their feathers were regularly replaced. Thus, it is conceivable that different individuals of *Confuciusornis*, both males and females, molted their pintail feathers, and that they did so in such a way that at a given time, the feathers of some individuals would be fully formed while those of others would just be beginning to be renewed.

In addition, the diversity of strategies of feather growth and molting in extant birds points to other equally likely alternatives for explaining the observed pattern of tail feathers in the *Confuciusornis* fossil sample. For example, modern birds do not develop their display plumage until they become sexually active adults, and in some species, males and females molt asynchronically (that is, at different times). With this array of molting strategies in living birds, we should be wary of assuming, even in those cases in which a pintailed *Confuciusornis*

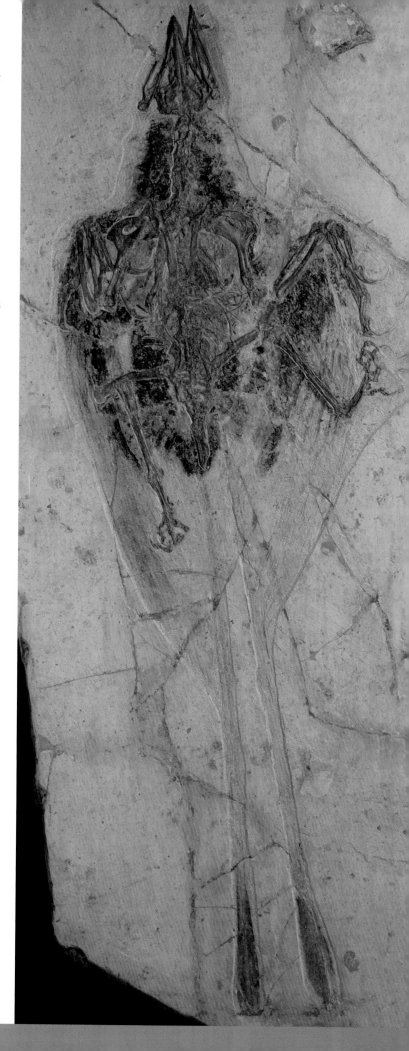

skeleton is preserved in the same slab as a "tailess" one, that they represent a male and a female. Even if we were to assume that because they are found in the same slab they lived at about the same time, two such individuals may well be of the same gender, one having not yet reached its sexual maturity. They could also be members of a molting population, either of the same sex or different sexes. Or it could be that *Confuciusornis* molted its feathers asynchronically, males before females or vice versa, in which case these two specimens might be of different genders but the long tail feathers would still not necessarily be exclusive to a particular sex.

To resolve the question of whether the observed differences in plumage can be ascribed to sexual differences, we may need to enlist the help of morphometrics, a mathematical discipline that studies changes in shape from a statistical point of view. By measuring the bones of a sufficient number of skeletons, morphometric analyses could tell us whether the studied sample of *Confuciusornis* fossils reflects size-dependent features that can be ascribed to one or another sex. In other words, by sifting these measurements into two groups, these studies could tell us whether there is statistical support to argue that the females of this ancient bird differed in some size-related feature from the males. If those differences were to be found, we could then examine whether the long pintail feathers occurred in only one of the two groups discerned in the morphometric study; and if these feathers were found to be restricted to one such group, we could argue in favor of their sexual interpretation. Taking advantage of the large number of well-preserved skeletons of *Confuciusornis* housed at public institutions, Jesús Marugán-Lobón (from the Universidad Autónoma of Madrid) and I have started to examine whether the interpretation of the long

tail feathers of these birds as sexual attributes can be supported statistically. We still need to collect more data, but at the time of writing this, the implication that the long-feathered tails of some specimens of *Confuciusornis* may be a feature restricted to one sex is not yet obvious.

Our statistical study is still preliminary, but even if the final results were to support the presence of a long-feathered tail as an attribute of a single sex, determining whether these feathers adorned the tail of males or females will not be as clear-cut as some have suggested. Although the males of most living birds are larger and have a more ostentatious plumage than females, exceptions are not uncommon. In addition, certain studies indicate that nonavian theropod females were larger than males and displayed horns, crests, and other structures absent or attenuated in the latter. The fact that more flamboyant females are known for a number of living birds, and the evidence that this may have also been the case for the forerunners of birds, suggests that even if we were to accept the proposal of sexual dimorphism, the identification of *Confuciusornis* sexes on the basis of the long feathers that garnish the tails of some fossils is not simple.

Confuciusornis' Place in the Tree of Birds

Early attempts to understand the position of *Confuciusornis* within the avian family tree supported a relationship with enantiornithines, a far-flung group of Cretaceous flying birds that will be discussed in the next chapter. Guided by the unusual beaked condition of *Confuciusornis*, Hou and his associates first proposed a close kinship between the Liaoning bird and the toothless enantiornithine *Gobipteryx minuta* from the Late Cretaceous of the Gobi Desert. However, support for this proposal was very weak and virtually limited to the

beaked nature of these two birds. Soon after, this genealogical interpretation was modified by removing *Confuciusornis* from within enantiornithines and instead regarding it as the closest relative of the entire enantiornithine ensemble. Yet evidence supporting this next conclusion continued to be weak, since the grouping was based on the presence of two features— a pygostyle and pneumatized trunk vertebrae—that are widely distributed among other birds. Additionally, the interpretation of *Confuciusornis* as either an enantiornithine or as their nearest kin had unfounded implications for our understanding of the evolutionary fine-tuning of flight, because enantiornithines share a suite of flight-correlated features with modern birds—flexible wishbones, keeled breastbones, and modern forelimb proportions, among others—that are absent in the confuciusornithids. In other words, by connecting the ancestry of these archaic beaked birds to that of enantiornithines, Hou's hypothesis implied that the more sophisticated flight of both enantiornithines and modern birds had evolved independently from more clumsily flying ancestors.

Although problems related to these initial interpretations involved the limited number of bird species and anatomical features included in the study, the most important caveat dealt with the methodological and philosophical approach. I highlighted earlier how genealogical relationships are nowadays inferred using parsimony—the principle that chooses the simplest explanation for the evidence at hand. Hou's proposed relationship between *Confuciusornis* and enantiornithines was a clear departure from this principle, because the support for this evolutionary link was overcome by the number of features suggesting a closer relationship between enantiornithines and modern birds. That is, a simpler alternative hypothesis was that the abovementioned flight-correlated characteristics had evolved only once,

in the common ancestor of these two groups. In fact, when *Confuciusornis* was considered within the context of studies that included many more primitive birds, compared many other anatomical features, and used cladistic methods such as the parsimony principle, the genealogical signal was clear. These studies demonstrated that *Confuciusornis* (and its sibling species, *Changchengornis*) did not share a most recent common ancestor with enantiornithines but with a much larger group that included not only these birds but also a number of other Cretaceous lineages and all their living relatives. Today, the beaked birds from Liaoning are widely accepted as the earliest known episode in the evolutionary history of short-tailed birds.

Ground Foragers and Tree Dwellers

Understanding the lifestyle of extinct organisms is always problematic. The fossil record rarely preserves any evidence of the functions performed by ancient organisms, and despite the hundreds of specimens discovered in the Yixian beds, *Confuciusornis* is not an exception. Since its discovery, some researchers have regarded this ancient bird as an enhanced climber that used the sharp claws of its hands and feet to ascend tree trunks, and its opposable hallux to perch high in the canopy. Although it is easy to see how the shape of the claws and feet of this animal could have led to such an interpretation, it is also clear that preconceptions about the origin of flight from arboreal animals played a role. In fact, only under these circumstances can some reconstructions of *Confuciusornis* in bizarre squirrel-like postures be explained.

Earlier in this book, we addressed the problems related to interpreting similar features as evidence of both tree-climbing abilities and arboreality. Just as the presence of sharp claws in *Archaeopteryx* need not imply these

behaviors—because these features were also present in its ground-dwelling predecessors—their occurrence among the confuciusornithids requires no special explanation beyond evolutionary legacy. A conclusion of this nature should only be constructed upon the discovery of features that are exclusively known among tree-climbers, but no evidence of this sort has yet been offered for either

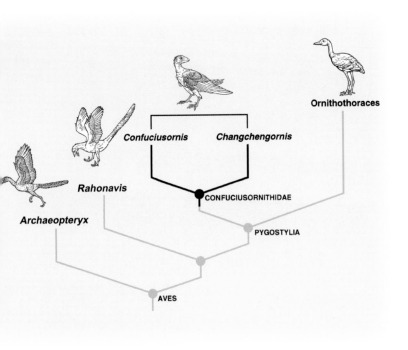

The evolutionary relationships of *Confuciusornis* and *Changchengornis* are well-understood. These confuciusornithids are amongst the most primitive pygostylians, the avian group that includes all short-tailed birds. (Image: E. Heck/S. Orell.)

Confuciusornis or *Changchengornis*.

Not only it is unlikely that the beaked birds from Liaoning were enhanced tree-climbers, but comparisons between their feet and those of extant perching birds suggest that they may have spent a fair amount of time foraging on the ground. The relative lengths of the first (hind) and second toes of these birds are dramatically different from those of extant perchers. While in *Confuciusornis* and *Changchengornis* the hind toe is shorter than the second toe (half, and two-thirds the length, respectively), in living perching birds it is usually longer than the second toe. Further support for a ground-foraging habit comes from Hopson's analysis of the relative lengths of the central phalanges of the third

toe of a wide range of living birds—a study mentioned earlier in the context of *Archaeopteryx*. This investigation showed that in species with grasping feet, the third phalanx of the third toe is longer than either the first or second phalanges of this toe. In *Confuciusornis*, however, the first phalanx is consistently the longest phalanx in the third toe. By observing the habits of many living birds included in his analysis, Hopson concluded that although the foot of *Confuciusornis* was somewhat closer to those of arboreal birds than to that of *Archaeopteryx*, the design of its foot approached that of birds that often forage on the ground. This relationship indicates that even if *Confuciusornis* and *Changchengornis* likely perched on trees, the anatomy of their feet does not support a primarily arboreal existence.

Our understanding of the flight capabilities of these birds also bears on our interpretation of their lifestyles. There is little doubt that with their long, narrow wings of high aspect ratio and their asymmetric flight feathers, both *Confuciusornis* and *Changchengornis* were capable of flying. In spite of the overall primitive characteristics of their skeletons, the large sternum, the elongate coracoid, and the pygostyle are some of the features typically correlated with flight capabilities not seen in *Archaeopteryx*, and streamer-like tails are present in many birds with substantial aerial maneuverability. It is conceivable that these birds flew in ways different to all living birds, but their capacity for aerial locomotion is unmistakable. In fact, if while sporting a much more primitive skeleton and lacking many of the flight-correlated characteristics present in confuciusornithids, *Archaeopteryx* may have been able to fly from the ground after a short take-off run, the aerodynamically better-equipped confuciusornithids may have even been capable of taking off from a standstill position on the ground. This inference bears on the issue of whether

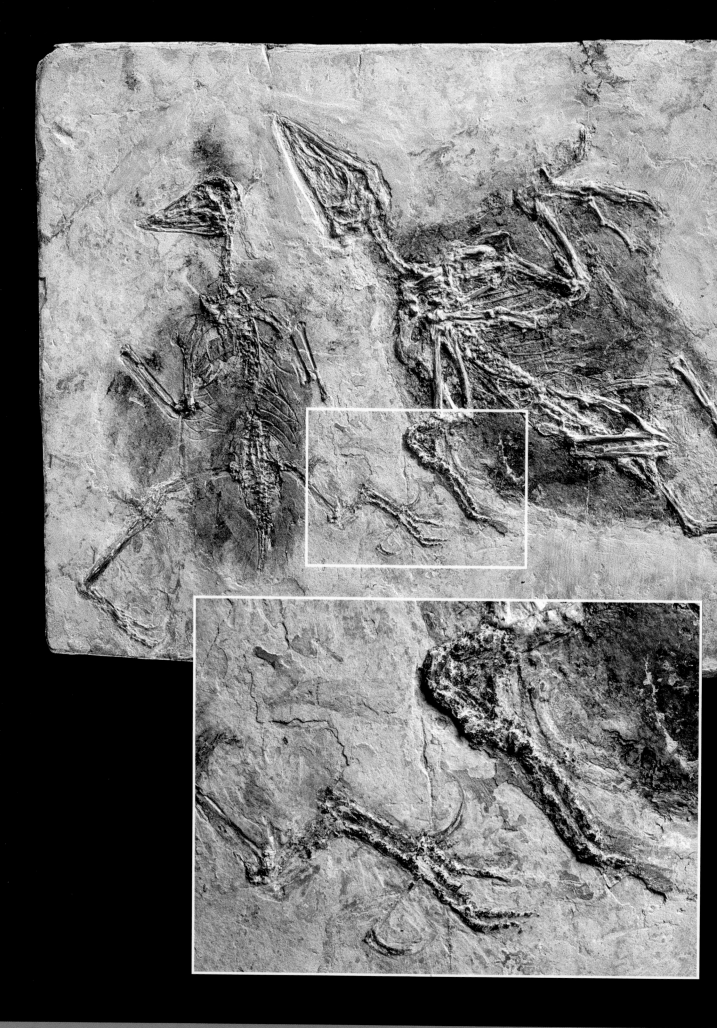

Confuciusornis or *Changchengornis* were tree-climbers. If they were capable of taking off from the ground, these birds are unlikely to have climbed trees in squirrel-like poses rather than taking flight and landing on them.

Ancient Eruptions and Mass Mortality

The extraordinary skeletons of birds and nonavian theropods that have put quiet Sihetun at the center of the paleontological community are not alone. The multicolored shales of the Yixian Formation reveal an environment teeming with life, one which flourished at a time when the geography of China was utterly different from today. In the Early Cretaceous, the Yangtze River and its Himalayan source were millions of years away and the area of its renowned gorges lay in the middle of a deep ocean. Emerged lands to the north were spotted by a network of shallow lakes and temperate forests that extended across Mongolia and as far as Kazakhstan.

But this verdant world was not immune to threats. During the Jurassic–Cretaceous transition, much of the eastern border of Asia experienced widespread volcanism, clues of which have been left in the abundant layers of ash and lava that alternate with Yixian's fossil-bearing deposits. From Santorini to Pompeii, Krakatoa, and Mount Saint Helens, our history is laden with events of comparable destruction. Emissions of poisonous gas and large clouds of ash may account for the large number and spectacular preservation of Yixian fossils, some of them even preserving snapshots of behavior. To be sure, the staggering number of fossils of *Confuciusornis* suggests episodes of mass mortality—Sihetun farmers once told me of more than 40 specimens contained in less than 100 square meters—and such profusion, along with the common occurrence of slabs containing two specimens, even hints at gregarious behavior. Catastrophic volcanic activity would also explain the extraordinary preservation—through rapid burial and limited bacterial activity—and the fact that the fossils show minimal disruption from scavenging. The definitive evidence of volcanic activity and the stunning preservation of the Yixian ecosystem document another example of devastation at the hands of nature. Here, the magnificent fossil ecosystem of Sihetun echoes the disaster of Pompeii, leaving behind a story many times more ancient.

The paleontological community is well aware of the vast amount of Chinese fossils that have been faked (completely or partially) to increase their market value. The inset shows the sculpted portions (a foot and a wing) of two *Confuciusornis* specimens. In this case, the artist candidly used a chicken wing as model, but other forgeries can be much more subtle. (Image: M. Ellison.)

CHAPTER SEVEN

The Great Mesozoic Radiation: The Flying Enantiornithes

In all likelihood, the profound locomotor transformations resulting from the abbreviation of the tail opened the door for a significant burst of diversity. With the tail being transformed into an independent locomotor module operating in concert with the wing locomotor module, the avian body acquired an essentially modern design. The fossil record tells us that an archaic lineage of short-tailed birds, the Enantiornithes, experienced a major evolutionary radiation at the beginning of the Cretaceous. Inland rocks formed between 130 and 120 million years ago preserve numerous fossils of these birds, whose skeletal anatomy and songbird sizes document a significant departure from more primitive fliers. Even in these early hours of the history of birds, the oldest enantiornithines had already evolved a series of important aerodynamic features that enabled them to fly at low speeds and to achieve a degree of maneuverability perhaps comparable to their living counterparts. Undoubtedly, these superior flying abilities facilitated the conquest of a wide range of environments and the development of a diversity of feeding specializations. The later evolution of enantiornithines was characterized by a distinct increment in size, a trend that highlights again the aerodynamic power achieved by this group.

The sparrow-sized,
120-million-year-old
Cathayornis yandica
illustrates the dramatic
reduction in size that
characterizes the
evolutionary divergence
of enantiornithines.
(Image: M. Ellison.)

Like no other ancient lineage of birds, the Enantiornithes embody the remarkable burst of fossil discoveries of the last two decades. Recorded throughout the Cretaceous and from every continent except Antarctica, half of all species of Mesozoic birds are members of this group. And yet its existence was first recognized just 25 years ago.

The First Hint

From penicillin to gunpowder, many scientific discoveries have been serendipitous. Avian paleontology is no exception. In 1974, Mesozoic paleontologist José Bonaparte organized a field party from Argentina's National University at Tucumán to explore the Late Cretaceous continental deposits that sparsely crop out in the thick, thorny acacia forest of the northwest corner of this country. In this nearly impenetrable forest, it is difficult, if not impossible, to see the small rock exposures from the air. Consequently, even wide-range exploration is essentially done on foot, by opening machete paths into the forest. It was during one of these harsh daily forays that technician Juan Carlos Leal lost his way in the woods. To his amazement, he came across a low hill from which several large bones were sticking out of the reddish sandstone. The excavations that followed his fortunate discovery led to several dinosaur skeletons being collected from this hill, which became known as El Brete after the ranch on which it is situated. Among these dinosaurs were the first specimens of the titanosaurid *Saltasaurus loricatus*, the first sauropod found with a protective armor of bony scutes. While collecting and preparing the remains of *Saltasaurus*, Bonaparte and his collaborators discovered a series of small bones buried amid the large sauropod remains. Bonaparte identified these as bones of unknown, primitive birds, although he passed them onto Cyril Walker, then a paleontologist at the

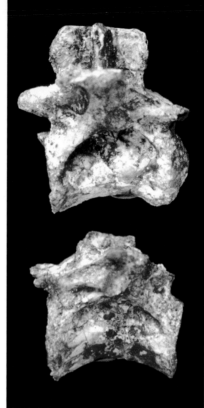

Natural History Museum in London, for formal study. In 1981, Walker published a brief report on these specimens, recognizing a new lineage of Mesozoic birds, which he named Enantiornithes; and, immortalizing Leal's serendipitous contribution to paleornithology, he named the first enantiornithine species, *Enantiornis leali*.

A Scientific Shift

José Bonaparte, however, was not the first to collect enantiornithine fossils. Bones of these ancient birds had been found decades earlier, during explorations of the Cretaceous outcrops of the American West, in the late 19th century. These specimens, however, had been mistakenly identified as nonavian theropods. Indeed, many Cretaceous birds discovered during the 19th and early 20th centuries were commonly interpreted either as remains of nonavian reptiles or as archaic representatives of extant lineages. Early studies of *Hesperornis* and its foot-propelled allies, for example, often regarded these toothed divers as aquatic relatives of ostriches and other paleognaths, or as early relatives of

A host of isolated bones of Late Cretaceous enantiornithines, like these trunk vertebrae, was unearthed from the El Brete locality in northwestern Argentina. This site produced the fossils that served as the basis for recognizing the existence of enantiornithines. (Image: L. Meeker/ L. Chiappe.)

loons and grebes. Likewise, the toothed *Ichthyornis* and other similar birds from the Late Cretaceous were interpreted as primitive members of charadriiforms, the group that traditionally includes gulls, terns, sandpipers, stilts, and other shorebirds. It seems as though, for most of the history of paleornithology, no one was ready to accept that the early phases of avian evolution involved the differentiation of several lineages that were more remotely related to the extant diversity of birds—and that, to a great extent, looked much more reptilian.

Walker's recognition of the Argentine assemblage as that of a primitive lineage of Cretaceous birds was a major benchmark in the history of Mesozoic avian paleontology. In retrospect, his short report was not only significant because it formally recognized an important chapter in the early evolution of birds, but also because it signaled the beginning of a major shift in our understanding of the first pages of avian evolution, one that would not fully occur until years later with the advent of more complete discoveries.

First Worldwide Discoveries

Soon after Walker's publication of the enantiornithine ensemble from El Brete, studies undertaken by University of Kansas paleontologist Larry Martin were the first to identify other Cretaceous species as members of the newly recognized group of early birds. In an influential 1983 publication, Martin argued that two previously described Late Cretaceous birds, *Alexornis antecedens* and *Gobipteryx minuta*, from Mexico and Mongolia respectively, were far-flung members of the Argentine Enantiornithes.

Expeditions from Occidental College and the Natural History Museum of Los Angeles County to the Late Cretaceous continental exposures of Baja California (Mexico) in the early 1970s collected an assortment of small bird bones in the vicinity of the village of El Rosario, on the Pacific coast. Expedition leader William Morris invited the late paleornithologist Pierce Brodkorb, from the University of Florida in Gainsville, to study these avian remains, which numbered among the handful of other Mesozoic land birds that had by then been discovered. In 1976, Brodkorb erected the new species *Alexornis antecedens* on the basis of this collection of bird bones. Following the tradition mentioned earlier, he classified *Alexornis* within modern birds, as the common ancestor of coraciiforms and piciforms, groups that include birds such as kingfishers, rollers, woodpeckers, and toucans. This initial identification was corrected in 1983, when Larry Martin demonstrated

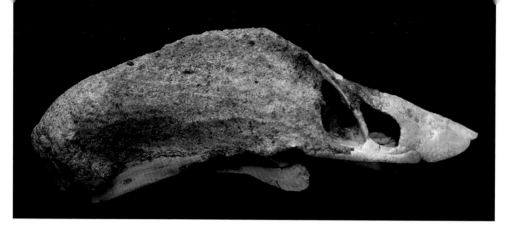

The toothless skull of the 75-million-year-old *Gobipteryx minuta*—the two windows on the beak represent the nostril (right) and the antorbital opening (left). This Mongolian species constitutes the only known beaked enantiornithine and one of the several lineages of early birds that evolved toothlessness. (Image: M. Ellison.)

the presence of several features shared between *Alexornis* and the assemblage of Argentine enantiornithine bones that Cyril Walker had described in 1981. My recent studies of the bones described by Brodkorb, now deposited at the Instituto de Geología in Mexico City, have confirmed Martin's identification. *Alexornis* was a sparrow-sized enantiornithine that inhabited the coastal regions of an ancient equatorial sea, for the Mexican peninsula was then far more to the south than it is today.

The Mongolian *Gobipteryx minuta* was recognized in 1974 on the basis of a fragmentary skull from the Nemegt Valley, an ample basin lined with jagged mountains that lies deep in the Gobi Desert. In his initial description, Polish paleontologist Andrzej Elzanowski regarded *Gobipteryx* as among the primitive paleognaths, the group of extant birds that includes all tinamous and ratites (ostriches, emus, and their kin). Yet Elzanowski's proposal was quickly challenged. Several researchers, Pierce Brodkorb among them, regarded the new skull as that of a Mesozoic reptile unrelated to birds. A second and better-preserved skull was found a few years later, but the disagreements about the avian relationship of *Gobipteryx* remained the same. In his 1983 article, Larry Martin defended the avian status of *Gobipteryx* and argued in favor of its classification within the enantiornithine birds Walker had described two years earlier. Yet because at that time *Gobipteryx* was known exclusively from skull remains that lacked comparable elements to those enantiornithines from Argentina,

Martin's comparisons were based on a series of embryonic skeletons that had been found in 1971 in the same Late Cretaceous rock layers of the Nemegt Valley that had yielded the fragmentary skulls of *Gobipteryx*. Elzanowski had published a detailed study of these exceptional embryos in 1981. However, and in spite of the extensive comparisons made between these embryos and *Gobipteryx*, Elzanowski's 1981 publication highlighted the limited data supporting

The falcon-sized skeleton of the 80-million-year-old *Neuquenornis volans* from Patagonia, Argentina. (Image: L. Chiappe.)

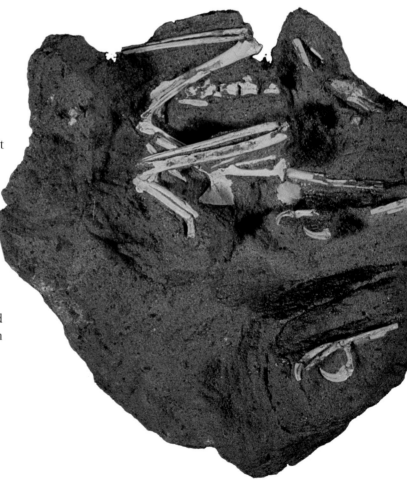

the identification of these embryos as those of the latter Gobi bird. Thus, Martin's identification of *Gobipteryx* as an enantiornithine remained unconvincing.

In recent years, however, Martin's conjecture of enantiornithine relationships has received support from the discovery of a much more complete, adult avian specimen from the Nemegt Valley. Combining elements from the skull and postcraneum, this specimen had been described in 1996 as a different enantiornithine species, "*Nanantius valifanovi*," by Russian paleornithologist Evgeny Kurochkin from the Paleontological Institute in Moscow. Nonetheless, research I conducted with Mark Norell and James Clark demonstrated that "*Nanantius valifanovi*" was nothing else than *Gobipteryx minuta*. Our investigation was based on a superbly preserved skull discovered in the mid-1990s by the Mongolian Academy of Sciences–American Museum of Natural History expeditions to the Nemegt Valley. In the course of this research, we realized that Kurochkin had misinterpreted several of the bones of the fragmentary skull of "*Nanantius valifanovi*." Namely, he had confused the lower jaw with the upper jaw, and vice versa, thus underestimating the resemblance between the skull of "*Nanantius valifanovi*" and that of *Gobipteryx*. Almost three decades after the species was first recognized, the enantiornithine relationship of *Gobipteryx* now became settled—the skeletal elements of "*Nanantius valifanovi*" showed many features indicative of Enantiornithes, which by the 1990s were starting to be found everywhere.

Enantiornithes Becomes a Natural Group

Although highly significant for the study of Mesozoic birds, the impressive collection of bird bones amassed by José Bonaparte and his collaborators, and later studied by Cyril Walker, was almost entirely composed of isolated elements.

Of the 60 or so bones found, only a handful were in articulation. Walker associated some of the bones by size, although this unwarranted approach left several researchers unsatisfied. The most important question at stake was whether this large bone assemblage represented a single taxonomic group, as claimed by Walker. Clearly, this issue could not be solved using the mostly disarticulated bones from El Brete—more complete specimens sharing characteristics with all these bones would have to be discovered. Such new evidence remained elusive for several years, but in 1988 it was unearthed from Late Cretaceous sandstones of northwestern Patagonia by Jorge Calvo, a paleontologist at the National University of Comahue in Argentina. This falcon-sized bird, which Calvo and I named *Neuquenornis volans* in 1994, was preserved with much of its skeleton still in articulation. Most importantly, its skeleton, from the forelimbs and hindlimbs to the vertebrae and girdles, exhibited numerous specialized features present in nearly all the isolated bones from El Brete. Thus the discovery of the 80-million-year-old *Neuquenornis* proved that most, if not all, of the El Brete avian bones were indeed part of a natural group of ancient birds—an evolutionary divergence of species sharing the same common ancestry.

An Unexpected Evolutionary Radiation

Like no other period in the history of paleornithology, the last two decades of the 20th century witnessed a spectacular and unexpected rate of fossil bird discoveries in Cretaceous beds worldwide. Most of these fossil remains are those of enantiornithines. Indeed, with more than 25 valid species and many more fossils either too fragmentary to be identifiable or representing early juveniles difficult to assign to particular species, the group that was modestly recognized at the

beginning of the 1980s rapidly became the most abundant and diverse lineage of Mesozoic birds. Today, after years of skepticism about the avian nature of these animals, the origin and evolution of enantiornithines is accepted as the earliest recognized evolutionary radiation of birds.

Yet the temporal divergence of this archaic radiation is not entirely clear. A number of well-preserved fossils reveal an evolutionary history that can be traced back almost 130 million years, but enantiornithine origins are likely to be much older. Some of the oldest known enantiornithines are represented by *Protopteryx fengningensis* and *Eoenantiornis buhleri*, toothed birds whose fossils have been collected in the Chinese provinces of Hebei and Liaoning, respectively. Even at such early stages of the enantiornithine radiation, these birds show many of the typical skeletal features of the group. *Protopteryx* is perhaps the most primitive known enantiornithine, but the features of its shoulder and hand that highlight its primitiveness with respect to other enantiornithines are poorly known in the tiny *Iberomesornis romerali*, a 115-million-year-old bird from central Spain, whose skeleton also displays features interpreted as primitive for the group. The abrupt appearance of enantiornithines in the fossil record, and the fact that the seemingly most primitive species are so widely distributed geographically, suggests that the origin of the group is significantly more ancient than these earliest records can convey, but we are baffled by the dearth of fossil birds from the beginning of the Cretaceous and earlier.

From these initial stages, enantiornithines show a tremendous diversity of shapes, likely reflecting a wide range of feeding behaviors and lifestyles. Their finch to robin-sized bodies also stand out when compared to the larger, more primitive birds that also lived in the Early Cretaceous. Indeed, enantiornithines are the oldest known birds to have developed sizes typical of modern songbirds. This distinct miniaturization of even the earliest members of the group is a clear indication of the superior aerodynamic abilities of these animals when compared to those of more archaic birds.

Lakebed sediments in northeastern China and Spain have been paramount in yielding the remains of the earliest enantiornithines. More than ten enantiornithine species, along with skeletons of early juveniles and a variety of fragmentary fossils that show the anatomical trademarks of the group, have been quarried from the 125–120-million-year-old shales that contain the Jehol Biota of northeastern China. A number of other species have also been named from these celebrated sites, but the validity of these species is problematic because they were based on either hatchlings (see *A Tale of Two Names*) or poorly preserved fossils. Other Early Cretaceous fossils of these birds have also been discovered in

Reconstruction of the Chinese *Protopteryx fengningensis*. This small enantiornithine is perhaps the most primitive and oldest member of the group. (Image: U. Kikutani.)

The finch-sized *Eoenantiornis buhleri* from the 125-million-year-old Yixian Formation of China. *Eoenantiornis* is one of the oldest known enantiornithines. The inset details the short snout and powerful teeth that characterize this species. (Image: L. Chiappe.)

Mongolian lake deposits and from rocks that formed near the sea shore of what is today Australia. Similarly, the fine 115-million-year-old lakebeds of Spain (see *The Real Gwangi*) have furnished the remains of at least four species of enantiornithines, along with a number of adult, juvenile, and even embryonic specimens that have not been named or that remain unstudied. Many of these specimens are exquisitely preserved, providing in some instances evidence of stomach contents. Thus, the fossil record tells us that an enormous diversity of enantiornithines had already diverged by the first half of the Cretaceous. Furthermore, comparisons between the Jehol Biota and the Early Cretaceous localities in Spain reveal a clear trend towards enantiornithine supremacy—the more primitive lineages that coexisted with the enantiornithines prior to 115 million years ago disappear from the fossil record at that time.

A wide range of evolutionary studies have documented how physical disparity within a lineage tends to increase over time. The evolution of enantiornithines throughout the Cretaceous is no exception. The diversification of these birds continued unabated until the very end of the period. Fossils of Late Cretaceous enantiornithines have been found in several localities of Canada, the United States, Mexico, Mongolia, Uzbekistan, Russia, Hungary, France, Lebanon, Madagascar, and Argentina, and some of these records extend to the last moments of the Cretaceous, geologic instants before the major mass extinction that defines the boundary between Mesozoic and Cenozoic eras. With a few exceptions, however, Late Cretaceous enantiornithine fossils are more incomplete than those recovered from the Early Cretaceous. Yet these fossils still document how late enantiornithines developed substantially larger sizes—some with wingspans reaching one-and-a-half meters—and a suite of specialized features, apparently associated with the evolution of novel lifestyles. Indeed, fossils of specialized swimmers, long-legged waders, near flightless species, and toothless lineages are all examples of the magnificent diversity that characterized the late evolution of enantiornithines.

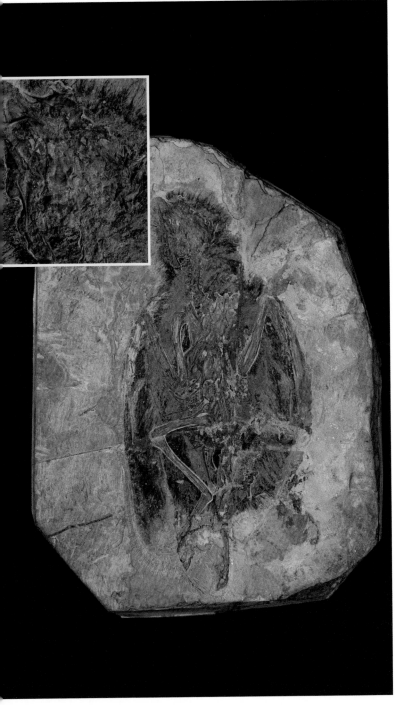

Many amateur fossil collectors have made great contributions to paleontology. Armando Díaz-Romeral, a relentless collector from Cuenca, a small town of Roman origin in central Spain, is one of them. In 1984, while exploring the Mesozoic rock layers surrounding his native town, Díaz-Romeral discovered the now renowned fossil site of Las Hoyas—a magnificent 115-million-year-old window on Cretaceous Europe. His discovery had been presaged years earlier, albeit in a very different fashion. In 1969, Hollywood released *Valley of Gwangi*, a rehash of Arthur Conan Doyle's *Lost World*, featuring Mexican cowboys battling vicious dinosaurs and flying reptiles. Ironically, *Gwangi* had been largely filmed near (and in) medieval Cuenca!

Las Hoyas is a small pocket of fossils in limestone some 20 meters deep, which once formed in the bottom of a shallow but permanent lake. Like the famous Solnhofen lagoons—the Late Jurassic coastal environments that entombed the remains of *Archaeopteryx*—the Las Hoyas lakes had thermally stratified waters and bottoms devoid of oxygen, thus providing exceptional conditions for the preservation of delicate skeletons and soft tissues. At that time, Spain and much of Western Europe formed a subtropical archipelago of large islands, located much closer to the equator than today. Swamps and lush vegetation, which developed not far from the shore of the tropical Tethys Sea, surrounded the ancient lake at Las Hoyas.

José L. Sanz and his associates from the Universidad Autónoma in Madrid have been excavating at Las Hoyas since its discovery. Through their efforts, this site has become one of the best-known European ecosystems of the Mesozoic Era. The systematic excavation of Las Hoyas has furnished thousands of fossils, from ferns and flowering plants to a myriad of insects and crustaceans, and numerous fish, frogs, salamanders, lizards, turtles, crocodiles, and dinosaurs. Among the most precious fossils from Las Hoyas are a variety of enantiornithine birds— *Iberomesornis*, *Concornis lacustris* and *Eoalulavis hoyasi* among others—which I have had the privilege of studying with my Spanish colleagues.

The information locked in the delicate bones of the ancient birds from Las Hoyas has been paramount in understanding the early chapters of birds' evolution. The spectacular preservation of fossils from this site has left a meticulous record of the anatomical and behavioral attributes of these ancient birds, to the point that *Eoalulavis* has an intact alula and even the remains of its last supper are preserved. The significance of the Las Hoyas birds goes beyond their bearing on the fine-tuning of flight and the clues they provide for understanding the physical transformations (and relative chronology) that occurred during the ancient history of birds. In documenting an intermediate morphology between their modern counterparts and their theropod forerunners, these archaic birds also claim their place within the family tree of dinosaurs. (Image: L. Chiappe.)

The tiny *Iberomesornis romerali* is a primitive enantiornithine from the 115-million-year-old limestones of Las Hoyas, Spain. (Image: R. Meier.)

The elongated jaws of *Longirostravis hani*, an enantiornithine from the Early Cretaceous of China, were specialized for capturing small prey that lived in soft substrate. The inset shows the tip of its toothed snout. (Image: L. Chiappe.)

An Aerodynamic Body

Fossils of enantiornithines highlight the extensive upgrading of their flight system, which is reflected directly or indirectly throughout the skeletal plan of these birds.

The majority of these birds retained the short snout with peg-like teeth and minimal degree of skull kinesis of *Archaeopteryx*, but new discoveries have started to reveal greater diversity of cranial architecture. Among these is the plover-sized, 125-million-year-old *Longirostravis hani*, which probably used its long, slender, and gently curved jaws to probe soft substrates. Also departing from the conservative skull design of

most enantiornithines is the slightly younger *Longipteryx chaoyangensis*, whose robust jaws, long snout, and massive teeth suggest fish-eating habits. The skull architecture of these enantiornithines was also strikingly different from that of the much younger, chicken-sized *Gobipteryx*, whose powerful jaws and sharply-edged beak became capable of crushing seeds. This significant diversity of skull designs suggests that the aerodynamic improvement of enantiornithines facilitated the invasion of new ecological niches.

Despite the diversity of cranial designs, the vertebral column of these birds remained virtually unchanged from that seen in more primitive short-tailed birds, although the elongation of the neck and abbreviation of the trunk moved a step further. A more significant difference involved the development of strong bony projections at the front of the trunk for the attachment of powerful neck muscles. This transformation very possibly reflects an enhanced precision of neck movement, perhaps in response to a skull that was becoming more specialized for particular behaviors—from feeding to nest construction.

More drastic transformations occurred in the shoulder and thorax. The elongation of the coracoid was one of these important transformations. By increasing the distance between the shoulder socket and the breastbone, the lengthening of the coracoid amplified the range of the wingstroke. Another key innovation involved the evolution of a passage delimited by joints of the coracoid, the scapula, and the furcula. Acting very much like a pulley, this triosseal canal funneled the supracoracoid tendon—a muscle largely responsible for the upstroke—from its origin in the sternum to its attachment on the top of the humerus. Such a rearrangement allowed this tendon to perform the rapid rotation of the humerus necessary for the elevation of the wing. Evolution also lightened the furcula and turned

it into a V-shaped bone able to spring sideways with each wingbeat, thus providing mechanical assistance to the wingstroke. At the same time, the furcula developed a long, downward projection called the hypocleideum, giving further strength to the ribcage. All these flight-related modifications were assisted by changes in the relative size and shape of the sternum, which grew larger in most enantiornithines and developed a deeper crest for the origin of larger flight muscles.

The 120-million-year-old *Longipteryx chaoyangensis* is another long-jawed enantiornithine from China. The large and stout teeth of this bird were suited for capturing fish. (Image: L. Chiappe.)

Reconstruction of the skeleton of *Longipteryx*. (Image: S. Abramowicz.)

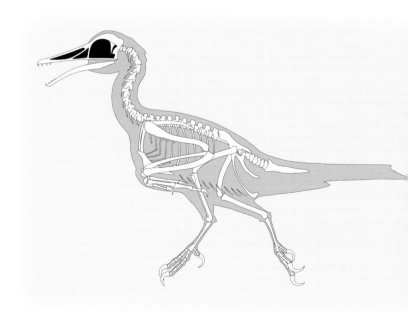

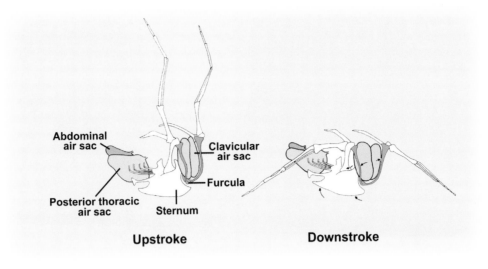

By storing elastic energy, the gracile, V-shaped furcula of enantiornithines provided mechanical assistance during the wingbeat and the breathing cycle. The black arrows (right) point out the sideways swinging of the V-shaped furcula of a modern starling which, in combination with the correlated ribcage movements, facilitates the aeration of the air sacs connected to the lungs. (Image after Dial et al, 1991.)

The shoulder girdle and sternum of the Spanish *Concornis lacustris*. Note the long coracoids, large sternum, and V-shaped furcula of this Early Cretaceous enantiornithine. (Image: L. Chiappe.)

The skeletal anatomy of the forelimbs of these birds also confirms a substantial upgrade in their flying capabilities. The wingbones evolved modern proportions, with the ulna and radius (the bones of the forearm) becoming longer than the humerus. This arrangement probably increased the amount of thrust produced during the flight stroke, because the outer portion of a wing generates most of the propulsive force. An enhanced flight performance can also be inferred from changes suggesting an overall strengthening of the wing. Now twice as thick as the radius, the ulna was a strong support for the attachment of the secondary flight feathers. Likewise, the metacarpals and carpal bones of the hand became interlocked and most commonly fused to each other, the outermost finger grew smaller, and while the remaining two fingers retained claws and the same number of phalanges as in more primitive birds, they also became proportionally shorter. The scant information available on the shape of the feathered wing also suggests aerodynamic improvements. The elliptical wings of the Spanish *Eoalulavis hoyasi* hint at enhanced maneuverability. Not surprisingly, other fossils suggest that enantiornithines also evolved a substantial range of wing designs—the longer forelimbs of *Longipteryx* must have supported a wing of greater aspect ratio, and the pheasant-sized *Elsornis keni* developed a stout and shortened wing of limited aerodynamic performance.

Unlike their highly transformed wing module, few changes are evident in the landing gear of the group. Indeed, enantiornithines retained many primitive features in their pelvis and hindlimbs. The bones of the pelvis show a general design that greatly resembles those of *Archaeopteryx* and *Confuciusornis*. The triangular ilium has a short, prong-like rear half, the hip socket or acetabulum

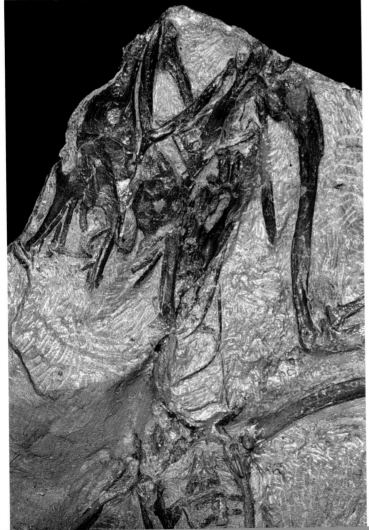

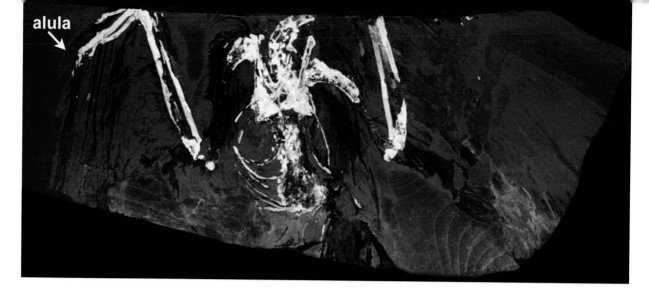

alula

is proportionally large, and both pubes meet each other at their tips. As with these more primitive birds, the legs of enantiornithines were slender and their feet anisodactyl—three main toes faced forward and the first one was oriented to the rear. Yet the diversity of foot designs of enantiornithines stands out among their Mesozoic counterparts. In most cases, the opposable toe grew longer and the foot became more suitable for perching—but in other instances, the feet became specialized for newly invaded ecological niches.

A Major Improvement in Flight Performance

We have seen how flight could have originated as nonavian theropods with feathered forelimbs evolved smaller sizes that allowed them to become airborne. Over the subsequent course of avian evolution, the refinement of flight entailed the appearance of a complex suite of physical specializations. Although it is unquestionably interpreted as a flying bird, the absence of a number of important aerodynamic structures suggests that *Archaeopteryx* was probably weaker and less able to maneuver in flight than most of its extant relatives. The primitive wing configuration of *Archaeopteryx* was essentially retained in the more derived confuciusornithids, but the long bony tail was shortened into a few free vertebrae and a pygostyle. The confuciusornithid wing also evolved

a propatagium, the skin fold joining the shoulder and wrist, a feature that would have increased the area of the lift-generating airfoil.

Estimates of important aerodynamic parameters such as wing loading (the ratio between the weight and the wing surface) also hint at refinements in the

The 115-million-year-old *Eoalulavis hoyasi* photographed under ultraviolet light. Note the preservation of the alula or bastard wing still attached to the innermost finger of the hand of this Spanish enantiornithine. (Image: J. Sanz.)

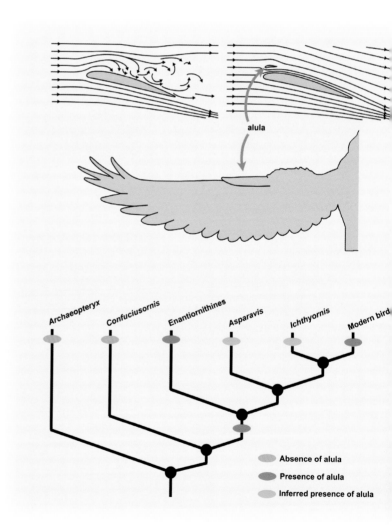

alula

Archaeopteryx · Confuciusornis · Enantiornithines · Asparavis · Ichthyornis · Modern bird

- Absence of alula
- Presence of alula
- Inferred presence of alula

flight capabilities of these birds—the wing loading of *Confuciusornis* was lower than in *Archaeopteryx* and less than half that calculated for *Caudipteryx*. The body miniaturization experienced during the transition from nonavian maniraptorans to the most primitive birds is further evidenced by the earliest enantiornithines, whose sizes are comparable to modern sparrows (*Iberomesornis*, *Sinornis santensis*) and thrushes (*Concornis*, *Eoenantiornis*). General flight performance is often correlated with size reduction—smaller birds are more maneuverable and their flight is energetically less expensive. The estimated wing loading for some enantiornithines is four or more times smaller than that of *Confuciusornis*. In spite of this, the clearest evidence for the enhanced flight ability achieved by the enantiornithines is documented by the evolution of a "bastard wing" or alula, a short tuft of feathers attached to the innermost finger of these birds. Operated by a complex system of nerves and muscles, and varying greatly in size from one bird to the other, this feathered tuft is a high-lift and safety device of important aerodynamic significance. During landing or braking, birds position their wings in such a way that the airfoil is angled with respect to the horizon. Such a position increases the animal's drag and reduces its speed, but beyond a certain angle, air turbulence generated over the wing could lead to stalling. Under these circumstances, the bastard wing acts like the wing flap on an airplane, producing a slot over the main wing. This slot helps to streamline the upper surface of the wing, minimizing turbulence and allowing braking without stalling. It is likely that the function of the well-developed alula of enantiornithines was assisted by other anti-stalling feather slots in the tip of their wings, although these are difficult to observe in fossils.

The advanced skeletal features of enantiornithines, coupled with the presence of an alula and their small size, suggest that even the earliest of these birds had aerodynamic abilities approaching those seen in extant forms. This evidence also indicates that just like most of their living counterparts, these ancient birds managed to take off from a standstill. With little doubt, enantiornithine fossils drive home the notion that slightly more than 20 million years after *Archaeopteryx*, some birds had already evolved flight capabilities comparable to those of their living relatives.

Enantiornithine Genealogy

Earlier we mentioned that enantiornithines were not universally accepted as birds until the more complete discoveries of the late 1980s and early 1990s. However, the precise placement of the group within the genealogy of early birds remained controversial for much longer.

In his initial study, Cyril Walker concluded that the enantiornithines occupied an intermediate position between *Archaeopteryx* and the Late Cretaceous hesperornithiforms, a group consistently considered to be more closely related to extant birds. This proposal was soon modified by Larry Martin, who in 1983 argued that enantiornithines and *Archaeopteryx* had a common ancestor shared neither by the Mesozoic hesperornithiforms and *Ichthyornis* nor living birds. Martin envisioned a fundamental evolutionary dichotomy at the onset of birds—*Archaeopteryx* and the enantiornithines on the one hand, and everything else on the other hand. A different view was subsequently proposed by Joel Cracraft, an ornithologist from the American Museum of Natural History in New York City. In 1986, Cracraft published a pioneering cladistic analysis of Mesozoic birds. Comparing more than 80 skeletal features among the best-represented lineages of early birds known at the time, he proposed three alternative genealogical hypotheses of enantiornithines. One of these regarded these birds as an early and

Aerodynamic function and evolution of the alula or bastard wing. This small tuft of feathers attached to the innermost finger creates a slot along the wing's leading edge that minimizes the turbulence generated when a bird flies at low speed. This device enhances control and maneuverability during take-off and landing—experimental studies on pigeon wings show that the deflection of the alula generates more than 20 percent of the lift provided by the wing. Current evidence indicates that the alula originated in the common ancestor of enantiornithines and ornithuromorphs. (Image: S. Abramowicz.)

extinct lineage of neornithines—the group that includes all living birds and their common ancestor. Another hypothesis considered enantiornithines to be the closest relatives of the ancestor of all living birds; and the third hypothesis placed them in a position slightly more primitive than *Ichthyornis*. Yet Cracraft's three hypotheses all supported the idea that enantiornithines were not especially related to *Archaeopteryx* but rather an offshoot of the avian line that led to modern birds.

Building on the work of Cracraft, in the early 1990s I conducted a series of cladistic analyses of Mesozoic birds—each one including more features and a larger number of species. My studies agreed with his results in that they showed enantiornithines as an offshoot of the line towards living birds; but my results were more consistent with Walker's early view in that they placed enantiornithines as a much more primitive branch than *Ichthyornis*. Since then, these results have been confirmed by several independent studies, all of which have rejected Larry Martin's proposed basal dichotomy. Martin and a few others have continued to argue that enantiornithines branched off from an evolutionary line separated from the line leading to extant birds, but their views

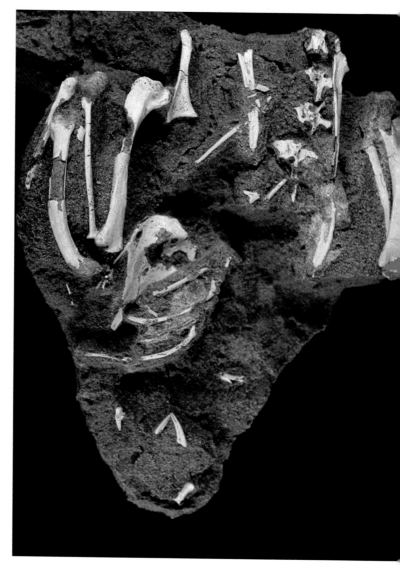

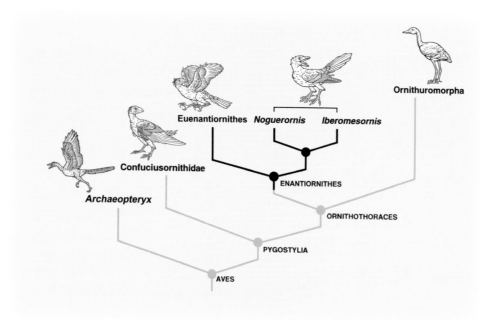

Cladogram showing the evolutionary relationships of enantiornithines to other groups of early birds. (Image: E. Heck/S. Orell.)

have been substantiated by a handful of
features allegedly shared by *Archaeopteryx*
and enantiornithines. For example, in a
1996 cladistic analysis of 36 bone variables
led by Chinese paleontologist Hou Lianhai,
the kinship between these birds was
supported by five characteristics that have
either been found in several other lineages
or are simply misinterpretations of the
anatomy of the archaic birds in question.
Such studies can hardly overturn the
evidence indicating that enantiornithines
are evolutionarily closer to living birds.
Part of this evidence includes many
attributes correlated with the enhanced
aerodynamic skills of these birds (the alula,
the similar wing proportions, and the
pygostyle, to mention a few), which, if
Martin's hypothesis
were endorsed, must have evolved
independently. Thus, accepting a basal
dichotomy of avian evolutionary
divergence would require accepting that
flight was independently fine-tuned in
both enantiornithine and neornithine birds.

Although the intermediate placement of
enantiornithines between *Archaeopteryx*
and neornithine birds is well-supported
today, the interrelationships among
different enantiornithine species have
been far less explored. Clearly, the
incompleteness of several species—
sometimes known on the basis of a single
bone—has hampered our understanding
of these evolutionary relationships, but
the great similarity in the anatomy of
many of these birds has also complicated
the picture. More detailed studies of
the anatomy of enantiornithines are
clearly needed. The pace at which their
diversity is being discovered is so rapid
that many of the known species have
been reported in preliminary studies
only. Deeper comparative studies of the
known enantiornithine species, as well
as the discovery of additional material,
are necessary for a better knowledge
of enantiornithine genealogy. The
interrelationships of these Cretaceous
birds remain one of the most challenging
issues in the early history of birds.

Their Diverse Environments

For decades, the early evolution of birds
was thought to have been mainly played
out in the marine realm. Most fossils
unearthed during the 19th and early 20th
centuries were collected from seabeds, and
only recently have the many discoveries
of Cretaceous birds in inland deposits
prompted reconsideration of this classic
notion. The discovery of enantiornithines
has greatly contributed to this conceptual
shift. Although these ancient birds have
been found in a variety of Cretaceous
deposits worldwide, most of them come
from rocks formed in terrestrial or
freshwater environments.

The earliest known members of this
group—those from the Early Cretaceous
of China and Spain—are recorded in
sediments deposited in calm, subtropical
lakes, which also entombed a variety
of plants, crustaceans and insects, fish,
amphibians, reptiles, and mammals. Late
Cretaceous enantiornithines are much
more widely distributed—but just as their
older counterparts, the majority of these
fossils occur in sedimentary rocks formed
inland. For the most part, fossils of Late
Cretaceous enantiornithines are contained
in sediments accumulated in the flood
plains and channels of ancient rivers that
either crossed the well-vegetated lowlands
of North America and Europe or the more
subhumid landscapes of southern South
America. Enantiornithines shared these
environments with a diversity of other
animals, including many large dinosaurs.
Some Late Cretaceous enantiornithines,
such as *Gobipteryx* and *Elsornis*, also
inhabited the arid landscapes of central
Asia. Together with an enormous
menagerie of ancient animals, fossils of
these birds have been recovered from
rocks formed in and around bodies of
water that spotted large fields of dunes
that developed in this part of the world
during the Late Cretaceous.

Although the bulk of the enantiornithine
fossil record—and thus, our interpretation
of their preferred environments—comes

from rocks formed in these diverse inland settings, discoveries of fossils contained in seabeds hint at an even greater ecological milieu. A handful of 100-million-year-old bones from Australia, including one found in the gut of a large ichthyosaur, are the earliest known enantiornithines from marine deposits. The occurrence of abundant logs within the rock layers containing these fossils suggests that these birds lived on a forested coast and foraged either on the seashore or in shallow waters. Interestingly, a host of geophysical evidence indicates that during the Early Cretaceous this part of Australia was very close to, if not within, the Antarctic Circle. Australia was then joined to Antarctica and located far to the south of its current position. Even if ice caps did not exist in the warm climate of the Mesozoic, studies of Australian Early Cretaceous leaves point at mean annual temperatures of about 10°C—as cold as it gets today in many parts of the American Midwest—and latitudinal estimates suggest long months of winter darkness. The fact that these enantiornithines would have had to cope with a rather cold and dark winter suggests that they could have already evolved a migratory behavior.

A few younger enantiornithines have also been discovered in marine rocks. One of these is *Halimornis thompsoni*, a small 80-million-year-old enantiornithine unearthed from the blackish chalks of the Mooreville Formation of western Alabama. *Halimornis* became buried some 50 kilometers offshore, in the tropical waters of the Pierre Seaway, which at that time cut across North America. More fossils of oceanic enantiornithines have been recently collected from offshore deposits of Late Cretaceous age in British Columbia, Canada. Together with *Halimornis*, these fossils represent the best case for inferring the oceanic and littoral habits of some of these birds—and they suggest that enantiornithines may have played a more important role in these ecosystems than previously thought.

Lifestyles and Foraging Preferences

The large number of enantiornithine species hints at a multitude of lineages specialized for quite different lifestyles. However, inferring these ecological habits is not simple. Most reconstructions have focused on anatomical comparisons with extant birds, but extrapolating the lifestyle of living organisms to animals that lived millions of years ago is problematic. The anatomy of both the skull and the foot has played a principal role in these reconstructions. On the one hand, there is a well-documented correlation between the structure of the skull, above all the snout, and the foraging specializations of living birds. On the other hand, when a bird is not flying, its feet interact with the substrate more than any other part of its body, and thus their anatomy is often indicative of the bird's lifestyle. Certainly, birds have evolved a variety of foot designs to provide solutions to the problems they face while interacting with a range of substrates, and these solutions are likely to have been similar for the extinct members of the group.

Numerous enantiornithines were small, similar in size to many modern songbirds, and the design of their feet, with a long first toe clasping against the remaining three toes, was well-suited for grasping. This suggests that, as do scores of their living counterparts, many of these archaic birds perched on branches. In some of these birds, evolution may have even taken this specialization a step further by rotating another of their toes backwards, a design typical of sophisticated perchers and tree-climbers such as parrots, woodpeckers, and nuthatches, although the fossil evidence suggesting this is not entirely conclusive. The grasping feet of many enantiornithines could have also been used for seizing and slaying prey. This is especially apparent in species whose opposable first toe bears a large, powerful

claw and whose kestrel-sized bodies are substantially bigger than those of many of their relatives. The radically different design of the feet of other enantiornithines indicates that some of these birds evolved quite different ecological adaptations. The impressive range of these specializations is best exemplified by the notable differences in the feet of three enantiornithines from the Late Cretaceous of El Brete in Argentina: the grasping foot of *Soroavisaurus australis* suggests perching and possibly predatorial activities, the long and slender footbones of *Lectavis brevipedalis* point at specializations ideal for wading in waterlogged environments, and the short, broad, asymmetrical feet of *Yungavolucris bretincola* are best suited for swimming.

Enantiornithines also exhibit substantial variety in the structure of their skulls. While most of the fossils exhibit triangular skulls with short, toothed jaws that are well-suited for pecking on a range of foods, others greatly broadened the spectrum of enantiornithine skull designs. Most notable among these are the stout, toothless jaws of the Mongolian *Gobipteryx*, a bird that probably fed on seeds and other hard items, and the long and slender snouts of the Chinese *Longipteryx* and *Longirostravis*. Yet the skull design of the latter two birds suggests very different foraging behaviors—the massive teeth of *Longipteryx* hint at fish-eating specializations kingfisher-style and the gently curved jaws of *Longirostravis*, which end in a handful of tiny teeth, suggest a mud-probing behavior similar to today's sandpipers.

Inferences about the lifestyles of enantiornithines have not been limited to the design of their feet and skulls. Based on observations of their wishbones and breastbones, paleornithologist Alan Feduccia has argued that the digestive

system of these birds could have included a large crop—an expansion of the esophagus to store food. Specifically, Feduccia compared the digestive tracts of the enantiornithine *Sinornis santensis* to that of the hoatzin, a peculiar tree-dwelling bird that lives in the rain forests of South America. Like those of most enantiornithines, the wishbone and breastbone of the hoatzin bear a long hypocleideum and a short rear keel, respectively. In the hoatzin, these features make room for a large and muscular crop that houses bacteria capable of digesting the cellulose fibers of the leaves this bird almost exclusively feeds on. Unfortunately, finding food items in the digestive tracts of Mesozoic birds is extremely rare, and neither leaves nor any other food items have been preserved in association with the known specimens of *Sinornis*. Regardless, Feduccia's suggestion seems at odds with the enhanced aerodynamic capabilities of enantiornithine birds, because leaves are poor in nutrients and thus rarely constitute an important part of the high-energy diet of small, active, warm-blooded animals.

Evidence suggesting that enantiornithines preferred nutrient-rich food items has recently become available. One of the discoveries involves a fragmentary Late Cretaceous specimen unearthed from marine limestones near Beirut, Lebanon (see *An Ancient Sap-eater*). This fossil is remarkable because several tiny pieces of amber—fossilized sap—lay muddled between its broken bones and feathers. Being the only fossil, among thousands discovered at the same site, in association with amber suggests that sap could have been part of the nutritious repertoire of this bird. Although inferring a sap-eating diet for a bird unearthed from marine rocks may appear unreasonable, the abundance of plants and land-dwelling organisms in these seabeds indicates the presence of forested neighboring islands, and the preservational characteristics of the fossil suggest that it was preyed upon

and probably regurgitated before sinking to the bottom of the sea. However, the most compelling evidence indicating an enantiornithine preference for highly nutritious meals comes from stomach contents discovered in the visceral cavity of *Eoalulavis hoyasi*, a finch-sized bird from the Early Cretaceous of Spain (see *The Real Gwangi*). Mixed between the ribs of this tiny fossil were several segments of the chitinous exoskeleton of a crustacean. This unusual discovery documents a high-energy diet in an enantiornithine whose wishbone shows the elongate hypocleideum used to infer a folivorous diet for the similarly sized *Sinornis*.

Sluggish Growth Rates

Beyond their diversity of shapes and their significance in understanding the fine-tuning of flight, fossils of enantiornithines have provided insights into several other aspects of the biology of ancient birds. One of these is the rate at which they grew.

Much of what we know about the growth rates of extinct animals comes from detailed examination of the structure of fossilized bone tissues. Bone is made up of specialized cells embedded in a matrix of the protein collagen with associated crystals of a phosphatic mineral called hydroxyapatite. In most cases, the process of mineral replacement that fossilizes a given bone also preserves the microstructure of its tissue. This phenomenon has allowed paleontologists to study aspects of the bone tissue of extinct animals and to infer a whole suite of features from these studies.

In 1994, I teamed up with Anusuya Chinsamy and Peter Dodson of the South African Museum and the University of Pennsylvania, respectively, to study the bone tissue of enantiornithines. We cross-sectioned the femurs of a couple of 70-million-year-old fossils from Argentina and prepared thin sections for microscopic studies. Our study revealed up to five growth rings demarcating concentric

AN ANCIENT SAP-EATER

Nearly 95 million years ago, a diversity of aquatic reptiles, fish, and invertebrates thrived in the tropical Tethys Ocean. A fraction of this exuberant marine community became exquisitely preserved in the fine-grained limestones that formed underneath the shallow waters that washed the ancient Near East. A few years ago, Lebanese stonemasons at Nammoûra, near Beirut, came across a peculiar fossil from these ancient rocks. Split into two small slabs, the new fossil contained the remains of a small bird still surrounded by portions of its feathers. Far from the fine preservation of other Nammoûra fossils, this small bird was preserved as a ball-like mass of powdered bone, but nonetheless, it was one of a kind.

Fossil birds from the Mesozoic of Gondwana are extremely rare, and although today we do not consider Lebanon to be part of the Southern Hemisphere, nearly 100 million years ago, it was. Abundant geological evidence indicates that many of the northern landmasses of the Mediterranean, as well as the Levant, have been accreted to the European continent as slices of continental crust became detached from the northern border of Gondwana and what is today Africa. Indeed, the eastern Mediterranean and the Arabian Peninsula remained part of a large Afro-Arabian continent until about 24 million years ago. Thus, although fragmentary, the new fossil was very significant because it represented—then and still now—the only known avian skeleton from the Mesozoic of northern Gondwana. Equally important was its unusual

preservation, because the pattern of its bones suggest that the animal was fed on by scavengers before its final entombment. But overall, the most interesting fact about this fossil was what was found in between its bones— several tiny balls of yellowish amber.

Although amber had previously been found in Nammoûra, no other fossil skeleton had been discovered in association with particles of fossilized resin. Certainly, this association had to mean something unique, something that probably pertained more to the behavior of the animal than to the characteristics of its fossilization. The interpretation given by Italian paleontologist Fabio Dalla Vecchia and me was that this small enantiornithine had been washed to its marine burial from a nearby terrestrial environment, and that the tiny amber balls mixed in with its skeleton were the remains of its last meal. (Image: F. Dalla Vecchia.)

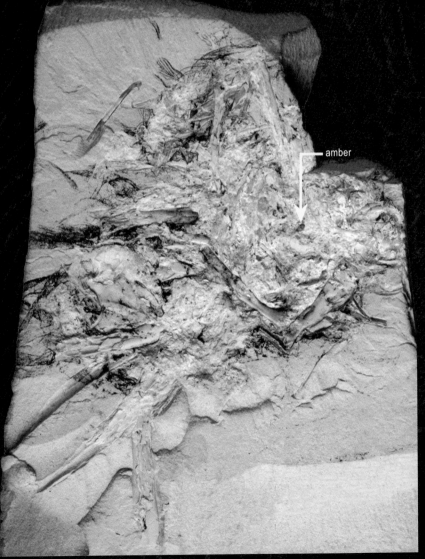

amber

A TALE OF TWO NAMES

Many of the recently discovered specimens from the Early Cretaceous of China are divided in two slabs, a consequence of the fact that the rocks formed by the slow accumulation of sediments at the bottom of a body of water, and they have zones of weakness right where they enclose a fossil body. As a result of the intense commercialization of Chinese fossils over the last few years, many farmers and fossil dealers are using the fact that most Liaoning fossils are preserved in two slabs to their great advantage—because each find potentially means two sales. Thus, pairs of slabs containing single individuals have found separate repositories and have been studied by separate teams of paleontologists. The naming saga of a tiny enantiornithine bird from the Early Cretaceous of Liaoning illustrates this quite well.

After discovery in 1997, one of the two slabs of this bird found repository at the Nanjing Institute of Geology and Paleontology; the other was housed at Beijing's National Geological Museum of China. The naming of *Liaoxiornis* is one of those cases in which teams of paleontologists compete to be the first to publish a name. Hou Lianhai and Chen Pei-Ji of the Institute of Vertebrate Paleontology and Paleoanthropology (Beijing) and the Nanjing institute, respectively, studied one slab and submitted for publication a paper with the name *Liaoxiornis delicatus*. Meanwhile, Ji Qiang and Ji Shu'an of Beijing's Geological Museum worked on the other "hemi-specimen" and submitted another paper, this time with the name *Lingyuanornis parvus*. In the end, Hou and Chen's paper was published a few weeks earlier, so *Liaoxiornis delicatus* gained priority: *Lingyuanornis parvus* went into the recycling bin. Although Hou and Chen won the naming competition, their study was less accurate than that of Ji and Ji. While Hou and Chen claimed the specimen to be that of an adult bird, the several immature features of *Liaoxiornis* did not escape the notice

of the Ji-team. Perhaps the most salient of these are the very large orbit and short snout, the large skull relative to the skeleton, and the loose connections between skull bones. For Ji and Ji the specimen's juvenility was self-evident, which was why they named their proposed species "*parvus*," which in Latin means "child." In the end, neither *Liaoxiornis* nor *Lingyuanornis* will survive. Because of the dramatic transformations that often occur during the development of a skeleton, taxonomists avoid using very immature specimens as holotypes for new species. The vast majority of species, both extinct and extant, are erected based on adult specimens, so morphological comparisons with neonates may highlight differences that are more related to their immaturity. Regardless of its naming contest, this hatchling is an important specimen, constituting one of the few Mesozoic birds known from a very young stage of skeletal development. (Image: M. Ellison.)

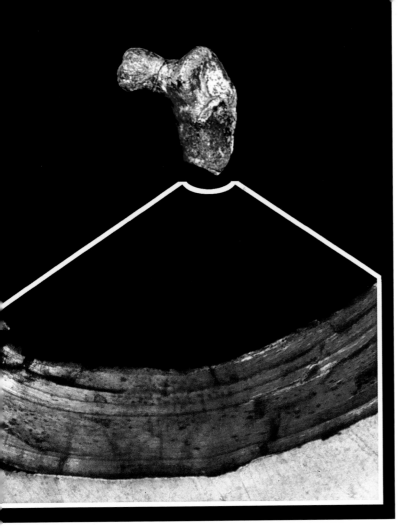

Bone microstructure of the femur of a small Late Cretaceous enantiornithine from El Brete, Argentina. The reduced number of blood canals and the presence of growth rings (thin, concentric lines) point at relatively slow rates of growth. Growth rings are believed to be formed annually, so the existence of several suggests that it took several years for this bird to reach its adult size. (Image: L. Meeker/A. Chinsamy.)

circles in the section of the shafts of these bones. Like the growth rings of a tree, the growth rings of bones reflect a reduction in the rate of bone formation, or even a complete pause in bone growth. These drops in the rate of bone formation appear to be correlated to periods of stress, which might result from adverse environmental conditions or limitations in food supply. Because in extant animals these times of stress (and their correlated growth rings) are often cyclical, occurring every year, the number of growth rings in the bones of extinct organisms has frequently been used in estimating the ages of individuals.

Recent studies of multiple bones from single animals, however, have shown that the number of growth rings may actually vary from bone to bone. In a study of the bone microstructure of the duck-billed dinosaur *Hypacrosaurus* led by Jack Horner from the Museum of the Rockies, the number of growth rings in the fossil skeleton varied from none in a toe

phalanx to eight in the tibia and femur. Thus although the number of growth rings could sometimes give us an idea of the individual age of an extinct animal, it should not just be taken at face value.

One clear outcome of my study with Chinsamy and Dodson was that the analyzed enantiornithines had experienced periods of substantial reductions in their rate of growth, if not complete cessations of growth. This unavoidable conclusion highlighted the remarkable differences between the pattern of bone formation of enantiornithines and that of living birds, which acquire adult size within a single year and typically do not experience interruptions or reductions in their rate of bone generation. The discovery of growth rings in the bones of enantiornithines suggested that these birds did *not* reach full size within the first year of their life.

The studies discussed here have coupled information derived from growth series of *Archaeopteryx* and *Confuciusornis* to indicate that early birds had rates of growth that greatly differed from their living counterparts. We saw earlier how some already fledged individuals—those endowed with flight feathers—of *Archaeopteryx* and *Confuciusornis* are 50 and 60 percent, respectively, the size of others. The fact that these small specimens are already fledged and that they show no obvious features indicating they died as neonates also suggests a dramatic difference in the rate of growth of these archaic birds and those of today.

Hatchling Strategies and Development

Fossils of enantiornithines have also cast light on aspects of the early development of ancient birds and helped us to understand the appearance and behaviors of their chicks. Living birds exhibit a vast spectrum of hatchling strategies in terms of behavior, ability to move, and appearance. Ornithologists

usually subdivide this spectrum into five developmental categories, whose boundaries are often poorly demarcated. At one end of the spectrum are the superprecocial megapods, plant mound-building gallinaceous birds that live in Australia, whose hatchlings are completely independent of their parents. The plumage of megapod chicks is already fledged when they hatch, and in some instances, these young birds are able to fly soon after they come out of the egg. The precocial chicks of ratites (ostriches and their flightless kin), pheasants and turkeys, ducks and geese, button-quails, cranes and rails follow the extreme precocial behavior of megapod hatchlings in the spectrum of avian developmental strategies. The precocial hatchlings of these birds also hatch feathered and capable of walking. Although the hatchlings of precocial birds are unable to fly and they are usually less autonomous than the young megapods, precocial hatchlings can search for food independently and they abandon the nest soon after they emerge from the egg. The other end of the spectrum of hatchling development is occupied by the chicks of parrots and songbirds, which are altricial. Altricial chicks hatch naked (without plumage) and with closed eyes. These chicks are incapable of locomotion and are completely dependent on their parents to provide for them, remaining in the nest until an advanced stage of development. Between these extremes of the precocial–altricial spectrum are the semiprecocial hatchlings of gulls and skuas, and the semialtricial hatchlings of storks and herons, whose behavioral, locomotor, and physical characteristics are intermediate between those of precocial and altricial hatchlings.

Rare discoveries of the embryos and chicks of enantiornithines have indicated that their hatchlings had physical appearances, behaviors, and walking abilities comparable to those of living precocial birds. Fossils from the Early Cretaceous of China and Spain

Modern birds exhibit a wide range of hatchling strategies, from the precocial chicks that hatch feathered and able to find their own food, to naked altricial hatchlings that are completely dependent on their parents. (Image: S. Abramowicz.)

have shown that these hatchlings were feathered and that the downy plumage of their wings was quickly fledged into fully formed flight feathers. Genealogical inference also points at the same conclusion: evolutionarily speaking, enantiornithines fall between a variety of precocial living birds and crocodiles, whose babies can also be described as precocial. Thus fossil evidence, as well as evolutionary bracketing near living animals with precocial hatchlings, strongly suggests that enantiornithine chicks hatched feathered, able to walk, and with the ability to find food by themselves.

Some researchers, however, have gone further, suggesting that baby enantiornithines had capabilities comparable to those of the chicks of megapods, thus classifying them within the superprecocial portion of the spectrum of avian hatchling strategies. One of these researchers is Polish paleontologist Andrzej Elzanowski, whose 1980s studies of well-preserved enantiornithine embryos from the Late Cretaceous of Mongolia highlighted features in common with the chicks and embryos of the Australian mound-builders. Elzanowski noticed that the wingbones of these embryos are proportionally longer than those of most extant hatchlings and that their bone formation is also more advanced than that exhibited by precocial chicks. These observations notwithstanding,

These two enantiorni-
thines from the Early
Cretaceous of China show
a number of features
indicating that they died
at a very young age. The
inset shows the wrist of
one of these juveniles
and how its semilunate
carpal is very similar
to the correspond-
ing bone in nonavian
maniraptoran theropods.
(Image: R. Meier.)

embryologist Matthias Starck (of Ludwig-Maximillians University in Munich) has shown how difficult it is to infer the developmental strategy of a hatchling on the basis of the degree of bone formation in an embryo or a chick. Even though precocial and altricial hatchlings often show significant differences in the ossification of the vertebral column, the rib segments joined to the breastbones, and the terminal finger bones of the wing and feet—these are usually ossified in precocial hatchlings but remain largely cartilaginous in altricial hatchlings—the degree of skeletal formation of these hatchlings may differ minimally. Starck calculated that while hatchling skeletons of the altricial budgerigar—an Australian parrot—were 7–13 percent ossified, those of precocial button-quails were only 15–25 percent ossified. If the difference in the amount of ossification of hatchlings with greatly different strategies is rather minimal, it is hard to imagine that we

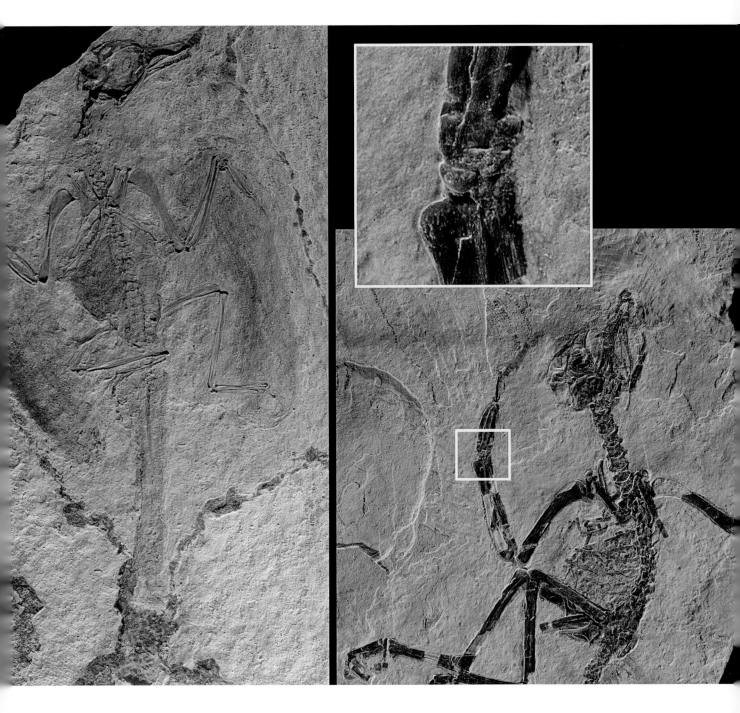

would be able to discriminate between precocial and superprecocial modes of development by simply observing the degree of ossification of a given fossil embryo or hatchling.

The handful of fossil enantiornithine babies has also provided a glimpse of the pathways taken by the skeletons of these birds during development. One important case involves the semilunate carpal, a crescent-shaped bone of the wrist that caps the upper ends of the metacarpals. Earlier on we highlighted the remarkable correspondence in the shape of the semilunate carpal of nonavian maniraptoran theropods and *Archaeopteryx*. In these fossils, this crescent-shaped bone is easily identifiable, because even in adults it remains separated from the metacarpals. However, comparisons with later birds—including enantiornithines—have been hampered by the fact that in the adults of these birds, all these bones are incorporated into a compound bone called the carpometacarpus. The discovery of enantiornithine fossils showing early stages of wrist development has demonstrated that the semilunate carpal of these birds remained virtually unchanged from the design present in *Archaeopteryx* and nonavian maniraptorans. Comparable correspondences in the structure of the ankle, whose compound bones are also obscured by fusion in adults, have also been detected through the discovery of enantiornithine hatchlings. By providing snapshots of the development of the primitive avian wrist and ankle, these exceptional fossils furnish unprecedented evidence of the remarkable similarity between the formation of the limbs of nonavian theropods and that of Mesozoic birds. Thus, this rare window on the early development of primitive birds not only illuminates aspects of the skeletal formation of enantiornithines but also highlights once more the dinosaurian ancestry of birds.

THE OLDEST PELLET

Raptor pellets are very rare in the fossil record. Recently, I teamed up with my colleague José Luis Sanz to study a curious bone assemblage collected from the Early Cretaceous Spanish locality of Las Hoyas, which showed distinct signs of having been digested. The 115-million-year-old limestones of Las Hoyas formed in the bottom of a very quiet lake. Fossil aggregates like this one are extraordinary for this site, where hundreds of fossils of land-dwelling vertebrates consisting of single individuals have been found. The new fossil aggregate contained the partial skeletons of four neonate birds, whose general characteristics identified them as enantiornithines. The four partial skeletons retain some of their original joints, although they are somewhat mixed up and overlapped with one another across a surface area of less than 25 square centimeters. Although two of these birds are nearly twice as large as the other two, none of them show significant differences in their degree of bone development. Anatomical differences in the tail vertebrae and the proportions of the hindlimbs indicate that the two larger individuals belonged to two different species. Furthermore, the notable size differences between these two and the smaller two individuals made us wonder about the possibility of there even being a third species in this peculiar bone assemblage.

predators that inhabited Las Hoyas' ancient ecosystem. Naturally, we could not rule out simple chance—the always-possible scenario that the four neonates had died far from one another but been buried together. But squaring this with what we know about the calm shallow waters of this ancient subtropical lake was not easy. A non-random alternative to our pellet interpretation was that this unique association of baby birds constituted a nest that was washed into the lake, sinking to the bottom with its four dwellers. Because the four individuals belonged to at least two, and possibly three species of enantiornithines, this scenario would require the extra assumption of nest parasitism. Some extant bird species (cuckoos, blackbirds, and others) do lay eggs in the nests of other birds. This complex behavior, however, has not been documented for any Mesozoic bird, nor is there any reason to believe that it existed in their nonavian ancestors. Given all this and also the fact that the aggregate showed features typical of pellets, we concluded that we were studying the oldest known mass of undigested remains regurgitated by a raptor.

Identifying the specific predator that had produced it was not easy, however, because pellets are common products of the behavioral repertoire of several of the animals recorded in Las Hoyas, including birds, a variety of nonavian reptiles, and fish. Comparisons to extant pellets allowed us to rule out crocodiles being responsible for this pellet; their gastric acids are so strong that their pellets contain nearly entirely digested bone, which was not the case in the Las Hoyas pellet. We were also able to rule out some of the large fish that are known from Las Hoyas, because the regurgitated items of fish are also very different from the Spanish pellet. The known lizards and birds of Las Hoyas were too small to be candidates. In the end, the animal responsible for this pellet is unknown—but the only predators known from Las Hoyas that cannot be ruled out are small nonavian theropods, such as the ornithomimid *Pelecanimimus*. This evidence indicates that the common behavior of birds in swallowing their prey whole and regurgitating the undigested remains may have been inherited from their dinosaurian ancestors. (Image: J. Sanz/R. Urabe.)

Because the bones had corroded ends, a surface pattern of pitting typical for digested bones of extant birds and mammals, and feather impressions surrounding the periphery of the bone aggregate much in the fashion of modern pellets, everything pointed to this being the regurgitated pellet of one of the

CHAPTER EIGHT

Early Predecessors of Modern Birds

The diverse enantiornithines that dominated the inland ecosystems of the Cretaceous shared an ancient common ancestor with a lineage that in time would evolve into all living birds. Together with their extant counterparts, these early predecessors of modern birds are grouped within the Ornithuromorpha. Despite our poor understanding of this critical phase of the evolutionary history of birds, the fossil record tells us that early ornithuromorphs evolved a number of modern features that distinctly differentiate them from their enantiornithine cousins. A handful of fossil discoveries of the last two decades have given us a glimpse of the exuberant diversity

that predated the differentiation of the closest relatives of modern birds. Known from the Early as well as the Late Cretaceous, these archaic ornithuromorphs document the initial steps leading to the evolution of the skeletal design that characterizes living birds and controls the way they move about.

Exposures of the Bajo de la Carpa Formation at the Patagonian locality of Boca del Sapo. All known specimens of the flightless *Patagopteryx deferrariisi* have been unearthed from these 80-million-year-old rocks. (Image: L. Chiappe.)

THE EVOLUTION OF
SECONDARY FLIGHTLESSNESS

Secondary flightlessness, the evolutionary process by which a species loses its capacity to fly, has been a common theme in the history of birds. Many lineages of birds—both extant and extinct—have flightless representatives. The large majority of these flightless birds are land dwellers and their anatomy as well as genealogical relationships indicate they evolved from relatives that once flew.

Regardless of the group in question and its lifestyle, secondary flightlessness is usually associated with a number of distinct physical attributes. Most evident among these is the abbreviation of the forelimb—sometimes as extreme as in the photographed Moa—and the changes in the proportions of its bones. For example, the bones of the hand and forearm of flightless birds are typically shorter than those of their flying relatives. In addition, the wishbones of flightless birds are often completely reduced and so is the keel of their breastbones. Not surprisingly, the flight musculature of these birds is

also greatly reduced. In the brown kiwi of New Zealand's North Island, the flight muscles barely make up 13 percent of body mass, but in most flying birds, the weight of these muscles is double that. Furthermore, in flightless birds, the hindlimbs are stout and proportionately larger than those of their flying relatives.

Attempts to explain the causes and processes that led to secondary flightlessness have focused on oceanic islands, where this phenomenon is common. Most islands have limited geographic barriers, minimal environmental diversity, and few, if any, indigenous land predators (only recently have humans introduced the predatory mammals that populate most of today's islands). With no major geographic barriers to cross or land predators to avoid, it is widely believed that evolution favored the loss of flight by relocating the energy necessary to maintain large flight muscles into activities such as feeding and breeding. In this context, flightlessness is interpreted as a by-product of reducing the size of the metabolically expensive flight muscles. While such interpretation appears reasonable for insular settings, it cannot explain the loss of flight in continental birds, where geographic barriers and land predators were not infrequent. In these settings, flightlessness has often been linked to an increment in size and/or speed, but the real causes underlying the evolution of flightlessness in continental lineages remain much more mysterious than those at play on islands. Indeed, in *Patagopteryx*—a continental bird that lived far from the ocean—flightlessness led neither to large size nor to specializations for running. The ancient Patagonian bird reminds us that the processes involved in the evolution of flightlessness are likely to be much more complex than those outlined above. (Image: American Museum of Natural History.)

Terrestrial Flightlessness Makes its Debut

One of the most primitive known ornithuromorphs is *Patagopteryx deferrariisi*, a hen-sized bird discovered in the Late Cretaceous of northwestern Patagonia, whose reduced wings and stout hindlimbs highlight its flightless nature. In the early 1980s, contractors working for the Comahue National University, in the Argentine town of Neuquén, leveled a large hill adjacent to the university's main building. The university was expanding—more classrooms, a library, a new museum, and other buildings had been planned for the leveled area. The project took longer than expected,

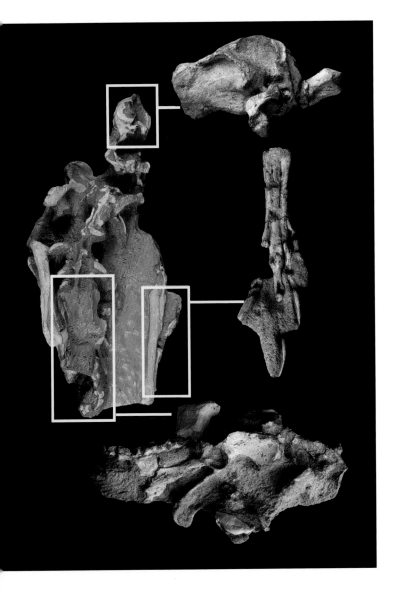

An articulated specimen of hen-sized *Patagopteryx* (left; cast made prior to complete preparation and disarticulation of the specimen). Details show the skull in left view (top), the right foot in front view (middle), and the pelvis in right view (bottom). (Image: L. Meeker.)

however, and with time, the leveled surface was eroded by rains and gusts of Patagonian wind. It did not take long for the first fossils to appear—after all, the hill was part of the 80-million-year-old Bajo de la Carpa Formation, one of Patagonia's richest dinosaur rock layers.

Between 1984 and 1985, Oscar de Ferrariis, at that time the Director of the university's budding Natural History Museum, discovered several specimens of a hen-sized bird—two of them preserving numerous portions of the skeleton. These were the first Mesozoic birds ever found in Patagonia and some of the very few known for the entire South American continent. Their body plan was in many respects unique among birds, and their obvious incapacity for flight made them not just one of the earliest examples of terrestrial flightlessness but also a suggestive discovery in a continent that had formed part of a landmass with plenty of other lineages of flightless birds—the rhea, ostrich, and kiwi among others.

De Ferrariis passed the precious fossils to José Bonaparte, Argentina's foremost authority on Mesozoic vertebrates. This led to my first contact with the terrestrial *Patagopteryx deferrariisi*, a relationship that would last several years, since, as a graduate student under the supervision of José Bonaparte, these skeletons as well as others subsequently discovered would become an integral part of my doctoral dissertation.

A Primitive Ornithuromorph: *Patagopteryx*'s Skeleton

Although a number of transformations towards the modern design characterize the skeleton of *Patagopteryx*, this bird is in many ways utterly unique. Some of these distinctive features—its short forelimbs and massive legs—are clearly related to its flightless nature, but others allude to the existence of a highly specialized lineage of ancient Patagonian birds for which *Patagopteryx* is our sole clue.

Even though half a dozen specimens have been discovered, little is known about the skull of this Patagonian bird. It was likely toothed, as were other early ornithuromorphs, but we know neither this nor the relative length of its jaws. However, evidence from the rear portion of the skull shows a significant departure from the primitive design present in *Archaeopteryx*, *Confuciusornis*, and enantiornithines. One of these developments relates to the extensive pneumatization of the skull bones, which in degree approaches that characterizing living birds. Another noticeable transformation involves the incorporation of the squamosal bone into an expanded braincase, a change that was probably connected to the enlargement of the brain, and therefore to the fine-tuning of sensorial capabilities. In this and other ornithuromorphs, the primitive diapsid configuration of more primitive birds had finally been modified into the derived diapsid condition of modern birds. This last transformation suggests that *Patagopteryx*, and presumably other ancient ornithuromorphs, saw a significant reduction in the musculature responsible for the powerful biting force of their forerunners.

Another noticeable transformation towards the modern design is found in the long, S-shaped neck of this early ornithuromorph, whose vertebrae have saddle-shaped joints. This advanced skeletal plan suggests that the neck of *Patagopteryx* functioned similarly to that of its extant counterparts. In modern birds, the neck is subdivided into three sections that act as separate functional units. While the front and rear sections move primarily downward from the normal position of the neck, the main range of movements of the middle section is upwards. This structural and functional design is fundamental to the pecking action we so often see in living birds. Other transformations of *Patagopteryx* that approach the skeletal

design of its living relatives are found in its hindquarters. For example, the pelvis is very wide, with each pubis and ischium well-separated from their counterparts (as opposed to contacting each other at their tips as in more primitive birds). Such a configuration suggests that *Patagopteryx* had evolved an ample pelvic canal, specialized for laying large eggs.

Beyond the features that place *Patagopteryx* evolutionarily closer to modern birds, it is the abbreviated wings and massively built legs of this fossil that provide the best testimony of its inability to fly. The wings are very small in comparison to its powerful hindlimbs—the length of the wing is less than half the length of the hindlimb. Clearly, such a short wing would have formed an airfoil inadequate for lifting the hen-sized body of this ancient bird. Furthermore, the stout humerus, longer than the bones of the forearm, has comparable proportions to most flightless birds, and the same is true of the short, clawed hand. Additional evidence supporting the flightless nature of *Patagopteryx* is found in its sternum, which is flat and unsuited for anchoring powerful flight muscles; the apparent absence of a wishbone; and the wide spacing of the joints between coracoids and sternum, all features typical of flightless birds.

Patagopteryx's Terrestrial Specializations

The stout legs of *Patagopteryx*, with their prominent structures for the attachment of strong muscles, indicate that the locomotion of this ancient Patagonian bird was governed by its hindlimbs. Powerful legs, however, are not only common to birds of an obligate terrestrial existence— they are also typical of flightless birds that have become specialized for foot-propelled swimming. Clearly, the robustness of the hindlimbs of *Patagopteryx* alone cannot discriminate between these very different lifestyles. Yet *Patagopteryx* does not exhibit any of the skeletal features that are usually associated with foot-

propelled divers and swimmers—such as the compression and elongation of the rear portion of the pelvis and sacrum, the hypertrophy of the upper end of the tibiotarsus, and the side compression of the tarsometatarsus. While *Patagopteryx* lacks these skeletal specializations of foot-propelled aquatic birds, it does display bony structures that strongly support an obligate terrestrial lifestyle. Indeed, it was because of these features and its aerial incapacity that previous researchers argued for an evolutionary connection between the ancient Patagonian bird and living ratites, the group that includes ostriches, rheas, emus, and other flightless, land-dwelling birds.

The skeletal anatomy of *Patagopteryx* leaves little doubt about its running abilities, but its hindlimbs do not show the specializations for high speed present in other land birds. Its tarsometatarsus is significantly shorter than the tibiotarsus, and its toes are large—the middle one is as long as the tarsometatarsus. In fast-running birds, however, the tarsometatarsus is almost as long as the tibiotarsus and the toes are short (and often fewer in number), minimizing their friction against the ground. The absence of these running specializations suggests that *Patagopteryx* was not able to achieve the speeds of ostriches and other fast-running, extant birds.

New Fossils from China

In the last few years, the Chinese Early Cretaceous lakebeds that buried much of the fantastic diversity of birds reviewed earlier in the book have also yielded evidence of a much older segment of the evolutionary history of ornithuromorphs than the one illustrated by *Patagopteryx*. Several fossils from Liaoning Province are likely to be members of the archaic radiation of ornithuromorphs, and among these, the 120-million-year-old *Yixianornis grabaui* and *Yanornis martini* are vivid testimonies of this portion of avian history.

Both these birds are toothed but they greatly differ in the design of their skulls. While the head of the smaller *Yixianornis*—a bird the size of a large nightjar—is short and broad, that of tern-sized *Yanornis* has a distinctly longer and narrower snout. Although the overall design of the wings of these birds—slightly longer than the hindlimbs, with the ulna and the radius longer than the humerus, and with compact hands carrying small claws on the two most internal fingers—resembles that of enantiornithines in a number of details, the wingbones of these archaic ornithuromorphs look much more like those

< The 120-million-year-old *Yixianornis grabaui*. The anatomy of the wing and shoulder of this Chinese bird closely approach the modern condition. (Image: Z. Zhou.)

Fleshed out reconstruction of *Yanornis*. (Image: U. Kikutani.)

< The tern-sized *Yanornis martini* is a larger contemporary of *Yixianornis*. Stomach contents in some specimens of *Yanornis* indicate that this bird was a fish-eater. (Image: Z. Zhou.)

MESOZOIC BIRDS
FROM THE SOUTHERN HEMISPHERE

For decades, the early stages of avian evolution were reconstructed almost exclusively on the basis of fossil evidence unearthed in the Mesozoic rocks of the Northern Hemisphere. In the wake of the recent surge in primitive avian discoveries, a number of fossils from Argentina, Madagascar, Antarctica, and other austral landmasses have started to illuminate the early history of birds in the Southern Hemisphere. These fossils, including skeletons, footprints, and feathers presumed to be avian, have made a significant addition to our knowledge about the archaic evolution of birds.

All indisputable Mesozoic avian remains from the Southern Hemisphere are of Cretaceous age. Fragmentary bones of enantiornithines from 100-million-year-old beds in east-central and southern Australia, and a handful of asymmetrically vaned feathers from northeastern Brazil, are the oldest testimonies of birds from the southern landmasses. However, most avian records from the Mesozoic of Gondwana come from the Late Cretaceous. In

South America, Argentina has yielded important remains of enantiornithines, early ornithuromorphs, and other fossils more closely related to modern birds—the illustration shows the falcon-sized enantiornithine *Nequenornis* perching on a branch in front of a pair of *Patagopteryx*. In Africa, Madagascar has contributed with the discovery of the long-tailed *Rahonavis*, the archaic ornithuromorph *Vorona*, and a number of fossils of enantiornithines. Very little is known of the early history of African birds outside Madagascar. The most notable occurrences include a handful of footprints from the Late Cretaceous of Morocco and a 90-million-year-old enantiornithine from Lebanon, which was then part of North Africa. Several frigid islands east of the Antarctic Peninsula have yielded a handful of 70-million-year-old fossils that appear to be closely related to, if not part of, several extant groups.

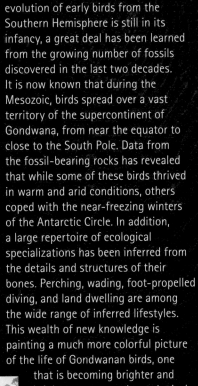

Although our knowledge of the evolution of early birds from the Southern Hemisphere is still in its infancy, a great deal has been learned from the growing number of fossils discovered in the last two decades. It is now known that during the Mesozoic, birds spread over a vast territory of the supercontinent of Gondwana, from near the equator to close to the South Pole. Data from the fossil-bearing rocks has revealed that while some of these birds thrived in warm and arid conditions, others coped with the near-freezing winters of the Antarctic Circle. In addition, a large repertoire of ecological specializations has been inferred from the details and structures of their bones. Perching, wading, foot-propelled diving, and land dwelling are among the wide range of inferred lifestyles. This wealth of new knowledge is painting a much more colorful picture of the life of Gondwanan birds, one that is becoming brighter and brighter as more paleontological expeditions scout the Mesozoic rock exposures of the austral continents. (Image: E. Heck.)

of living birds. Their shoulder also has an essentially modern configuration, and the sternum carries a large and strong keel for the attachment of powerful flight muscles. Other differences with respect to more primitive birds are evidenced in the tail, particularly the pygostyle (the rump bone at the end of the tail), which is much shorter and compressed sideways. The significance of this transformation is not entirely clear. The abbreviation of the tail of these early ornithuromorphs is probably related to the evolution of the fan-like organization of the tail feathers (rectrices), a design typical of many modern birds that maximizes lift. However, very little is known about the plumage of early ornithuromorphs, and further fossil discoveries may need to be conducted prior to considering this line of thought further.

Other extraordinarily well-preserved, archaic ornithuromorphs have also been recently unearthed in China. One of these fossils is *Hongshanornis longicresta*, a bird from Inner Mongolia that is much smaller and somewhat older than both *Yixianornis* and *Yanornis*. The seemingly beaked skull of *Hongshanornis* has been described as evidence that this early ornithuromorph lacked teeth—however, remnants of what appear to be tooth sockets in the lower jaw of this small bird make such an interpretation questionable. A wealth of Early Cretaceous birds has also been recently quarried in the northwestern province of Gansu, far to the west of Liaoning. Thus far, most of these fossils seem to be new specimens of *Gansus yumenensis*, a grebe-sized bird previously known by a fragmentary foot. Being the first Mesozoic bird ever discovered in China, this historically important specimen had been interpreted as the remains of an ancient wader and an early relative of modern shorebirds. The new fossils, however, demonstrate the primitive ornithuromorph nature of *Gansus*, thus indicating that although this bird was evolutionarily close to modern birds, it was very distant from any living

shorebird. Just like *Yanornis*, *Yixianornis*, and *Hongshanornis*, *Gansus* possessed a flight system of modern design, with many features in the wing, shoulder, and sternum to highlight its aerodynamic proficiency.

Given the little that is still known about these archaic ornithuromorphs, it is hard to agree on whether they represent evolutionary stages more primitive than the much younger *Patagopteryx*. The specialized nature of the latter, whose reduced flight system can be hardly compared to those of its Chinese counterparts, further complicates this issue. Some features, however, do suggest that *Hongshanornis*, *Yanornis*, and *Yixianornis* are among the most primitively known ornithuromorphs and could well illustrate earlier stages in the evolution of the group than that represented by the Late Cretaceous Patagonian bird. One such feature lies in the configuration of the pelvis, which in the Chinese birds is characterized by a narrow pelvic canal (both pubes contact each other over a long distance), a primitive feature absent in *Patagopteryx*. However, the evolutionary divergence of ornithuromorphs was likely characterized by as much experimentation as other chapters of the early evolution of birds, so further discoveries may be needed before we can decipher the relationships among these and other archaic ornithuromorphs.

An Exquisite Ornithuromorph from Mongolia

Also representing a very early phase of the evolution of ornithuromorphs is the 70-million-year-old *Apsaravis ukhaana*, whose beautifully preserved skeleton was discovered in the late 1990s in the renowned locality of Ukhaa Tolgod (see *Ukhaa Tolgod: An Extraordinary Cretaceous Ecosystem*). *Apsaravis* was a medium-sized bird—slightly smaller than a mallard—with a sharp beak, toothless jaws, well-developed wings, and a foot

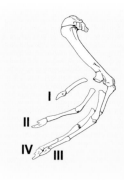

The long and slender toes of *Gansus yumenensis* and the overall anatomy of its foot suggest that this bird was largely aquatic. This interpretation is confirmed by much better preserved specimens discovered recently. (Image after Rich et al., 1986.)

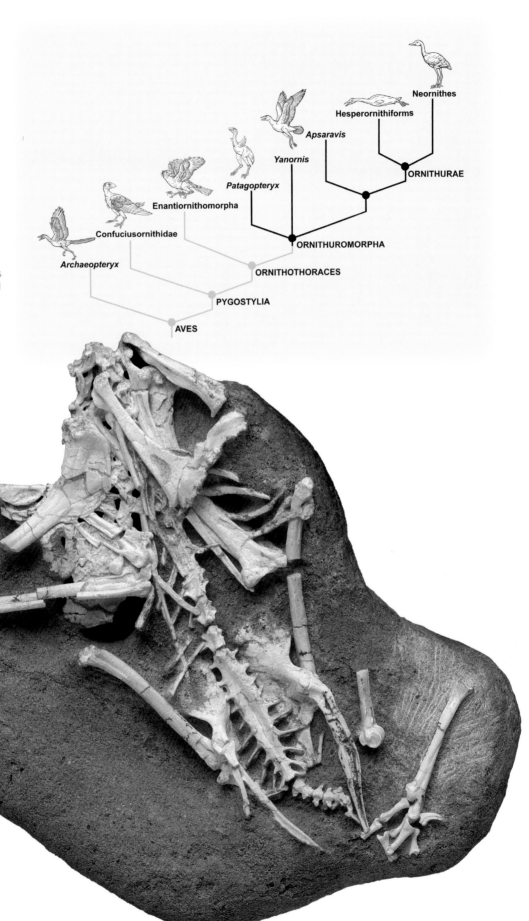

Cladogram of relationships of primitive ornithuromorphs. The relationships between *Patagopteryx* and some of the newly discovered primitive ornithuromorphs—*Yanornis*, *Yixianornis*, and *Hongshanornis*—still need to be analyzed in further detail. (Image: E. Heck/S. Orell.)

The 75-million-year-old *Apsaravis ukhaana* is one of the best preserved primitive ornithuromorphs. This Mongolian fossil provides evidence that the skeletal features allowing the automatic unfolding of the modern wing evolved early on in ornithuromorph history. (Image: M. Ellison.)

Neornithes

Hesperornithiforms

Apsaravis

Yanornis

ORNITHURAE

Patagopteryx

Enantiornithomorpha

ORNITHUROMORPHA

Confuciusornithidae

ORNITHOTHORACES

Archaeopteryx

PYGOSTYLIA

AVES

with only three toes. Even though its skull is largely incomplete, the lack of teeth in the lower jaw suggests that *Apsaravis* could represent another instance of tooth loss among birds, a recurrent theme in the evolution of birds that has also occurred in the confuciusornithids, the enantiornithine *Gobipteryx*, and modern, neornithine birds.

Despite the limitations that obscure our understanding of the sequence of evolutionary transformations that occurred at the onset of the ornithuromorph divergence, many features of this Mongolian fossil confirm the anatomical makeover from early ornithuromorphs. *Apsaravis* tells us again that these early relatives of modern birds developed a longer neck with a full set of saddle-shaped vertebrae, a modern design of the shoulder girdle, and hindlimb transformations that had postural and locomotory significance. *Apsaravis* also shows that at some point in the early history of ornithuromorphs, birds evolved a broad pelvis of modern configuration, an important transformation likely connected to the evolution of larger eggs, and possibly a more substantial investment in parental care. Although parental involvement in raising chicks must have been extensive in all primitive birds, it is possible that early in their evolutionary history, ornithuromorphs developed levels of offspring attendance comparable to those of many living birds.

Yet perhaps the most significant innovation of the skeleton of *Apsaravis* is found in its hand, particularly in the design of the innermost metacarpal—the bone that supports the finger operating the alula. In *Apsaravis* this bone has developed a prominent projection called the extensor process, which in modern birds receives the attachment of the extensor ligament. This strong ligament plays a key role in flight: it supports the propatagium—the airfoil between the wrist and the shoulder—and operates the unfolding mechanism of the wing's tip.

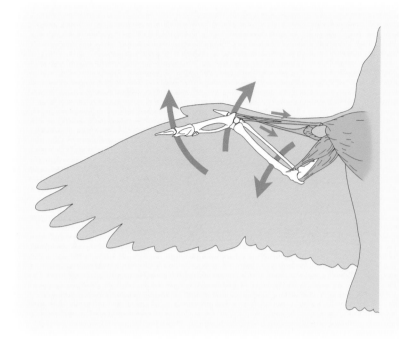

In living birds, the wing is automatically unfolded by the combined actions of the innermost metacarpal and this ligament. When the extensor ligament is tensed, the extensor process of the innermost metacarpal works like a lever, unfolding the end of the wing, and triggering a series of movements of the wrist and forearm that spread out the rest of the wing. The energetic advantages of this mechanism over one operated by individual muscles become obvious when considering the frequency at which the wing is spread out during flight—10 to 20 times per second in small birds flying at cruising speeds. Even if an incipient version of this mechanism could have evolved among *Apsaravis*' predecessors, the modern design seen in this bird is a clear breakthrough in the refinement of avian flight.

In modern birds, the wing is unfolded by the combination of ligamental tension and the lever-like action of the extensor process of the hand. Arrows show how the tension of the extensor ligament (small arrows) generates the unfolding of the wing (large arrows). (Image: S. Abramowicz.)

Early Ornithuromorphs' Lifestyles

Fossils of archaic ornithuromorphs are still too few to provide a clear picture of the lifestyles of these birds, but fortuitous discoveries have given us an idea of the feeding preferences of some

of these ancient birds. Vertebrae, fin rays, and other bones of a ray-finned fish found inside the visceral cavity of a specimen of *Yanornis* have documented the fish-eating habits of this animal and provided the earliest direct evidence of such a behavior among birds. At the same time, the discovery of large numbers of gastroliths or gizzard stones in another fossil of *Yanornis* suggests that this bird could have switched its main diet in response to seasonal changes in food supplies. Examples of living birds that switch their main diet from one season to another are well-known. These changes often go from diets that are low in non-digestible fibers—fish, insects, crustaceans, and other animals—to seeds, grains, and other high-fiber food items. These dietary changes are typically accompanied by the ingestion of gravel and grit, which by remaining inside the gizzard help to macerate and thus digest the high-fiber food.

The foot structure of *Gansus* and the abundance of webbed footprints possibly made by archaic ornithuromorphs in several Early Cretaceous localities suggest that a number of these birds also foraged near bodies of water. Despite the terrestrial specializations of *Patagopteryx* and our limited knowledge of the early phases of ornithuromorph evolution, there is clear evidence that many of these birds inhabited near-water environments. Perhaps this existence heralded the specializations to marine environments developed by more advanced ornithuro-morphs, such as the familiar hesper-ornithiforms of the Late Cretaceous of North America. Our story now turns to these fossils, which illustrate a step even closer to the evolutionary origin of living birds.

These specimens of *Gansus yumenensis* are two of the many fossils of this bird that have been recently found in the Early Cretaceous of Gansu Province. The anatomy of the feet of *Gansus* suggests an aquatic existence. (Image: H. You.)

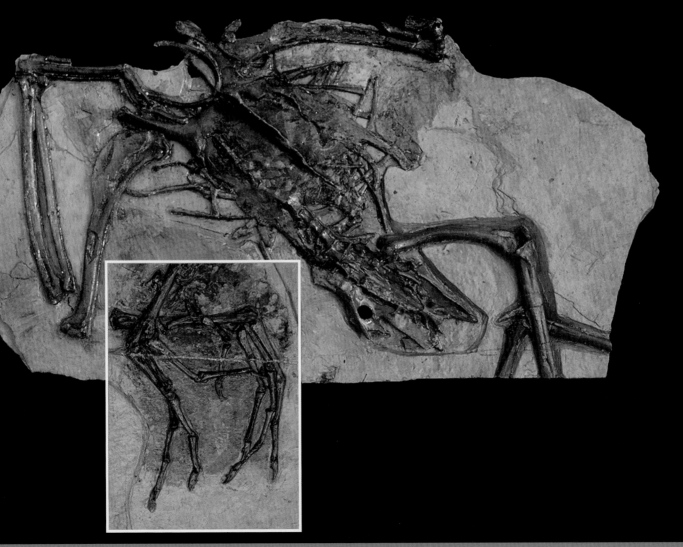

CHAPTER NINE

Hesperornis and its Relatives: Divers of the Cretaceous Seas

In the Late Cretaceous, shallow seas flooded vast areas of the northern continents. Evidence of this has remained in the rocks and fossils of central North America and the boundaries of Europe and Asia, where tongues of water connected the ancient equatorial Tethys to the predecessor of the Artic Ocean. These North American and Eurasian seas, the Pierre Seaway and Turgai Strait, respectively, brought a more littoral climate to the already subtropical weather of these continental lands and hosted a flourishing life among which we found the remains of a peculiar lineage of toothed birds, the hesperornithiforms, large flightless, foot-propelled divers of the Cretaceous seas.

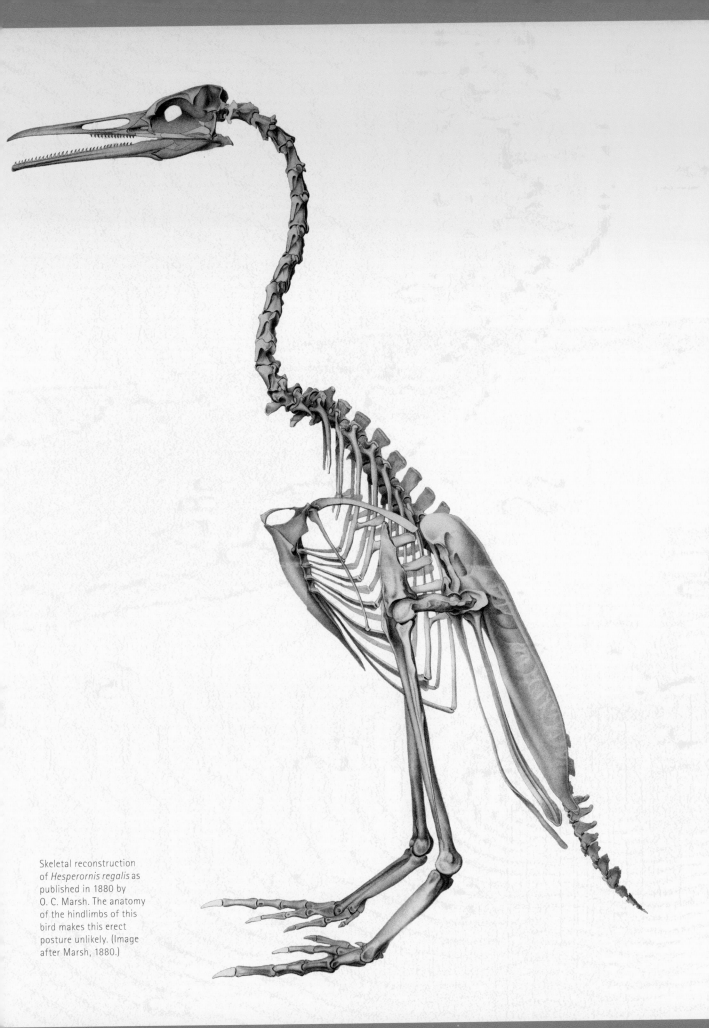

Skeletal reconstruction of *Hesperornis regalis* as published in 1880 by O. C. Marsh. The anatomy of the hindlimbs of this bird makes this erect posture unlikely. (Image after Marsh, 1880.)

First Mesozoic Birds in America

Fossil bones of hesperornithiforms were the first Mesozoic avian remains to be found in the New World. These early discoveries came at a time when parties of paleontologists from the emerging museums of the eastern United States began to prospect for dinosaurs and other ancient creatures in the Mesozoic rocks that crop out east of the Rocky Mountains. The pioneer expeditions had not only to face the bitter cold of the winter and the extreme heat of the summer, but also the danger of working in hostile territories, for the paleontological exploration of these regions began during the American Indian Wars.

The first evidence of these birds was unearthed in 1870, when Yale University paleontologist Othniel Charles Marsh collected a fragment of a large tibia near the Smoky Hill River in western Kansas. Facing menacing natives, the party had to cut short their prospecting season, but Marsh and his fossil hunters returned to this area the following year, this time with a large escort of the United States army. The 1871 Yale expedition discovered several additional specimens of the same bird, which Marsh used to establish the new species *Hesperornis regalis*—the "royal bird of the west." It was not until the following year, however, that a much better picture of this bizarre bird came to light. In that year, the Yale fossil collectors stumbled upon an exquisitely preserved skeleton near the Smoky Hill River, one that included a skull and lower jaws. Measuring nearly 2 meters from the end of its bill to the tip of its toes, and with a host of highly specialized diving features, *Hesperornis* had the size of the largest penguin and skeletal features like no other Mesozoic bird that had been found before. That said, by the early 1870s there was not much to compare it with, for only a handful of other Mesozoic birds had then been discovered, or at least, recognized. These included a single skeletal specimen of *Archaeopteryx* and

some isolated bones from the Cretaceous of New Jersey and England.

Interestingly, one of the most remarkable features of *Hesperornis* evaded Marsh until 1875, when he realized that the bird he had found a few years earlier was toothed. At the time, this was an unexpected conclusion, because the idea that birds had teeth early in their evolutionary history was not an entirely clear concept. After a decade of collecting in the rich fossiliferous deposits of the ancient Pierre Seaway, Marsh amassed a collection of dozens of specimens of *Hesperornis regalis*, which with the exception of a few toebones and the very end of the tail provided an accurate picture of the anatomy of this bird. In 1880, the renowned paleontologist wrote a seminal monograph—*Odontornithes: Extinct Toothed Birds of North America*—in which, on the basis of a collection of some 50 specimens, he carefully described the anatomy of *Hesperornis* and provided exquisite illustrations and comprehensive measurements of all its bones. This volume still stands as the most important source for the study of this peculiar lineage of ancient birds—the first ones to be found in the Americas.

Diversity of Hesperornithiforms and Their Fossil Record

Not much scientific research has been conducted on hesperornithiforms, a fascinating early chapter of the evolutionary history of birds that is

Fossils of hesperornithiforms have been collected from many marine and brackish sediments that span the last 35 million years of the Mesozoic Era. Uncontroversial records of these birds are limited to North America and Eurasia. (Image: R. Urabe.)

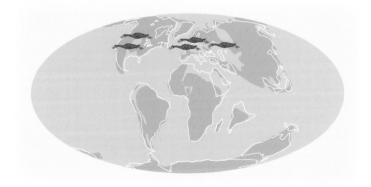

in great need of revision. Nonetheless, a steady number of discoveries since Marsh's first expeditions to the Smoky Hill River have documented the existence of a large diversity of these aquatic birds. In all, about a dozen hesperornithiform species have been named, but because some of these are based on single elements and other largely fragmentary fossils, such a count likely represents an overestimation of the known diversity of these birds.

Although primarily known from Late Cretaceous deposits, there is little doubt that the origin of the hesperornithiforms is to be found in the Early Cretaceous or even earlier. The fossil bones of the 100-million-year-old *Enaliornis* from the Cambridge Greensand of England may represent the first known record of this lineage of seabirds. The first remains of birds from the Cambridge Greensand appear to have been unearthed as early as 1858, even before the first recognized specimen of *Archaeopteryx*. These fossils were mentioned the following year by British geologist Charles Lyell, the architect of "uniformitarianism"—the fundamental concept that postulates that

the same geologic processes at play today operated also in the past. The fate (and identity) of these avian bones remains a mystery, however. In 1876, when paleontologist Harry Seeley reviewed the Cretaceous birds from the British Isles, they had already disappeared. In his report, Seeley described a large assemblage of other bird bones he had collected from the Cambridge Greensand. Although he named two new species of birds, *Enaliornis barretti* and *Enaliornis sedgwicki*, the validity of these still remains controversial. Undoubtedly, however, the bones he found did belong to a loon-sized, foot-propelled diving bird known nowhere else.

Just as with its putative relative *Hesperornis*, little work followed the 19th century description of these bones by Seeley, until paleontologist Peter Galton from the University of Bridgeport (Connecticut) and his co-workers restudied the specimens in the 1990s. Although initially regarded as an archaic loon, *Enaliornis* was first identified as a hesperornithiform by University of Kansas paleontologist Larry Martin in 1976. This view has

AVIAN SPOILS OF THE BONE WARS

Although the vitriolic confrontation of Othniel Charles Marsh with his foe, American paleontologist Edward Drinker Cope, has been recounted countless times, Marsh's legacy to the study of Mesozoic birds is often remembered in much less detail. However, the contributions of this pioneer fossil hunter, whose teams relentlessly collected fossils in the wake of colonization of the American West and the creation of a system of railroads, deserve a very special place in the annals of paleornithology. With the London specimen of *Archaeopteryx* as the only known Mesozoic-aged bird,

the discovery of *Hesperornis* (whose tooth is illustrated here) and a variety of other ancient North American birds in the early 1870s provided an enormous wealth of information about the deep history of birds. Even though these archaic avians were just a handful of species among the hundreds of new vertebrate species that resulted from the adventurous exploration of the fossil-rich badlands of the American West, they made a remarkable contribution to our understanding of the history of life. Not only did these birds exercise a profound influence on the outcome of the 19th century

evolutionary debate, but for over a hundred years *Hesperornis* and its allies shaped our perception of the early evolution of birds. (Image: O. C. Marsh.)

prevailed today, and the foot-propelled diver *Enaliornis* is regarded as a close relative of other primitive members of the hesperornithiforms, such as the 90-million-year-old *Pasquiaornis tankei* and *Pasquiaornis hardiei* from nearshore deposits of the Canadian Late Cretaceous. The remains of these primitive hesperornithiforms were briefly reported in 1997 by a Canadian team led by Tim Tokaryk from Saskatchewan's Eastend Fossil Research Station as part of a large collection of disarticulated bird bones from the Carrot River in central-eastern Saskatchewan. Although these two species mostly differ from one another in size—*Pasquiaornis hardiei* being roughly two-thirds the size of *Pasquiaornis tankei*—they show weaker hindlimbs and fewer foot-propelled diving specializations than any other hesperornithiform. Several portions of the humerus, tentatively identified as belonging to these species, suggest that these birds may have had much less atrophied wings than other hesperornithiforms. *Pasquiaornis* is not only the least-specialized hesperornithiform but also the oldest known North American bird.

From slightly younger deposits of the Pierre Seaway of western Kansas are the remains of *Baptornis advenus* and *Parahesperornis alexi*, both seemingly restricted to rock layers formed between 88 and 86 million years ago. Marsh discovered the loon-sized *Baptornis* in 1877, only five years after his initial report of *Hesperornis*. The famed Yale paleontologist regarded these birds as close relatives, although he emphasized the more specialized foot-propelled apparatus of *Hesperornis*. Nearly a century later, in 1976, Larry Martin and James Tate Jr. published a revision of this and subsequently discovered specimens of *Baptornis*, ratifying Marsh's hypothesis of a hesperornithiform relationship. Since Martin and Tate's publication, a fragmentary vertebra collected from the approximately

80-million-year-old marine sandstones of southern Saskatchewan (Canada) has been assigned to *Baptornis*. Interestingly, the occurrence of numerous bones of horned dinosaurs as well as some remains of duckbills and theropods suggest these sandstones formed much closer to the shore than the offshore chalk of western Kansas, where all other specimens of *Baptornis* had been found. If correct, this discovery would not only indicate a greater temporal range for this lineage but also a larger geographic distribution and perhaps a more plastic ecological specialization. Nonetheless, given that this record is based on a single vertebra, the alleged presence of *Baptornis* in nearshore deposits of Saskatchewan should be treated with caution. A contemporary of *Baptornis*, *Parahesperornis alexi* approaches more the highly specialized diving skeletal design of *Hesperornis*. Nonetheless, *Parahesperornis* was substantially smaller than *Hesperornis regalis* and its feet were distinctly less specialized for foot propelling.

A number of species much like *Hesperornis regalis* have been named since the discovery of this bird in the 19th century. Most of these are based on minor differences of foot structure and could well be synonyms of the latter. Others, like *Coniornis altus* from the Claggett Formation of Montana, may indeed be different species. Fossils very similar in general anatomy to *Hesperornis regalis* have been found in marine rocks from the Late Cretaceous of Eastern Europe and central Asia. Remains of the large *Asiahesperornis bashanovi*—a bird perhaps larger than *Hesperornis regalis*—are relatively abundant in near-shore deposits of the Kostanai Oblast of northern Kazakhstan. This large hesperornithiform swam and dived in the ancient Turgai Strait, a shallow marine tongue that began to connect the Tethys Sea, the ancient predecessor of the Mediterranean, and the Arctic Ocean some 75 million years ago. Fragmentary and slightly older

fossils from Sweden and Russia show that a diversity of these birds thrived in the flooded realm of the Late Cretaceous of Europe. Some of these fossils belong to a bird similar to *Baptornis* and others indicate the existence of a much larger hesperornithiform, *Hesperornis rossicus*, which was perhaps even larger than *Hesperornis regalis*. With powerful, recurved teeth on long and robust jaws, and sophisticated foot-propelled diving, *Hesperornis rossicus* was likely the ecological equivalent of its North American cousin. Nonetheless, more investigations on these fossils will need to be conducted before we can determine their genealogical relationships to other hesperornithiforms, and their precise functional specializations.

Neogaeornis wetzeli, a bird collected at the beginning of the 20th century from the Late Cretaceous marine deposits of central Chile, known from an isolated foot-bone, stands as the only putative hesperornithiform from the austral supercontinent of Gondwana. Although *Neogaeornis* was long accepted as a member of this archaic lineage of foot-propelled divers, a recent review of its only known bone—a tarsometatarsus—by Smithsonian Institution paleornithologist Storrs Olson recognized it as an early member of the loon family. Indeed, the design of the foot of *Neogaeornis* approaches that of modern foot-propelled divers such as loons and lacks some of the typical features of hesperornithiforms. Olson's identification of this bird as a loon, however, should be treated with care. Deciphering the genealogical relationships of birds known from single bones is often difficult, especially at a time when many bird lineages were experimenting with various skeletal designs. By removing *Neogaeornis* from the roster of hesperornithiforms, this peculiar lineage reveals itself as a radiation of birds that evolved exclusively in the Northern Hemisphere, in the oceans and waterways of the ancient supercontinent of Laurasia.

The Marine Realm of the Pierre Seaway

The Cretaceous was a period of great oceanic transgressions, in which sea levels may have been as much as 200 meters higher than today. To a great extent, this remarkable change in sea level may be correlated with the intense volcanism of the period, with mid-ocean ridges being formed as a result of the breakup of the large Mesozoic supercontinents of Laurasia and Gondwana. In North America, the lowlands that during the Late Jurassic had prevailed west of the Appalachian Mountains began to be flooded early in the Cretaceous. Shallow waters advanced to the north from the Gulf of Mexico and to the south from the Arctic Ocean, slowly flooding the states of Texas, Oklahoma, Kansas, Nebraska, Colorado, the Dakotas, Saskatchewan and the Northern Territories, creating a vast sea bordered by the Appalachians to the east and another range formed by the folded continental crust of eastern California and Oregon, Arizona, Nevada, Idaho, Washington, and Alberta.

This expansion of the Gulf of Mexico and the Arctic, however, was not uniform. The Middle Cretaceous witnessed a retreat of the ocean, even though much larger floodings occurred in the Late Cretaceous. As in other parts of the world at that time, the warm waters of this shallow Late Cretaceous sea hosted extensive coral reefs. Most of the rocks that formed on the sea bottom were made by the calcareous bodies of microscopic algae, which produced thick, nearly horizontal layers of a soft, yellowish-white variety of limestone known as *chalk*. Large accumulations of other microorganisms at the bottom of this tropical sea led to the formation of hydrocarbons, which in time formed much of today's oil. This vast mantle of salt water at the center of the continent, which at times could have been as wide as 1,600 kilometers, hosted a wide range of creatures whose abundant fossil remains are preserved in badlands

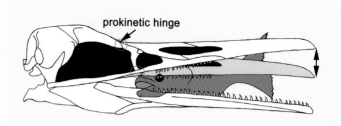

prokinetic hinge

widely distributed throughout central North America. In addition to a host of invertebrates, the tropical waters of this ancient sea teemed with fish of all sizes and the large plesiosaurs and mosasaurs that likely fed on them.

A Diving Body

Much like grebes and loons, *Hesperornis* and its relatives had compressed feet and torpedo-like bodies, features that helped them reduce their friction against the water. Unlike these modern analogs, however, the hesperornithiforms had toothed skulls with braincases that were in proportion much smaller than those of any of their living counterparts. The skull of *Hesperornis* and other hesperornithiforms was long and narrow, and its snout ended in a sharp tip. Evidence from some specimens indicates that a horny and weakly hooked bill covered the toothless front portion of the skull.

The upper and lower teeth of these birds had broad roots and much narrower crowns, with sharp, backward-pointing tips. Such a design was aptly suited for grasping the fish they probably ate. Evidence of resorption pits in the roots of their teeth indicates that these birds, like other ancient toothed birds and their nonavian theropod relatives, shed their teeth regularly. These teeth, however, were set in a common groove, at least until the birds became full-grown adults, when their tooth sockets developed discrete walls.

The toothed mandible of these peculiar birds bears a mobile joint at the center of each lower jaw that could have helped widen the mouth, a functionally useful specialization for predatory birds that may have had to swallow large food items. This was not the only area of mobility of the skull, however. A number of special types of skull kinesis or regions of movement between skull bones have been proposed for *Hesperornis* and its allies, but research led by the late Paul Bühler, a functional morphologist from the University of Hohenheim in

Germany, demonstrated that the skull of hesperornithiforms had a type of prokinesis not much different from that of most living birds. In a prokinetic skull, the upper jaw moves up and down as a single unit, independent of the rest of the skull. This movement is permitted because there is an area of bending in front of the orbit, and because a system of struts formed by several bones of the palate produces the forces to lift the upper jaw up and down. The result is that the mouth can open wider than it would if the skull were devoid of kinesis. Clearly, with a large and long skull able to perform this kind of movement and with the lower jaws being able to expand thanks to their internal joints, *Hesperornis* and its relatives must have been able to catch relatively large fish. This inferred diet is congruent with the presence of large glandular depressions

The long jaws and strong, pointy teeth of *Hesperornis* and other hesperornithiforms were specialized for capturing fish. Movable connections between certain bones of the skull, particularly in front of the orbit (red circles), facilitated the independent movement of the upper jaw with respect to the rest of the skull (arrow), thus allowing these birds to ingest large fish. (Image after Bühler et al., 1988.)

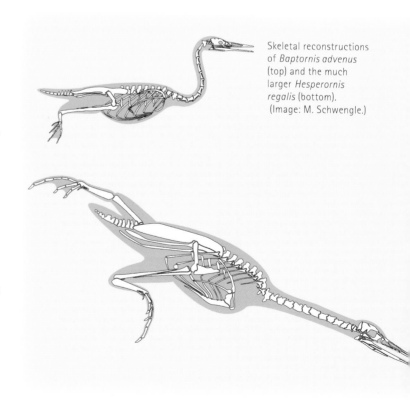

Skeletal reconstructions of *Baptornis advenus* (top) and the much larger *Hesperornis regalis* (bottom). (Image: M. Schwengle.)

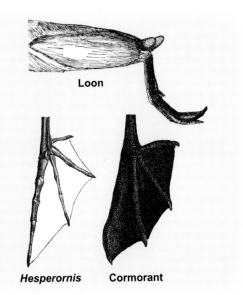

Loon

Hesperornis   **Cormorant**

The anatomy of the pelvis and hindlimbs of hesperornithiforms indicates that the upper legs of these birds were contained within their torpedo-like bodies, very much like in loons (top). Although the foot of hesperornithiforms is often reconstructed as webbed and similar to the feet of cormorants (bottom), some believe that the ancient seabirds had lobed feet similar to those of grebes. (Image after Heilmann, 1926.)

on the roof of the orbit, above the eyes. Similar depressions in extant oceanic birds such as gannets, pelicans, albatrosses, and penguins, house "salt glands" to help these birds process and secrete the large amount of salt they ingest along with fish.

The skull of *Hesperornis* was attached to a long, slender, S-curved neck, whose number of vertebrae falls within the range typical of modern birds. *Hesperornis* and its relatives were the first birds (and the most primitive) to show the fully formed saddle-shaped vertebral joints that characterize much of the spine of extant birds. The presence of this condition suggests that these birds had highly movable necks, capable of performing the precise movements required to capture fast-swimming fish. The number of trunk vertebrae of *Hesperornis* also approaches that typical of living birds, leading to an elongation of the neck and a strengthening of the trunk, which likely evolved in response to aerodynamic pressures in the flighted ancestors common to hesperornithiforms and their living counterparts. These ancient birds, however, lacked the fusion of the trunk vertebrae into a single element—the notarium of most extant birds—which adds strength to the trunk and, as such, to the whole ribcage.

The tiny forelimbs of the hesperornithiforms are the best confirmation of their incapacity to fly. Only a rod-like, rudimentary humerus is known for *Hesperornis*, and it is likely that this was the only bone of its highly atrophiated forelimb. The arms of *Baptornis* were somewhat less reduced, still bearing remnants of the ulna and radius, the two bones of the forearm. Undoubtedly, the aerodynamic incapacity of hesperornithiforms evolved secondarily, from ancestors that were fliers, as hesperornithiforms have a long history of flying predecessors. Furthermore, those hesperornithiforms believed to be the most primitive members of the group, such as *Pasquiaornis*, had much larger forelimbs than the highly specialized *Hesperornis*. *Baptornis* was somewhat in between these two. Although useless for flying, the abbreviated forelimbs of hesperornithiforms may have helped them steer while diving and swimming under the water. Not surprisingly, the entire flight apparatus of these birds was atrophiated. The shoulder girdle was small, displaying the wide angle between the scapula and coracoid typical of other flightless birds, and the sternum lacked any evidence of a keel for anchoring the large breast muscles of flying birds.

As with most of the remaining skeleton, the pelvis and hindlimbs of these birds display a host of unique features, most of which can be easily correlated to specializations for their aquatic existence and their foot-propelled locomotion. Some of these features are already developed in the earliest known members of the group, such as *Enaliornis* and *Pasquiaornis*, which exhibit a shortened femur, a crest-like projection on the upper end of the tibia, highly compressed metatarsals, and thick, solid bones. The general morphology and their resemblance to extant foot-propelled divers suggest that hesperornithiforms held their hindlimbs sideways, with the femur projecting

at an angle of 90 degrees from the pelvis, the tibia parallel to the latter, and the foot pointing backwards. Other noticeable diving specializations include a long pelvis, increasing the surface for the attachment of large swimming muscles; a partial closure of the hip socket that would have strengthened this area against the potent movements of the femur; and a large kneecap or patella that, together with the projected upper end of the tibia, provided a lever aiding the powerful movements of the hindlimbs. Perhaps the most interesting of these diving specializations, however, is the peculiar rotation of the toes evident in advanced hesperornithiforms such as *Hesperornis* and *Parahesperornis*. The highly compressed feet of these birds bore toes with special joints that allowed them to turn sideways during the recovery phase of the propulsive stroke of the hindlimb. This movement, similar to that of an oar feathered on its return, must have substantially minimized the friction of their large feet against the water. Although not yet fully developed, an incipient degree of this sophisticated energy-saving mechanism can be seen in the feet of *Baptornis* and other more primitive hesperornithiforms.

In spite of the absence of direct evidence, some researchers believe that the toes of hesperornithiforms were lobed, much like those of today's grebes, although others have argued that they were webbed. It is also possible that both the femur and tibia were incorporated into the muscular body, with the feet being directed to either side. This limb arrangement would have given their bodies a torpedo-like appearance, with the legs placed as side propellers—a design typical of modern divers. While of great advantage in the water, the backward position of the legs and their possible incorporation into the body suggests that these birds could not have placed their hindlimbs under their bodies. With their legs located behind their center of gravity, it is unlikely that they were

able to adopt an erect walking posture. Most likely, *Hesperornis* and its relatives pushed themselves seal-like on their stomachs when they had to venture onto the shore.

Further swimming specializations are apparent in the unique design of the tail of these birds. In *Hesperornis*, the number of tail vertebrae retaining mobile joints is greater than those of other short-tailed birds—the pygostyle or rump bone of the tailend of modern birds is restricted to the last two vertebrae. The depressed outline and paddle shape of all these vertebrae suggests restrictions for lateral motion but greater up and down mobility. It is likely that the unique design of the tail of *Hesperornis* and its kin facilitated underwater steering.

The discovery of some exquisitely preserved specimens has shed light on the external appearance of the feet and plumage of hesperornithiforms. A specimen collected in western Kansas as early as 1894 preserved evidence of the smooth (rather than imbricated) scutes that lined the feet of these ancient birds. This same specimen, later used as the holotype of *Parahesperornis alexi*, has also provided evidence of the long semiplumes, feathers of downy appearance but with a more distinct shaft, that presumably covered much of its body.

Aquatic Ostriches to Loons' Cousins: Genealogical Relationships of *Hesperornis*

Perhaps because of the specialized skeletal anatomy of *Hesperornis* and its relatives, hypotheses of the genealogical relationships of these birds have differed greatly. In the late 19th century, superficial similarities in the palate and the weak union between some skull bones led O. C. Marsh to believe that *Hesperornis* and ratites—ostriches and their allies—had evolved from a remote common ancestor. According to this view, both lineages had reduced

forelimbs due to disuse, either from the aquatic specialization of *Hesperornis* and its relatives or the terrestrial specialization of ratites. Marsh believed that *Hesperornis'* ancestors had to pass through flighted phases comparable to modern loons. However, he appears to prefer the view that *Hesperornis* and ratites shared a much closer relationship, one in which *Hesperornis* was essentially regarded as a weird, aquatic ostrich. Following on with this interpretation, Marsh envisioned that *Hesperornis'* wings were first reduced during its adaptation for a ground-dwelling existence, and that it was only after this terrestrial phase that the bird conquered the aquatic environment.

Several subsequent researchers seriously criticized Marsh's interpretation of *Hesperornis* as an aquatic ostrich. Renowned 20th century paleornithologists Kalman Lambrecht from Germany and Pierce Brodkorb from the University of Florida at Gainsville argued for a close relationship between hesperornithiforms and modern foot-propelled divers such as loons and grebes. Some authors classified fossils previously thought to be members of the same hesperornithiform lineage

within different groups of foot-propelled divers. Brodkorb, for example, regarded *Enaliornis* as an Early Cretaceous loon and *Baptornis* as a grebe of the terminal Cretaceous. Features from the hindlimbs, such as a large triangular patella and compressed foot bones, however, unite the many species of hesperornithiforms into an evolutionary lineage, one that must have evolved from a single ancestor. Indeed, during the second part of the 20th century, more and more workers began interpreting these birds as an extinct primitive lineage with no particularly close relationships to any extant group of birds. A notable exception, however, was Polish paleontologist Andrzej Elzanowski from the University of Wroclaw, who, based on his reinterpretation of several cranial features of *Hesperornis*, regarded this bird as the earliest branch of neognaths—the group that includes all living birds other than ratites and their tinamou relatives. This view, however, is not endorsed by most students of avian evolution, who interpret the hesperornithiform radiation as an early divergence of ornithurines, the ornithuromorph group that also includes modern birds and their closest kin.

GROWTH SPEEDS UP

Structural details of bone tissue are often preserved in fossil skeletons. Studies of the bone microstructure in cross-sections of long bones and other skeletal elements have revealed the presence of growth rings in nearly all primitive lineages of birds. These concentric growth rings, usually interpreted as the result of yearly cycles, provide evidence of a significant slowdown in the rate of growth of an individual—or even a complete cessation of its growth. The most direct conclusion in light of this evidence is that most

primitive birds grew at rates that were much slower than their living relatives, which typically reach their maximum size within a year. Aside from the specializations related to low buoyancy, the dense bone of *Hesperornis* and its relatives is essentially indistinguishable from that of modern birds. The tissue of the bones of these birds, often named *fibro-lamellar* in reference to its woven-looking, fibrous appearance, is a fast-growing tissue that lacks growth rings or any sign of interruption that may suggest a pause in an otherwise

continuous growth. This is also the case with a variety of other Mesozoic fossils that are habitually interpreted as more closely related to extant birds than *Hesperornis*. The presence of uninterrupted growth in hesperornithiforms and in birds genealogically closer to those of today suggests that a fundamental change in the pattern of growth must have occurred during the evolutionary origin of the ornithurines, the large group of ornithuromorphs that encompasses hesperornithiforms, modern birds, and their closest relatives.

A Life at Sea: Daily Existence for the Hesperornithiforms

Without a doubt, *Hesperornis* was incapable of flying—along perhaps with all of its known close relatives—a functional inference already made explicit in the original work by O. C. Marsh. For a long time, Marsh and other researchers perceived these birds as *primarily* flightless, having their origin in an ancestor devoid of aerodynamic capabilities—but the discovery of a large diversity of more primitive and yet unquestionably flighted birds in the last two decades has permanently knocked down this early view. Hesperornithiforms are nowadays universally accepted as *secondarily* flightless—birds that, like all other known flightless avian lineages, evolved from ancestral forms that *were* able to take to the air. Confirming this current view is the fact that a trend toward simplification of the forelimb and increment in size is seen from the most primitive hesperornithiforms through to the highly derived forms such as *Hesperornis regalis.*

As in the case of their aerodynamic capabilities, the skeletons of *Hesperornis* and its relatives provide undisputable evidence for their foot-propelled diving specializations. The very nature of their bone tissue attests to this functional inference. Just as with their more noticeable specializations (flippers, torpedo-like bodies, and others), the bone tissue of aquatic tetrapods is characterized by a remarkable degree of evolutionary convergence. From manatees to penguins to a variety of extinct aquatic reptiles, all exhibit a highly compact bone tissue that is much denser than that of their terrestrial relatives. The bone tissue of *Hesperornis* is also much denser than that of non-aquatic birds. Such a specialization endows these animals with a compact skeletal framework of low buoyancy, which is highly beneficial for their aquatic lifestyle. In addition, the abundance of hesperornithiform fossils in rocks formed far from the shore of the Pierre Seaway is another clear indication of their lifestyle.

There is little doubt that *Hesperornis* and its relatives ate fish. Their long necks were specialized for rapid flexure and their long, slender jaws and sharp teeth appear to have been well-designed for catching fish. A *Baptornis* skeleton associated with fossilized dung containing bones of the small fish *Encodus*, ubiquitous in the marine deposits of the Pierre Seaway, provides direct evidence of this, and the presence of several points of intramandibular mobility suggest that this and other hesperornithiforms were able to ingest very large prey. Some of the large forms,

Cladogram showing the evolutionary relationships of hesperornithiforms. These birds are largely regarded as the earliest divergence of ornithurines. (Image: E. Heck/S. Orell.)

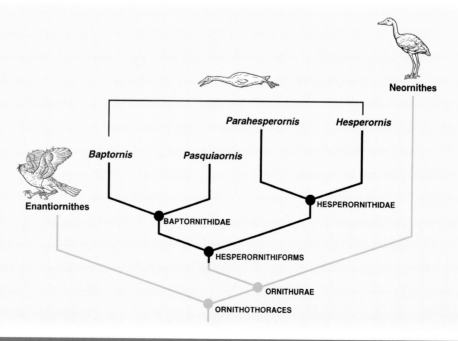

Fleshed out reconstruction of *Hesperornis regalis*. (Image: S. Abramowicz.)

such as *Hesperornis* and *Asiahesperornis*, probably spent much of their time at sea; circumstantial evidence in support of this comes from the fact that the smaller and presumably less pelagic species are more abundant in nearshore environments.

Several paleontologists have argued that all these birds nested in rookeries on shallow coasts, or perhaps on offshore islands. The discovery of bones of young individuals in Arctic rocks has led some to believe that these colonies could have existed within the Arctic circle. University of North Carolina paleontologist Alan Feduccia speculated that *Hesperornis* could have given birth to live young, an idea perhaps originated in the fact that most fossils of this large bird have been found in offshore rocks, apparently very far from any suitable coastline where they could have nested. In the end, nothing is really known about the nesting habits of these birds. If they indeed gave birth to live young, *Hesperornis* and its kin would be the only group of birds—and the only dinosaurs—to have evolved such a reproductive strategy.

CHAPTER TEN

The Beginnings of Modern Birds

The long Mesozoic saga of avian evolution ends with the rise of modern birds, Neornithes, the group that includes the 10,000 living species and all Cenozoic-aged fossil birds. Despite centuries of scientific inquiry, the origin and early diversification of neornithines continues to be the focus of much research. Who are the closest relatives of these birds? How are the major neornithine groups related to one another? Which of these lineages, if any, evolved during the Mesozoic Era?

When and where did these modern bird groups diverge? What was their pattern of diversification? These are just some of the major questions that current ornithological research is trying to answer, an endeavor that has brought together a diverse array of scientific disciplines. Before we review these major issues, let us look at some of the fossils that, albeit not included within neornithines, are generally accepted as being part of the immediate evolutionary history of this group.

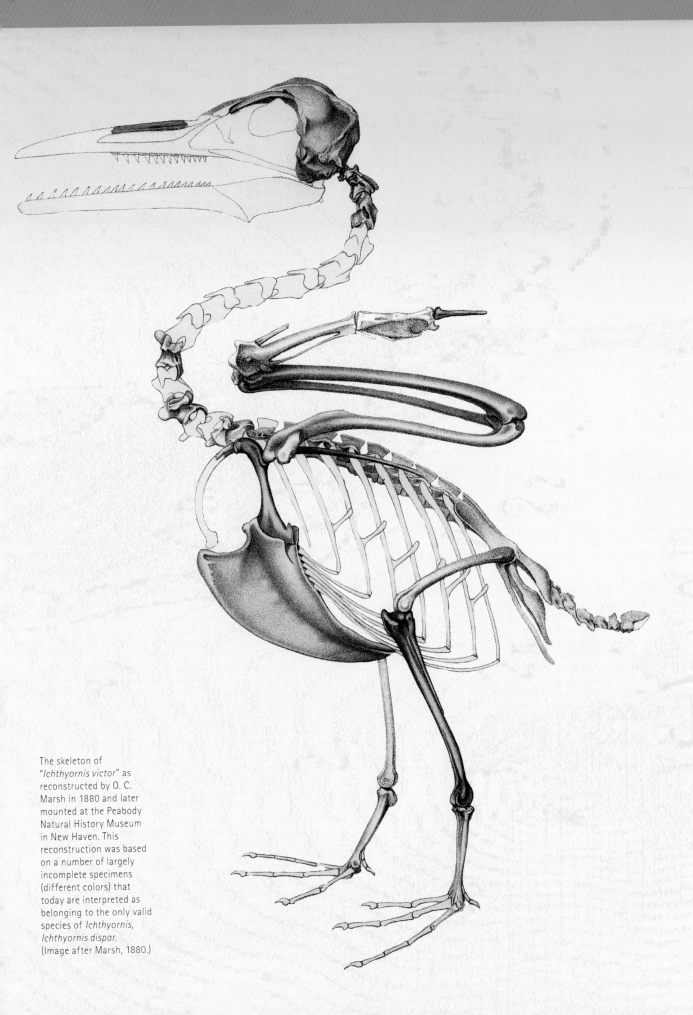

The skeleton of "*Ichthyornis victor*" as reconstructed by O. C. Marsh in 1880 and later mounted at the Peabody Natural History Museum in New Haven. This reconstruction was based on a number of largely incomplete specimens (different colors) that today are interpreted as belonging to the only valid species of *Ichthyornis*, *Ichthyornis dispar*. (Image after Marsh, 1880.)

Ichthyornis: An Ongoing Puzzle

Historically linked to the hesperornithiforms is *Ichthyornis*, a bird that also lived in the marine environments of the Late Cretaceous Pierre Seaway of North America. Indeed, *Ichthyornis* is one of the most familiar fossil birds from the Mesozoic—and like *Hesperornis* it also played a key role in the evolutionary debate of the 19th century. The publication of the seminal work of O. C. Marsh, *Odontornithes*, in which *Ichthyornis* and *Hesperornis* were thoroughly described and illustrated, provided Charles Darwin with a cornerstone to support his ideas about evolution and to keep at bay the vitriolic debates he had become involved in.

In spite of the aura of familiarity that crowns *Ichthyornis*, very little research on this ancient bird followed Marsh's contributions. Until recently, *Ichthyornis* ranked as one of the least understood Mesozoic birds (see *The Mistaken Tooth*), even though its remains are relatively abundant (nearly 80 specimens, albeit largely represented by single bones, are found at the Peabody Museum of Natural History alone). For nearly a century, most of the fossils of this bird remained mounted on plaster bases in displays at the Peabody Museum of Natural History in New Haven, and it was not until the late 1990s that they were removed from their plaster enclosures and used as the focus of a doctoral dissertation, that of Julia Clarke. A great deal of new and sorely needed information on *Ichthyornis* came as a result of her thorough study. By research in the archives of the Peabody Museum and through careful analysis of the newly prepared bones, Clarke was able to demonstrate that much of what was thought to belong to the holotype of *Ichthyornis victor*, one of the species Marsh copiously featured in *Odontornithes*, was in fact a paleontological chimera. Not only was the mount of this species composed of more than a dozen individuals, but none of them included the holotype and some belonged to different species, closer to the ancestry of modern birds. Another important conclusion drawn by Clarke was that the several species of *Ichthyornis* that had been named since Marsh's initial studies could not be justified on the basis of bone anatomy. She concluded that of the eight previously named species of *Ichthyornis*, only *Ichthyornis dispar* was valid.

Ichthyornis dispar was a flying bird with powerful wings and a large keel in its breast bone. In many other respects, the skeleton of this ancient bird is also fully modern. This overall similarity with the basic structure of modern birds highlights the evolutionary proximity of *Ichthyornis* to its living counterparts, even if details of its skull, wings, vertebral column, and pelvis indicate that this ancient bird was not immediately close to the ancestry of modern birds. For example, the head still bears sharp teeth on the lower jaws and the rear half of the upper jaw (the premaxilla was toothless), a snout of ancestral configuration (the maxilla is larger than the premaxilla), and a primitive design of the joints between the two lower jaws and between these and the skull. Furthermore, the skeleton exhibits a lesser degree of pneumatization and a more primitive architecture of the pelvis and synsacrum when compared to those of living birds.

Skeletal reconstruction of the 80–85-million-year-old *Ichthyornis*. (Image after Rowe et al., 1998.)

Although most fossils of *Ichthyornis dispar* at the Peabody Museum of Natural History are from rocks formed between 80 and 85 million years ago at the bottom of the Pierre Seaway in what is now Kansas, a handful of other remains from the same region, together with incomplete and often fragmentary fossils from California, Alabama, New Mexico, Texas, and several Canadian localities indicate that this bird could have evolved as much as ten million years earlier. However, the incomplete nature of these fossils prevents reliable identification. This is also the case for a series of fragmentary bones from several localities in central Asia that have been thought to be from either *Ichthyornis* or a close relative of this bird. Certainly, all these incomplete fossils need to be studied in light of Clarke's recent revision and the many other developments in the field of Mesozoic birds. For example, a cursory examination of these purported "*Ichthyornis*" fossils indicates that some are likely to be the remains of near-shore enantiornithines.

The various fossils of *Ichthyornis dispar* indicate that this ancient bird ranged in size from that of a small gull or tern to that of a prion. While most specimens in the collection from the Peabody Museum belong to birds whose body size differed within a 20 percent range, some bones are more than 30 percent smaller than others. Such diversity in size could perhaps be attributable to sexual differences—males larger than females, or vice versa—or it could reflect an evolutionary trend during the apparently long evolutionary history of this species. Alternatively, the observed size range in fossils that otherwise seem to belong to either adults or to individuals reaching adulthood may indicate differences in the way these animals grew compared to their living relatives. Despite the fact that the bone tissue characteristics of *Ichthyornis* are much closer to those of living birds than those of enantiornithines or their more archaic equivalents, as-yet undetected differences in the growth patterns of *Ichthyornis dispar* may well account for the presence of such a size range. Much more needs to be learned about the growth patterns and physiology of early birds. In the case of *Ichthyornis*, sexual differences, changes through time, and differences in the way these birds grew could have all contributed to the diversity of sizes represented in the available fossil collection. Furthermore, there is the possibility that this size range may be highlighting the existence of more species than those inferred from anatomy alone. However, considerations of whether size differences between anatomically equivalent fossils underlie a still-undetected evolutionary diversification will have to await a greater understanding of the previously mentioned biological factors.

Other Immediate Relatives of Modern Birds

Although *Ichthyornis* is the most familiar close relative of modern birds, several recently described fossils, together with others previously misidentified as either *Ichthyornis* or its kin, could represent evolutionary branches nearer to the origin of neornithines. One of these fossils is *Apatornis celer*, a bird long considered to be closely related to *Ichthyornis* and from the same marine environments. Very little, however, is known about *Apatornis*; fossils of this small bird are extremely fragmentary. Despite *Apatornis* having often been interpreted as evolutionarily connected to *Ichthyornis*, Clarke's recent revision of these fossils has supported a closer link between *Apatornis* and living birds.

Another Late Cretaceous bird with close relationships to its extant counterparts is the Argentine *Limenavis patagonica*, whose 70-million-year-old remains show many detailed similarities to the basic structure of the modern avian wing. Unfortunately, we also know very

little about this ancient Patagonian bird. Although it is one of the most complete Cretaceous birds interpreted as very near to the ancestry of neornithines, the remains of *Limenavis* include only portions of a wing. Such a fossil is too incomplete to paint an accurate picture of the appearance and lifestyle of this bird, even though the shape of the bones and the geologic context in which they were found indicate that *Limenavis* had flight capabilities comparable to those of many living birds and that it inhabited the warm, coastal plains that lined the ancient South Atlantic Ocean. The only known remains of *Limenavis* are also too incomplete to allow us to judge its exact placement within the family tree of birds, even though several features of its wing indicate it was closer to the ancestry of neornithines than *Ichthyornis*.

More complete, although not necessarily more closely related to living birds, is a new discovery from the Late Cretaceous of Belgium. Including portions of skull and mandible, wing, shoulder, vertebral column, and feet, this fossil is likely the youngest known pre-modern bird—its remains were found in rocks that predate the Cretaceous–Tertiary boundary by 800,000 years. With the size of a large seagull, long wings and

toothed jaws, this bird inhabited the warm waters and coastal environments of a shallow sea that at the time flooded much of Europe. Most important, its existence less than a million years before the end of the Cretaceous indicates that archaic, pre-modern birds lived until the very end of the Mesozoic and posits the possibility that some of them survived the mass extinction that marked the end of this Era.

Other fossils thought to be among the closest relatives of modern birds are even more incomplete—most of them are represented by single bones. Such a deficient record of the putative closest relatives of neornithines has historically

Skeletal reconstruction of a Late Cretaceous seabird from Belgium. This gull-sized bird is perhaps the youngest known fossil of a pre-modern lineage. (Image: G. Dyke.)

THE MISTAKEN TOOTH

Although most birds from the Mesozoic have been shown to have teeth, the idea of toothed birds was not always received so naturally. When O. C. Marsh described the first remains of *Ichthyornis*, in 1872, an incomplete toothed jaw found on the same slab was regarded as that of a small nonavian reptile. Further preparation of the slab the following year revealed another portion of a jaw along with an avian skull, thus forcing Marsh to reconsider his initial statement and to regard these toothed jaws as those of the

bird he had reported the previous year. More than seven decades after Marsh's studies of his outstanding collection of fossils from the Pierre Seaway of North America, American paleontologist Joseph T. Gregory proposed that the toothed jaws thought to be of *Ichthyornis* were not those of a bird but of a baby mosasaur, a group of marine reptiles related to monitor lizards that lived at the end of the Cretaceous period. His was, however, not a lone voice— several other paleontologists joined the cause, plucking the teeth from

Ichthyornis' mouth. This idea gained such momentum that in the early 1970s renowned paleornithologist Pierce Brodkorb was able to refer to "the fable of the toothed birds," also including *Hesperornis* among the discredited. Brodkorb and others argued that the "unnatural" teeth of *Ichthyornis* were the result of O. C. Marsh, who was alleged to have tinkered with them. In time, Marsh was vindicated as new fossils proved that not only these birds, but also a plethora of their pre-modern predecessors, did in fact have teeth.

haunted our capacity for resolving the evolutionary origin of this important group of living vertebrates. The risks involved in drawing conclusions from fossils represented by single bones were highlighted by Clarke and me in our study of *Limenavis*. When we used all the characteristics observable in its wing remains, our cladistic analysis placed *Limenavis* as a close relative of neornithines (albeit not nested within them), but when we performed separate cladistic analyses using features from only its humerus, its ulna, or other single bones, the result of these analyses nested each of the bones within a different lineage of living birds. Clearly, the results of cladistic analyses based on such incomplete evidence are unreliable. Our experiment highlighted the dangers of positing the existence of modern lineages of birds in the Cretaceous when these arguments are based on isolated and often fragmentary fossil bones.

Even though most fossils thought to be very closely related to the origin of modern birds are of Late Cretaceous

Although long identified as a bird closely related to its modern counterparts, the Mongolian Early Cretaceous *Ambiortus dementjevi* is possibly a member of a more primitive ornithuromorph lineage. (Image: G. Dyke.)

age, a handful of birds from much older rocks have also been allied to their living counterparts. However, most these fossils are fragmentary and therefore difficult to interpret. One of them is *Gansus yumensis*, a plover-sized, long-toed bird that lived during the Early Cretaceous in what is today the Chinese northwestern province of Gansu. We saw earlier how, based on a study of the foot—which, for many years, was all that was known of this bird—Chinese paleontologist Hou Lianhai has argued that *Gansus* was an ancient near-relative of modern shorebirds. Although its foot superficially resembles that of some living shorebirds and its remains were unearthed from rocks formed on a lakebed, the absence of some notable features suggests that *Gansus* is neither so closely related to modern shorebirds as Hou suggests, nor very close to any other living lineage. Newly discovered fossils support the view that *Gansus* is part of a more primitive ornithuromorph radiation and that perhaps it is closer to the origin of ornithurines.

Another Mesozoic fossil bird thought to be closely allied to extant birds is *Ambiortus dementjevi*, also from the Early Cretaceous but found in central Mongolia. The only known fossil of *Ambiortus* preserves a good portion of the neck and trunk, a shoulder girdle, a segment of the sternum, and much of its clawed wings. Similar in size to *Apsaravis*, *Ambiortus* was a flying bird that lived in a lake region comparable to that hosting the temporally equivalent biota of Liaoning, although much farther away from the seashore and likely more arid. Russian paleontologist Evgeny Kurochkin of the Paleontological Institute in Moscow has long argued that *Ambiortus* is an archaic member of the paleognaths—neornithines that include the flying tinamous and the flightless ratites. More recently, Kurochkin has argued that *Otogornis genghisi* from the Early Cretaceous of Inner Mongolia (northern China) is a close relative of

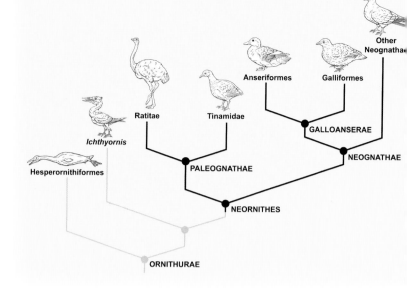

Ambiortus and also another example of an archaic flying paleognath. However, given the incomplete nature of *Otogornis*—the holotype consists of poorly preserved portions of its wings and shoulder girdles—such a claim remains largely unconvincing. In fact, several features of *Otogornis* suggest that it is not only unrelated to paleognaths, but that it is more likely an enantiornithine. Similarly, features of *Ambiortus* suggest that this bird is not an archaic paleognath but most likely a member of the early ornithuromorph radiation.

Despite the tantalizing clues these claims appear to provide to a much earlier origin of neornithines, nothing approaching the evolutionary level of *Ichthyornis*—let alone something that could be considered closer to living birds—exists in the diverse and exquisitely preserved Early Cretaceous avifauna from Liaoning Province (China). *Ambiortus* and *Gansus* may well be primitive ornithuromorphs, fossils of which are well-represented in the Liaoning deposits. These alleged Early Cretaceous near-relatives of modern birds need to be examined in the light of new discoveries such as *Yanornis* and *Yixianornis*, archaic ornithuromorphs whose wings and shoulders show a remarkably similarity to those of living neornithines.

Evolutionary Relationships of Modern Birds

Despite centuries of research, the genealogical interrelationships of many major lineages of living birds are still far from being settled. Nonetheless, the convergence of classic morphological investigations (which study the form and structure of birds) with molecular techniques for devising genetic relationships designed in the last 40 years have provided us with some important points of consensus about how neornithine lineages are related to one another (see *Molecules and Family Trees)*. Most researchers overwhelmingly accept a

primitive neornithine divergence. On one side of this basal divergence are the paleognaths—birds that include all ratites (ostriches, rheas, cassowaries, emus, kiwis, and their extinct flightless relatives) and the flighted tinamous from South and Central America. On the other side are the neognaths, which include all other lineages of neornithine birds. Numerous studies also support the view that among neognaths, galliforms (grouse, pheasants, and other landfowl) and anseriforms (screamers, ducks, geese, and other waterfowl) share a common heritage, and that all remaining neognaths evolved from a different ancestor. The group formed by galliforms and anseriforms is called Galloanserae; the other, larger group of neognaths is called Neoaves. Pelecaniforms (pelicans, boobies, cormorants, darters, and frigate birds) and gruiforms (cranes, rails, bustards, and their relatives) are considered among the most primitive neoavians, but the common ancestry of each of these large groups of birds is often disputed—seriamas, for example, are one of the groups traditionally included within gruiforms but that most likely represent an independent lineage of their own. All major lineages of songbirds or passeriforms, of which there are more than 5,000 living species altogether, are thought to have evolved from a common ancestor, and these diverse birds are

Although the evolutionary relationships of modern birds are far from settled, there is a great consensus about the sequence of early divergences of the group: ratites (ostriches, emus, and their kin) and tinamous constitute the first split, followed by an enormous group including waterfowl and landfowl, on the one hand, and all other modern lineages on the other hand. (Image: E. Heck/S. Orell.)

commonly regarded as the most advanced neoavians.

Beyond these points of consensus, there is far less agreement about which major neornithine lineage is more closely related to which. Some studies support the common ancestry of penguins, loons, and petrels, and a close relationship between swifts and nightjars. Other investigations suggest kinship links between New World vultures, storks, falcons and hawks, and shared evolutionary relationships between woodpeckers and songbirds. Yet evidence contradicting all these groupings has also been proposed. Our current understanding of the genealogical relationships of neoavians leaves many important branches of the evolutionary tree of modern birds unresolved—the

relationships of owls, pigeons, parrots, and cuckoos are among the uncertainties. Such a lack of resolution could well be the result of a bush-like pattern of divergence—a large number of lineage splits in a very short period of time—but research on the evolutionary relationships of birds is far from exhausted. We can only hope that as a growing number of genealogical studies based on the anatomy of living and fossil birds becomes integrated with investigations of their genetic make-up, better consensus on how to categorize the present diversity of birds will eventually emerge; but ornithologists may be facing a dimmer future, for some evidence suggests that the interrelationships among several large groups of neoavians may well

Fossils alleged to represent many lineages of modern birds have been found in Late Cretaceous rocks worldwide. These fossils, however, are largely represented by individual bones that are very difficult to identify with confidence. (Image: R. Urabe.)

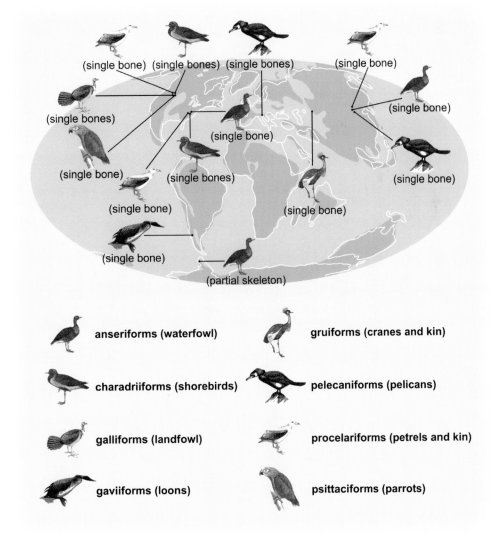

anseriforms (waterfowl)

charadriiforms (shorebirds)

galliforms (landfowl)

gaviiforms (loons)

gruiforms (cranes and kin)

pelecaniforms (pelicans)

procelariforms (petrels and kin)

psittaciforms (parrots)

remain obscure due to the characteristics of their early divergence—the nearly simultaneous evolution of their genomes.

Did Modern Lineages of Birds have a Mesozoic Origin?

Despite Late Cretaceous fossils such as *Ichthyornis*, *Limenavis*, and *Apatornis* documenting widespread differentiation close to the ancestry of neornithines, these fossils are of little use in deciding whether extant lineages of birds had their origin in the Mesozoic. Fossils seemingly pertaining to galliforms, anseriforms, pelecaniforms, gruiforms, charadriiforms (gulls, terns, sandpipers, and other shorebirds), procellariiforms (petrels, albatrosses, prions, and their kin), gaviiforms (loons), and psittaciforms (parrots) have been reported from the Late Cretaceous of nearly every continent—but the vast majority of these occurrences are represented by single bones. We have seen how attempts at placing isolated fossil bones within a family tree are often unwarranted, a caveat exacerbated by the fact that the genealogical relationships of neornithines are poorly understood on the basis of characteristics of their bones.

The earliest fossils represented by complete skeletons that can unambiguously be interpreted as members of modern avian lineages are approximately 55 million years old—these Early Tertiary birds manifest the divergence of a large number of neornithine groups, including paleognaths, galliforms, anseriforms, pelecaniforms, psittaciforms, and many others. For most researchers, the occurrence of these fossils only ten million years after the end of the Cretaceous hints at the Mesozoic differentiation of at least some neornithine lineages. This idea

MOLECULES AND FAMILY TREES

Over the last few decades, several different molecular techniques have taken a dominant role in research on the genealogical relationships among organisms. Today, the most important molecular approach to the study of evolutionary relatedness is DNA sequencing. This approach entails isolating and amplifying homologous genes of the species under consideration through using techniques such as degenerate polymerase chain reaction (PCR) to decipher the sequence of nucleotide acids that make up the selected genes. Modern technology has made this process a fast and inexpensive one, readily available at universities and museums.

Genes are formed by four different nucleotides—adenine, thymine, cytosine, and guanine—which over time and as a result of mutation have replaced each other or been deleted from their initial position. This process of nucleotide

swapping and deletion has created the wide diversity of genes that form the genome of organisms. DNA sequencing has endowed researchers with important means for understanding genealogical patterns. Because the genetic code of every organism includes millions of nucleotide molecules, DNA sequencing allows an immense number of characteristics to be compared, even if these are restricted to the presence or absence of the four nucleotides. Furthermore, because PCR techniques can greatly amplify even the smallest portion of DNA, molecular researchers can obtain their samples from a wide variety of sources, from blood to feathers to eggshell membranes, and even from some fossils. By comparing the homologous genes of the species under study and detecting the changes or deletions of their nucleotide acids, molecular researchers produce genealogical trees very much in the same

way cladograms are produced using anatomical information.

Today, the complete sequences of many genes are freely available on large Internet databases—and the entire genome of several organisms (humans, yeast, chicken, fruit flies, bacteria, and others) is on hand for sequence comparison. Although many molecular genealogies are consistent with traditional classifications derived from anatomy, others have forced established classifications to be reexamined. Yet molecular genealogies are not exempt from constraints, and such studies often contradict each other. Regardless, gene-based approaches have become a powerful tool for inferring evolutionary relatedness, and it is likely that a greater understanding of the biological pattern of inheritance will emerge as more and more DNA studies are combined with traditional morphology.

The 70-million-year-old *Vegavis iaai* is one of the few examples of Late Cretaceous fossil birds that can be confidently identified as belonging to a modern lineage. This Antarctic bird has been recognized as a waterfowl closely related to ducks. (Image: L. Chiappe.)

is also supported by the presence of a few anatomically modern fossils of Late Cretaceous age, whose identification as early representatives of extant lineages has gained acceptance. One of these is *Vegavis iaai*, a fossil known by a partial skeleton (missing its skull) from the Late Cretaceous marine rocks of Vega Island, near the Antarctic Peninsula. Several details of its skeleton not only support the identification of this fossil as a neornithine but also as a close relative of living ducks. A holdover from the slow fragmentation of the supercontinent of Gondwana, the Antarctic Peninsula and its nearby islands joined the White Continent with South America for many millions of years, even after the end of the Mesozoic. Free from ice caps and enjoying the temperate climates that characterized high-latitude regions of the Late Cretaceous, the ancient Antarctic waterfowl *Vegavis* shared its coastal environment with a diverse fauna and flora.

Another likely representative of the Cretaceous divergence of neornithines is *Teviornis gobiensis*, a fossil represented by portions of the wing of a bird that lived in the Mongolian Gobi Desert some 75 million years ago. A few features from the hand support the placement of *Teviornis* within the presbyornithids, birds that like *Vegavis* shared a common ancestry with living ducks. Remains

of presbyornithids have largely been recovered from lacustrine, riverine, and coastal environments (habitats which are typical for extant anseriforms). *Teviornis* probably frequented the ponds and streams that spotted the warm and semi-arid environments of the Cretaceous Gobi Desert, and its occurrence in these environments indicates that the ecological preferences of waterfowl have remained essentially unchanged throughout their long evolutionary history.

The identification of these Late Cretaceous fossils as anseriforms hints at a broader differentiation of Mesozoic neornithines. Although the fossil record can hardly ever be taken at face value, fossils can provide a clearer account of temporal patterns when calibrated on the basis of genealogical interpretations. Given two closely related lineages, the oldest known record of them provides information about the minimum age of their divergence from a common ancestor. If anseriforms had already evolved by the Late Cretaceous, it is logical to infer a Late Cretaceous divergence of galliforms, for these two lineages originated from the common ancestor of all galloanserines. But, following the same reasoning, the presence of anseriforms in the Late Cretaceous also implies the pre-Tertiary differentiation of both neoavians and paleognaths— because the former is accepted as the closest group to galloanserines, and the latter is considered to be the earliest split of the neornithine family tree. Thus, the interpretation of some of the best preserved anatomically modern fossils of Late Cretaceous age as members of the early anseriform radiation has profound implications for understanding the temporal divergences of the main lineages of neornithines. In spite of these developments and the nearly universal acceptance that the origin of neornithines dates back to the Cretaceous, the rate of differentiation and the precise chronology of the ancient radiation of modern avian lineages remains contentious.

Temporal Origin of Extant Lineages

Although the idea that some neornithine lineages originated during the Mesozoic is also supported by genealogical studies based on the genetic make-up of living birds, these molecular investigations have often provided dates of divergence that are significantly earlier than what the fossil record tells us about those evolutionary events. Known as "molecular clocks," these techniques use the age of fossils identified as members of extant lineages and the differences in the sequence of nucleotides—the structural units of nucleic acids such as DNA—of homologous genes to estimate the temporal split among neornithines. Because mutations have altered the sequence of nucleotides throughout the history of any two lineages, the differences between two genes are a function of the time these lineages have been separated since they evolved from their common ancestor. The oldest fossil occurrence of any two lineages provides a minimal age for their split and a basis for inferring the rate at which their genes have changed since they diverged from the gene carried by their common ancestor. Assuming that homologous genes in other lineages mutated at the same rate as those of these two lineages, the temporal divergence of more distantly related lineages can also be calculated. Some molecular estimates have placed the divergence of paleognaths and neognaths as long ago as 120 million years, roughly the time of the Jehol Biota of northeastern China (Liaoning), and indicated that most lineages of neornithines split from one another prior to 100 million years ago. As in this case, molecular clocks based on other groups of organisms have also resulted in temporal estimates that are usually older than those obtained from the direct reading of the fossil record. For example, molecular estimates have dated the origin of multicellular organisms to between 700 and 1,000 million years ago,

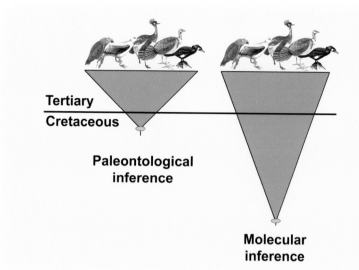

Tertiary
Cretaceous

Paleontological inference

Molecular inference

Inferences about the temporal divergence of modern birds based on molecular genealogies have consistently resulted in substantially older dates. This discrepancy, however, is beginning to wane as more complete and better-dated fossils become available. (Image after Dyke and van Tuinen, 2004.)

but the earliest representatives of these organisms are about 600 million years old; and the divergence of flowering plants has been estimated to be 20 to 70 million years older than the oldest known fossils of these plants, which are approximately 120–130 million years old.

There are several explanations for the fact that molecular estimates are typically older than the origination dates provided by the earliest known fossils of the groups being considered. A commonly cited concern is that molecular approaches require genes to evolve at a constant rate, both across time and between species, something known to be generally false. If the rates at which genes evolve vary among organisms, molecular clocks based on the genetic change measured for two closely related lineages may either underestimate or overestimate the time of divergence of more remotely related lineages. Likewise, if the rates of genetic evolution vary across time, calibrations based on a handful of fossils are likely to be at best gross estimations. These important caveats have been mitigated by developing tests that detect variation in the rate of molecular evolution, but these tests are not very strong and they still allow a significant amount of variation to go undetected, especially when only small portions of genes are examined.

In addition, molecular biologists have not always relied on fossils that are both confidently classified and well-dated for the temporal calibrations they have proposed. In some instances, these calibrations have used dates taken from previous molecular clocks, a procedure that copies (and even multiplies) whatever error may have been present in the original estimate. However, molecular estimates are clearly useful, and the discrepancy they often exhibit when compared to the ages provided by fossils is not only a result of the problems inherent to these clock-like approaches. If molecular clocks render origination dates that are typically considered to be too old, the dates provided by the fossil record are often deemed as too young. In the end, paleontological estimates are likely detecting a different type of event than the one measured by molecular clocks. While fossil occurrences may be detecting the origin of features that identify the fossils as members of one or another lineage, molecular estimates may be measuring the genetic divergence that

predated the anatomical differentiation of these lineages. Furthermore, the earliest known fossil of a given lineage is very unlikely the first member of the lineage. The fossil record thus underestimates the time at which two lineages diverged anatomically, let alone genetically.

We have discussed how the Cretaceous fossil record of birds includes a few occurrences that could well represent the earliest known neornithines—but these Late Cretaceous fossils postdate molecular estimates for the origin of neornithines by more than 30 million years. Could the Cretaceous record of birds be underestimating the origin of modern lineages so deeply? The absence of any Early Cretaceous fossil that could be entertained as a member of a modern group makes it difficult to embrace the explosive Early Cretaceous radiation of neornithines proposed by some molecular advocates. When several new species of Early Cretaceous birds are recognized every year, such an argument cannot even find refuge in the incompleteness of the fossil record.

Genealogical relationships of pre-modern lineages of birds, with their known fossil record highlighted. All these primitive birds vanished before the Cenozoic Era, more than 65 million years ago, but how this disappearance relates to the mass extinction of the end of the Mesozoic remains unclear. (Image after Chiappe and Dyke, 2002.)

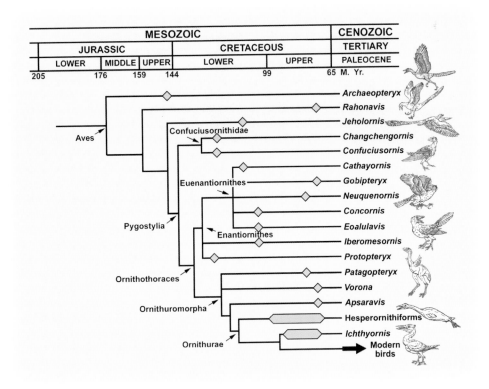

If neornithines did originate so early in time, their initial divergence must have been minimal and likely restricted ecologically or geographically. As some researchers have argued, evidence of this ancient radiation could still be hidden in rocks of largely unexplored regions, such as those that once formed the ancient supercontinent of Gondwana. This view has been recently spearheaded by Joel Cracraft, an ornithologist at the American Museum of Natural History in New York. Combining the research of other scientists with his own investigations, Cracraft has pointed out that many primitive lineages of neornithines—paleognaths, galloanserines, and gruiforms—as well as the most archaic songbirds, are predominantly distributed across the former Gondwanan landmasses. This view has generally been embraced by molecular biologists, who in some instances also see a causal correlation between the fragmentation of Gondwana and the divergence of groups distributed in the southern hemisphere. In contrast to views favoring the interpretation of such a geographic pattern as the result of independent colonization, Cracraft argues that the ancientmost representatives of these lineages dispersed across Antarctica and essentially acquired their modern geographic distribution prior to the break-up of Gondwana in the Early Cretaceous. Nonetheless, such a view receives no support from the paleontological record of today. Although only a handful of avian fossils have been recovered from the Early Cretaceous of Gondwana, those so far reported belong to enantiornithines or perhaps more primitive birds. Yet Cracraft remains hopeful that evidence in support of his views will be discovered with further paleontological exploration of the Cretaceous rocks of Gondwana. Indeed, minimal exploration of the Early Cretaceous deposits of Africa has been carried out, and many of the extensive Mesozoic outcrops of South America still remain to be surveyed paleontologically. With such a minute sample of Early Cretaceous fossils available from the Gondwanan landmasses, the door to future discoveries of this alleged ancient neornithine radiation remains open.

A "Big-bang" Evolutionary Radiation of Modern Birds?

Paleontologists and some molecular biologists generally agree with the view that neornithines originated in the Late Cretaceous and remained with relatively low diversity, at least in terms of ecological and anatomical divergence; but the perceived rate at which they evolved—the magnitude of the ancientmost radiation of modern birds—is still an area of contention.

Inspired by the apparent existence of Late Mesozoic anseriforms, one side of the debate argues in favor of a rather narrow divergence of Late Cretaceous neornithines—one mostly limited to the early members of paleognaths, galloanserines, and neoavians—followed by a larger diversification of modern lineages in the Early Tertiary. In contrast, and guided more by the sudden appearance of numerous fossils of modern birds some 55 million years ago, the other side of the debate supports a much more explosive rate of evolution—a "big bang" of neornithine divergence in the first million years of Tertiary history, preceded by minimal Cretaceous differentiation. The latter view is eminently defended by Alan Feduccia, who also envisions that the extinction of pre-modern birds during the Cretaceous allowed the newcomers to occupy ecological niches vacated by their avian predecessors.

The limitations of today's fossil record prevent us from entirely dismissing any of the main tenets of these views of modern avian diversification. Comparisons between Late Cretaceous fossil birds and their Early Tertiary equivalents highlight the disappearance of all pre-modern lineages—but whether the extinction of these archaic birds

Most lineages of modern birds, like this 50-million-year-old swift *Scaniacypselus szarskii* from Germany's Messel oil shales, are recorded 10–15 million years after the end of the Mesozoic. (Image: G. Mayr.)

happened abruptly at the Cretaceous–Tertiary boundary or gradually during millions of years of Mesozoic history remains unclear. Even if the occurrence of enantiornithines and other non-neornithine birds in strata very close to the Cretaceous–Tertiary boundary suggests that some primitive lineages could have shared the doom of many other organisms at the mass extinction that marks the end of the Mesozoic Era, the record of birds contiguous to this event remains largely incomplete, and the precise stratigraphic provenance of most fossils is poorly documented. Furthermore, most lineages of Late Cretaceous birds—*Patagopteryx*, *Limenavis*, and *Apsaravis* to mention a few—are recorded at a single locality

(and known only from one moment in time); and of those known by ranges, some have their last occurrence more than ten million years before the end of the Mesozoic. If, in the face of a far richer fossil record, the extinction rate of the last nonavian dinosaurs still lingers unclear, the question of how abruptly the archaic Mesozoic lineages of birds disappeared will likely remain unanswered for years to come.

Regardless of whether one envisages the extinction of the magnificent diversity of archaic birds that thrived during the Mesozoic as sporadic or *en masse*, it is reasonable to assume that the diversification of neornithines followed these extinctions, with the newly divergent lineages occupying

the available ecological spaces both on land and in the ocean. Ecological niches occupied by foot-propelled, fish-eating hesperornithiforms were subsequently invaded by the evolution of grebes and loons; the roles that flightless birds such as *Patagopteryx* played in the Late Cretaceous ecosystems were taken over by ratites and other ground-dwelling neornithines; and the ecological functions performed by arboreal and wading enantiornithines were seized by parrots, woodpeckers, and other tree-dwelling birds, and by long-legged presbyornithids and other early relatives of modern waterfowl and shorebirds. Undoubtedly, the Cretaceous–Tertiary transition was a critical moment in the evolutionary history of birds. Important levels of diversification become evident when fossil birds are compared on both sides of this boundary—but the extent of the Late Cretaceous radiation of neornithines and the pace it gathered during the first million years of the Tertiary remains unclear.

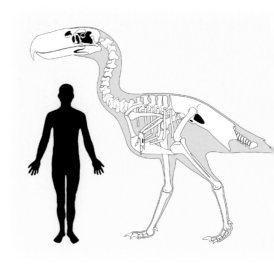

The phorusrhacids are one of the spectacular groups that radiated early in the Cenozoic after the demise of all the pre-modern lineages of birds and their large dinosaurian relatives. For tens of millions of years, these enormous "terror birds" occupied the top predatory niches of South America. (Image: S. Bertelli/ S. Abramowicz.)

The Early Cenozoic Record: The Modern Radiation of Birds

Even though the fossil record of Late Cretaceous birds does not support the sudden eradication of all pre-modern lineages at the Cretaceous–Tertiary boundary, it does show a distinct turnover in the composition of avian lineages when the Cretaceous is compared to the Early Tertiary. During the Late Paleocene and Eocene epochs, the fossil record of modern birds improves greatly, both in number of specimens and their completeness. Evidence of this significant turnover is beautifully preserved in a great number of fossils of birds that lived between 10 and 15 million years after the end of the Mesozoic. Rocks of the Green River Formation of North America and the Fur Formation, London Clay, and Messel oil shales of Europe contain remains of virtually all major lineages of modern birds except songbirds

(passerines) and perhaps hummingbirds (trochiliforms). Although many of these fossils belong to primitive representatives of living groups, as opposed to being nested within extant lineages, they remain critical for better understanding the genealogical relationships of neornithines, at least when these are analyzed from the osteological perspective. Furthermore, because these well-preserved fossils are complete enough to be used for reliable cladistic analysis, they provide critical evidence for estimating the time of divergence of extant lineages.

These Early Tertiary fossils show a great deal of diversity as well as evolutionary convergence, and when the history of neornithines is overviewed, one discovers that the spectacular variety of birds that surrounds us is overshadowed by those that evolved and became extinct during their long evolutionary past. Although, unlike mammals, most lineages of birds did not increase in size after the disappearance of the large nonavian dinosaurs—perhaps due to the constraints imposed by the physics of flight—several lineages evolved flightlessness and gigantism, both within the herbivore and carnivore guilds. The gigantic mihirungs evolved among the anseriforms. These enormous herbivorous ground birds lived in Australia and

reached sizes between 1.5 and 3 meters tall. A whole radiation of large predatory ground birds with massive hooked beaks evolved within the gruiforms. A well-known example of this radiation is the North American *Diatryma*, a colossal bird with a half-meter-long skull and diminutive forelimbs that was likely both a predator and a scavenger, although a recent claim has suggested that it could have been a herbivore. The most diverse group of this gruiform radiation of enormous ground birds includes the phorusrhacids—sometimes referred to as "terror birds"—which in the face of minimal competition with large carnivorous mammals ruled the ecosystems of South America during much of the Tertiary.

Other lineages of Tertiary birds also evolved into entire groups that are not represented today. Two peculiar lineages diverged from a pelecaniform stock: the penguin-like plotopterids and the awesome pseudodontornithids. The former were huge, wing-propelled divers that lived within the confines of the Pacific Rim. Some plotopterids stood nearly 2 meters tall, and their forelimbs developed into flippers similar to those of modern penguins. Also from the marine realm, the pseudodontornithids were superb predators with long jaws edged by prominent, tooth-like, bony protuberances. With a lightweight skeleton and wingspans reaching 6 meters, these long-range, pelagic birds must have flown over large oceanic distances in search of food.

Many other fascinating chapters of bird evolution are recounted by fossils of the last 65 million years. Just like its Mesozoic counterpart, Cenozoic avian paleontology is a dynamic field of discovery and scientific inquiry. The contributions of dozens of paleontologists working in this field have illuminated many aspects of the evolution of modern birds. Most unexpectedly, these investigations have shown that the conspicuous features of some extant birds appeared long after their divergence as discrete lineages—Eocene parrots lacked the short and curved beak of their living heirs, primitive flamingoes had straighter bills, and early hummingbirds had not yet evolved the specializations that would allow their descendants to hover. Even the last major radiation of birds, which has left us a legacy of more than 5,000 species of living songbirds and other passeriforms, did not occur until 20–25 million years ago.

With this last evolutionary radiation, the many varieties, ecological roles, and geographic distributions of the world's avifaunas became, at last, similar to those we see today. When surrounded by this splendid diversity, one cannot help but marvel that these feathered animals are the remnants of a long evolutionary saga, one that can be traced right back to the magnificent dinosaurs that ruled the world of the Mesozoic.

Epilogue

The last chapter brought us up to the rise of modern birds, and with this, to the final pages of the Mesozoic saga of avian evolution. But the end of this impressive story of biological diversification is in itself the beginning of another evolutionary epic, for the divergence of modern birds at the closure of the Cretaceous was the start of a radiation that has extended the dinosaurian legacy to the present day.

Throughout these pages, we have seen how in the last few years an unprecedented number of fossil discoveries concerning the origin and early evolution of birds have been made. In addition to impressive evidence documenting the great similarity between the skeletal design of birds and that of several lineages of theropod dinosaurs, key discoveries have revealed that many other features previously thought to be unique to birds also evolved among nonavian maniraptoran theropods. Indeed, a trove of new fossils has shown that a great deal of evolutionary experimentation preceded the origin of birds, and that the degree of birdness achieved by some nonavian maniraptorans has blurred the line distinguishing birds from their closest relatives. This large body of evidence has cemented the notion that birds are the living descendants of the nonavian maniraptoran theropod dinosaurs, thus settling a century-old controversy about one of the most significant events in the history of life.

Paramount among these new developments has been the discovery of an array of feathered nonavian maniraptorans, indicating that not just feathers but also flight evolved prior to the rise of birds. Regardless of the ecological milieu of the origin of avian flight, it is now apparent that such a feat originated as nonavian theropods with feathered forelimbs, and the capacity to flap them, evolved smaller sizes and larger wings that in time allowed them to become airborne. The wingstroke—the key to powered flight—thus preceded aerial locomotion, and flight itself most likely originated as the by-product of a behavior involving wing-assisted running.

Over the subsequent course of evolution, the refinement of flight involved numerous other transformations—including not just modifications in the skeleton and plumage but also changes in the physiology, the musculature, and the neural system. Fortunately, every major lineage of early birds provides evidence that allows us to piece together many of the critical transformations that led to the development of modern avian flight. Indeed, the fossil record tells us of a progressive enhancement of flight abilities throughout the Mesozoic evolution of birds, a step-by-step progression that involved not only the development of many aerodynamic features but also the disassociation of the tail from terrestrial locomotion and an overall escalation of investment in the forelimb at the expense of the hindlimb.

This understanding of the evolutionary fine-tuning of avian flight would not have been reached without the recent burst of discoveries of archaic, Mesozoic birds. Not only has this fossil menagerie documented an unexpected diversity of avian designs, but by filling the enormous anatomical and genealogical gap that crippled our earlier perception of the early history of birds, these new fossils have painted a much more complete picture of how the unique physical systems of modern birds came about. Although *Archaeopteryx* still stands alone in the Late Jurassic, the new discoveries tell us about a range of birds with long bony tails that thrived in the early part of the Cretaceous. This primitive aviary is certain to grow, as fossils from the 25 million years separating the 150-million-year-old *Archaeopteryx* from the oldest known Early Cretaceous birds come to light.

Yet our window on this fascinating chapter of the early history of birds is still disappointingly small—no long-tailed bird other than the island-relic *Rahonavis* is recorded beyond 120 million

years ago—and the evolutionary transition between long-tailed and short-tailed birds is still shrouded in mystery. The earliest short-tailed birds appear suddenly with the first records of Cretaceous birds. The aerodynamic significance attached to the reduction of the tail and the existence of a spectrum of wing designs suggests a substantial improvement in aerial proficiency with respect to their clumsy-flighted, long-tailed forerunners. However, these earliest short-tailed birds seem to have shared the fate of their long-tailed counterparts—none are recorded beyond 120 million years ago. From this time on, the evolutionary history of birds is largely recorded within two lineages: the enantiornithines and the ornithuromorphs. The earliest representatives of these birds already show a significant departure in the basic design of their flight gear when compared with their predecessors. Critical transformations in skeletal and muscular architecture were coupled with innovations in plumage and a substantial reduction in size, perhaps a consequence of important changes in growth rates and development. The superior aerodynamic capabilities of these birds must have unlocked a wide range of ecological and evolutionary opportunities that are mirrored in the geographic distribution, diversity of lifestyles, and trophic strategies inferred from their fossils. A wealth of recent enantiornithine fossils has highlighted their enormous adaptive radiation, and new discoveries of archaic ornithuromorphs have documented how, very early in their history, these birds evolved an essentially modern flight apparatus.

Despite huge gains in our knowledge about the origin and early evolution of birds, many important questions remain unanswered. Which lineage of nonavian maniraptorans includes the closest relatives of birds? How far can the dawn of birds and the divergence of their archaic groups be traced back in time? What was the extent of the Cretaceous radiation of modern birds? Will the discrepancy between the paleontological record and the timing of divergence derived from molecular studies of living birds ever be resolved? How devastating was the mass extinction at the end of the Mesozoic for the avifaunas of the Late Cretaceous? Did some of the archaic lineages of birds solve the physical challenges of flight in ways different from those devised by their living counterparts? When did birds acquire their existing high metabolic levels?

The answers to these and many other questions clearly lie in our ability to discover new fossils and our capacity to refine our analytical techniques. If the past is the key to the future, I believe that many of these issues will be largely answered in the years to come. In my two decades of research on these topics, I have had the privilege to see an enormous body of new evidence come to light. Fossils that I had never expected to see have become abundant, and a magnificent drama of diversification has been unfolded in ways no one expected. As a consequence of this unexpected bonanza, a field of research that was once largely limited to a small number of researchers working on a limited sample of fossils has accommodated an army of scientists working in a much greater number of scientific disciplines, and nothing seems to indicate that we are near the end of the rally.

In this book, I have tried to summarize the evidence locked in thousands of fossils that document the origin of birds from some of the most fearsome creatures ever to live on Earth, and the evolutionary Odyssey that predecessors of our playful companions endured throughout the second half of the Mesozoic Era. Undoubtedly, the most important outcome of recent years is the solution to a century-old problem. Perhaps next time you see crows roosting in a tree's canopy, you will remember that these familiar animals are the bearers of a glorious pedigree.

Acknowledgements

This book would be greatly incomplete if not for a number of people who generously created images and/or gave me permission to use their images. I am especially indebted to Stephanie Abramowicz, Mick Ellison, William Evans, Ed Heck, Utako Kikutani, Richard Meier, Stacie Orell, Timothy Rowe, Michelle Schwengle, Helmut Tischlinger, and Ryan Urabe, who created, designed, or granted most of the photographs and illustrations found in this book, and to Nicholas Frankfurt for creating the cover image. Other images were provided by Anusuya Chinsamy, Sankar Chatterjee, Julia Clarke, Philip Currie, Favio Dalla Vecchia, Kenneth Dial, Gareth Dyke, Alan Feduccia, Kimball Garrett, Steve Gatesy, Ted Goslow, James Hopson, Farish Jenkins, Hou Lianhai, Robert Loveridge, David Martill, Larry Martin, Gerald Mayr, Loraine Meeker, Ricardo Melchor, Mark Norell, Kevin Padian, Richard Prum, Ji Qiang, Silvio Renesto, Christopher Ries, Martin Röper, Scott Sampson, José Luis Sanz, Tamaki Sato, Michael Shapiro, Hans-Dieter Sues, Alexander Vargas, Günther Viohl, Peter Wellnhofer, Lawrence Witmer, Xu Xing, Hailu You Zhou Zhonghe, the American Museum of Natural History (New York), the Bayerische Staatssammlung für Palaeontologie und Geologie (Munich), the Geologisk Museum (Copenhagen), the Natural History Museum of Los Angeles County (Los Angeles), the Museo de Arte Hispanoamericano Isaac Fernández Blanco (Buenos Aires)—I extend my gratitude to all of them. Stacie Orell and William Evans provided instrumental editorial assistance and Stephanie Abramowicz and Ryan Urabe made extraordinary efforts during the final phases of this project. I acknowledge the careful comments of L. Dingus and J. Clark. I am also very grateful to my agent, Edite Kroll, and my editors, Luna Han, Robert Harington, and Thomas Moore, and to Catherine Page for her careful proofreading and final editing of the manuscript. This book also benefited from many years of discussions with a good number of colleagues and funding from the American Museum of Natural History, the Antorchas Foundation, the Discovery Channel, the John Simon Guggenheim Foundation, the National Geographic Society, the National Science Foundation, the Natural History Museum of Los Angeles County, and the Universidad Autónoma de Madrid—to all concerned, I express my sincere appreciation. Finally, I would like to extend my deepest gratitude to my wife, Megan Walsh, for her patience and unconditional support throughout the time it took for this book to be finished.

Glossary

Altricial One end of the range of hatchling strategies, referring to their physiology, locomotion, behavior, and external appearance. Altricial birds hatch naked and with the eyes closed; they are incapable of locomotion and completely dependent on their parents for food; and they do not leave the nest until an advanced stage of development.

Analogy A similar characteristic of two or more species or lineages is called analogous if its presence in these species is interpreted as the product of convergent evolution.

Archosaur A species that evolved from the common ancestor of living crocodiles and birds.

Aspect ratio The ratio between the span (distance between tips) and the average width (distance from leading to trailing edge) of a wing.

Biota The collection of organisms within a geographic area.

Cladistics A biological discipline that studies how species or lineages are related to one another based on their common ancestry.

Cladogram A branching diagram depicting the genealogical relationships among species or lineages, which are connected to one another by their common ancestry.

Collagen A fibrous connective protein found in many tissues, including skin, bone, cartilage, and tendons.

Compound bones Skeletal elements formed by several bones that fuse to one another during development.

Convergent evolution The evolution of similar characteristics by different species or lineages.

Developmental pathway The sequence of transformations during the development of an organ or a part of an organism.

Diapsid skull A skull that is perforated by two large openings behind the orbit: the supratemporal and infratemporal openings.

Gondwana Large Mesozoic continent formed by modern South America, Africa, Australia, Antarctica, Madagascar, and India, as well as portions of the Eastern Mediterranean landmasses.

Holotype A specimen designated as type specimen in the original description of a species.

Homology A characteristic of two or more species or lineages is called homologous if its presence in these species is interpreted as the product common descent.

Keratin A hard, protein substance that is the main component of hairs, feathers, scales, and nails.

Laurasia Mesozoic supercontinent that consisted of most landmasses of today's Northern Hemisphere.

Life history The sequence of changes experienced by an organism, from its earliest embryonic stages to the time of its death.

Lineage A group of organisms or species that descended from a single common ancestor.

Pentadactyl A hand or foot with five digits.

Precocial One end of the range of hatchling strategies, referring to their physiology, locomotion, behavior, and external appearance. Precocial birds hatch feathered and are able to walk and sometimes fly soon after hatching; they search for food on their own and do not remain in the nest.

Regulatory genes Genes that regulate the expression of one or more protein-coding genes.

Sexual dimorphism The existence of qualitative or quantitative differences between the genders of a species.

Sister groups Two groups of species that are interpreted as sharing a most recent common ancestor.

Sister species Two species that have evolved from an ancestor not shared by any other species.

Skull kinesis Relative motion between skull bones or parts within a skull. The skull kinesis of most birds falls within two categories: prokinesis (upper beak moves as a unit) and rhynchokinesis (portions of the upper beak move as units).

Synapomorphy A characteristic shared by two or more species or lineages that originated in their most recent common ancestor. Homologous features are synapomorphic at some level.

Tectonic processes Geologic mechanisms responsible for the large-scale movements of the Earth's crust.

Wing loading The ratio between the weight of a bird and its wing area.

Further Reading

Chapter 1

Bock, WJ (1964) Kinetics of the avian skull, *Journal of Morphology*, 114: 1–42.

Bühler, P (1981) Functional anatomy of the avian jaw apparatus. In AS King & J Mc-Lelland (eds.) *Form and Function in Birds*, vol. 2, Academic Press, New York, 439–468.

Busbey, AB, Coenraads, RR, Willis, P & Roots, D (1996) *Rocks & Fossils, The Nature Company Guides*, Time-Life Books, Sydney.

Chiappe, LM & Dyke, G (2002) The Mesozoic radiation of birds, *Annual Review of Ecology and Systematics*, 33: 91–124.

Chiappe, LM, Ji, S, Ji, Q, & Norell, MA (1999) Anatomy and systematics of the Confuciusornithidae from the late Mesozoic of northeastern China, *Bulletin of the American Museum of Natural History*, 242: 1–89.

Dingus, L & Rowe, T (1997) *The Mistaken Extinction*, WH Freeman & Co., New York.

Gee, H (1999) *In Search of Deep Time*, The Free Press, New York.

King, AS & McLelland, J (1984) *Birds: Their Structure and Function*, Baillière Tindall, London.

Martin, RA (2004) *Missing Links*, Jones & Bartlett Publishers, Massachusetts.

Palmer, DP (1999) *Atlas of the Prehistoric World*, Discovery Books, New York.

Paul, G (2000) *The Scientific American Book of Dinosaurs*, Byron Preiss Visual Publications, New York.

Proctor, NS & Lynch, PJ (1993) *Manual of Ornithology: Avian Structure and Function*, Yale University Press, New Haven.

Rowe, T, Kishi, K, Merck Jr, J, & Colbert, M (1998) *The Age of Dinosaurs*, 3rd edition, Educational Interactive Multimedia on CD-ROM for Macintosh and PC Computers, W. H. Freeman & Co., New York.

Weishampel, D, Dodson, P, & Osmólska, H (eds.) (2004) *The Dinosauria*, 2nd edition, University of California Press, Berkeley.

Chapter 2

Bundle, MW & Dial, KP (2003) Mechanics of wing-assisted incline running (WAIR), *Journal of Experimental Biology*, 206(24): 4,553–4,564.

Burgers, P & Chiappe, LM (1999) The wing of *Archaeopteryx* as a primary thrust generator, *Nature*, 399: 60–62. *Science*, 278: 666–668.

Chatterjee, S (1997) *The Rise of Birds*, Johns Hopkins University Press, Baltimore.

—— (1999) *Protoavis* and the early evolution of birds, *Palaeontographica*, 254: 1–100.

Chiappe LM (1997) Climbing *Archaeopteryx*? A response to Yalden, *Archaeopteryx*, 15: 109–112.

—— (2004) The closest relatives of birds, *Ornitología Neotropical*, 15 (Suppl.): 101–116.

Chiappe, LM & Dyke, G (2002) The Mesozoic radiation of birds, *Annual Review of Ecology and Systematics*, 33: 91–124.

Chiappe, LM & Witmer, L (eds.) (2002) *Mesozoic Birds: Above the Heads of Dinosaurs*, University of California Press, Berkeley.

Chinsamy, A (2005) *The Microstructure of Dinosaur Bone*, Johns Hopkins University Press, Baltimore.

Currie, PJ & Chen, PJ (2001) Anatomy of *Sinosauropteryx prima* from Liaoning, northeastern China, *Canadian Journal of Earth Sciences*, 38: 1,705–1,727.

Dalton, S (1999) *The Miracle of Flight*, Firefly Books, Buffalo.

Dial, KD (2003) Wing-assisted incline running and the evolution of flight, *Science*, 299: 402–404.

Feduccia, A (1999) *The Origin and Evolution of Birds*, 2nd Edition, Yale University Press, New Haven.

Feduccia, A & Nowicki, J (2002) The hand of birds revealed by early ostrich embryos, *Naturwissenschaften*, 89: 391–393.

Gauthier, JA & Gall, LF (eds.) (2001) New perspectives on the origin and early evolution of birds: proceedings of the international symposium in honor of John H. Ostrom, Peabody Museum of Natural History, New Haven, *Nature*, 440: 329–332.

Gauthier, JA & Padian, K (1985) Phylogenetic, Functional, and Aerodynamic Analyses of the Origin of Birds and their Flight. In MK Hecht, JH Ostrom, G Viohl, & P Wellnhofer (eds.), *The Beginnings of Birds*, Freunde des Jura-Museum, Eichstätt, 185–197.

Heilmann, G (1926) *Origin of Birds*, Witherby, London.

Larsson, HCE & Wagner, GP (2002) Pentadactyl Ground State of the Avian Wing, *Journal of Experimental Zoology*, 294: 146–151.

Lucas, AM & Stettenheim, PR (1972) *Avian Anatomy: Integument. I & II*. Agriculture Handbook 362, USDA, Washington, D.C.

Norell, MA, Clark, JM, Chiappe, LM, & Dashzeveg, D

(1995) A nesting dinosaur, *Nature*, 378: 774–776.

Norell, MA, Clark, JM, Dashzeveg, D, Barsbold, R, & Chiappe, LM (1994) A theropod dinosaur embryo, and the affinities of the Flaming Cliffs dinosaur eggs, *Science*, 266: 779–782.

Norell, MA & Xu, X (2005) Feathered dinosaurs. *Annual Review of Earth and Planetary Science*, 33: 277–299.

Osborn, HF (1924) Three new theropods, *Protoceratops* zone, central Mongolia, *American Museum Novitates*, 144: 1–12.

Ostrom, JH (1969) Osteology of *Deinonychus antirrhopus*, an unusual theropod from the lower Cretaceous of Montana, *Bulletin of the Peabody Museum of Natural History*, 30: 1–165.

—— (1976) *Archaeopteryx* and the origin of birds, *Biological Journal of the Linnaean Society*, 8: 91–182.

Padian, K & Chiappe, LM (1998) The origin of birds and their flight, *Scientific American*, 278: 38–47.

Padian, K, de Ricqlès, AJ, & Horner, JR (2001) Dinosaurian growth rates and bird origins, *Nature*, 412: 405–408.

Paul, PS (2002) *Dinosaurs of the Air*, John Hopkins University Press, Baltimore.

Prum, R (1999) Development and evolutionary origin of feathers, *Journal of Experimental Zoology*, 285: 291–306.

Rowe, T, Ketcham, R, Denison, C, Colbert, M, Xu, X, & Currie, PJ (2001) The *Archaeoraptor* Forgery, *Nature*, 410: 539–540.

Rowe, T, Kishi, K, Merck Jr, J, & Colbert, M (1998) *The Age of Dinosaurs*, 3rd edition, Educational Interactive Multimedia on CD-ROM for Macintosh and PC Computers, W. H. Freeman & Co., New York.

Sato, T, Cheng, Y, Wu, X, Zelenitsky, DK, & Hsiao, Y (2005) A pair of shelled eggs inside a female dinosaur, *Science*, 308: 375.

Shapiro, MD (2002) Developmental morphology of limb reduction in *Hemiergis* (Squamata: Scincidae): Chondrogenesis, osteogenesis, and heterochrony, *Journal of Morphology*, 254: 211–231.

Witmer, LM (1991) Perspectives on avian origins, In HP Schultze & L Trueb (eds.), *Origins of the Higher Groups of Tetrapods*, Comstock Publ. Association, Ithaca, 427–466.

Xu, X, Zhou, Z, Wang, X, Kuang, X, Zhang, F, & Du, X (2003) Four-winged dinosaurs from China, *Nature*, 421: 335–340.

Zhou, Z, Barrett, PM, & Hilton, J (2003) An exceptionally preserved Lower Cretaceous ecosystem, *Nature*, 421: 807–814.

Chapter 3

Chiappe, LM, Norell MA, & Clark JM (1996) Phylogenetic position of *Mononykus* from the Upper Cretaceous of the Gobi Desert, *Memoirs of the Queensland Museum*, 39: 557–582.

—— (1997) *Mononykus* and birds: methods and evidence, *Auk*, 14(2): 300–302.

—— (1998) The skull of a new relative of the stem-group bird *Mononykus*, *Nature*, 392: 272–278.

Gatesy, SM & Dial, KP (1996) From frond to fan: *Archaeopteryx* and the evolution of short tailed birds, *Evolution*, 50: 2,037–2,048.

Norell, MA, Chiappe, LM, & Clark, J (1993) New limb on the avian family tree, *Natural History*, 9: 38–43.

Novacek, M (1996) *Dinosaurs of the Flaming Cliffs*, Anchor Books, New York.

Novas, FE (1997) Anatomy of *Patagonykus puertai* (Theropoda, Maniraptora, Alvarezsauridae) from the Late Cretaceous of Patagonia, *Journal of Vertebrate Paleontology*, 17: 137–166.

Perle, A, Norell, MA, Chiappe, LM, & Clark, JM (1993) Flightless bird from the Cretaceous of Mongolia, *Nature*, 362: 623–626.

Senter, P (2005) Function in the stunted forelimbs of *Mononykus olecranus* (Theropoda), a dinosaur anteater, *Paleobiology*, 31(3): 373–381.

Chapter 4

Barthel, KW, Swinburne, NHM, & Conway Morris, S (1990) *Solnhofen: A Study in Mesozoic Palaeontology*, Cambridge University Press, Cambridge.

Chatterjee, S (1999) *Protoavis* and the early evolution of birds, *Palaeontographica*, 254: 1–100.

Gatesy, SM & Dial, KP (1996) From frond to fan: *Archaeopteryx* and the evolution of short-tailed birds, *Evolution*, 50: 2,037–2,048.

Hecht, MK, Ostrom, JH, Viohl, G, & Wellnhofer, P (eds.) (1985) *The Beginnings of Birds: Proceedings of the International* Archaeopteryx *Conference*, Jura Museum, Eichstätt.

Heilmann, G (1926) *Origin of Birds*, Witherby, London.

Hopson, J (2001) Ecomorphology of avian and nonavian theropod phalangeal proportions: implications for the arboreal versus terrestrial origin of bird flight. In JA Gauthier & LF Gall (eds.) *New perspectives on the origin and early evolution of birds: proceedings of the international symposium in honor of John H. Ostrom*, Peabody Museum of Natural History, New Haven, 211–235.

Jenkins, FA (1993) The evolution of the avian shoulder joint, *American Journal of Science*, 293: 253–267.

Martin, LD, Zhou, Z, Hou, L, & Feduccia, A (1998) *Confuciusornis sanctus* compared to *Archaeopteryx lithographica*, *Naturwissenschaften*, 85: 286–289.

O'Connor, PM & Claessens, LPA (2005) Basic avian pulmonary design ventilation in non-avian theropod dinosaurs, *Nature*, 436: 253–256.

Ostrom, JH (1976) *Archaeopteryx* and the origin of birds, *Biological Journal of the Linnaean Society*, 8: 91–182.

Rowe, T, Kishi, K, Merck Jr, J, & Colbert, M (1998) *The Age of Dinosaurs*, 3rd edition. Educational Interactive Multimedia on CD-ROM for Macintosh and PC Computers, W. H. Freeman & Co., New York.

Viohl, G (1998) *Jura-Museum Eichstätt Willibaldsburg: The fossils of the Solnhofen Lithographic Limestone*, Bioschöfliches Seminar, St. Willibald, Eichstätt.

Chapter 5

Forster, CA, Sampson, SD, Chiappe, LM & Krause DW (1998) The theropodan ancestry of birds: new evidence from the Late Cretaceous of Madagascar, *Science*, 279: 1,915–1,919.

Gatesy, SM (1990) Caudofemoral musculature and the evolution of theropod locomotion, *Paleobiology*, 16: 170–186.

Ji, Q, Ji, S, Lü, J, You, H, Chen, W, Liu, Y, & Liu, Y (2005) First avialian bird from China (*Jinfengopteryx elegans* gen. et sp. nov.). *Geological Bulletin of China*, 24(3): 197–203.

Stokstad, E (2002) Fossil Bird From China Turns Tail, Spills Guts, *Science*, 297: 495.

Wellnhofer, P (1974) Das fünfte Skelettexemplar von *Archaeopteryx*, *Palaeontographica*, 147: 169–216.

Zhou, Z & Zhang, F (2002) A long-tailed, seed-eating bird from the Early Cretaceous of China, *Nature*, 418: 405–409.

–– (2003) *Jeholornis* compared to *Archaeopteryx*, with a new understanding of the earliest avian evolution, *Naturwissenschaften*, 90: 220–225.

Chapter 6

Chiappe, LM, Ji, S, Ji, Q, & Norell, MA (1999) Anatomy and systematics of the Confuciusornithidae from the late Mesozoic of northeastern China, *Bulletin of the American Museum of Natural History*, 242: 1–89.

Gatesy, SM & Dial, KP (1996) Locomotor modules and the evolution of avian flight, *Evolution*, 50: 331–340.

Hou, L, Zhou, Z, Martin, LD, & Feduccia, A (1995) A beaked bird from the Jurassic of China, *Nature*, 377: 616–618.

Ji, Q, Chiappe, LM, & Ji, S (1999) A new Late Mesozoic confuciusornithid bird from China, *Journal of Vertebrate Paleontology*, 19(1): 1–7.

Stokstad, E (2001) A Peek at China's Paleontological Bounty, *Science*, 291: 234–235.

Zhou, Z & Zhang, F (2003) Anatomy of the primitive bird *Sapeornis chaoyangensis* from the Early Cretaceous of Liaoning, China, *Canadian Journal of Earth Sciences*, 40: 731–747.

Chapter 7

Chiappe, LM (1995) The first 85 million years of avian evolution, *Nature*, 378: 349–354.

–– (1998) Wings over Spain, *Natural History*, 9: 30–32.

Chinsamy, A, Chiappe, LM, & Dodson, P (1994) Growth rings in Mesozoic avian bones: physiological implications for basal birds, *Nature*, 368: 196–197.

Dial, KP, Goslow, GE, & Jenkins, FA (1991) The functional anatomy of the shoulder in the European starling (*Sturnus vulgaris*). *Journal of Morphology*, 207: 327–344.

Martin, LD (1983) The origin and early radiation of birds. In AH Brush & GA Clark (eds.), *Perspectives in ornithology*, Cambridge University Press, Cambridge, 291–338.

Sanz, JL, Chiappe, LM, Pérez-Moreno, BP, Buscalioni, AD, & Moratalla, J (1996) A new Lower Cretaceous bird from Spain: implications for the evolution of flight, *Nature*, 382: 442–445.

Walker, CA (1981) New subclass of birds from the Cretaceous of South America, *Nature*, 292: 51–53.

Chapter 8

Norell, MA & Clarke, JA (2001) Fossil that fills a critical gap in avian evolution, *Nature*, 409: 181–184.

Rich, PV, Hou, L, Ono, K, & Baird, RF (1986) A review of the fossil birds of China, Japan and southeast Asia, *Geobios*, 19(6): 755–772.

Zhou, ZH & Zhang, F (2001) Two new ornithurine birds from the Early Cretaceous of western Liaoning, China, *Chinese Science Bulletin*, 46: 1,258–1,264.

–– (2005) Discovery of an ornithurine bird and its implication for Early Cretaceous avian radiation, *Proceedings of the National Academy of Sciences*, 102(52): 18,998–19,002.

Chapter 9

Bühler, P, Martin, LD, & Witmer, LM (1988) Cranial kinesis in the Late Cretaceous birds *Hesperornis* and *Parahesperornis*, *Auk*, 105: 111–122.

Chinsamy, A, Martin, LD, & Dodson, P (1998) Bone microstructure of the diving *Hesperornis* and the volant *Ichthyornis* from the Niobrara Chalk of western Kansas, *Cretaceous Research*, 19: 225–235.

Marsh, OC (1880) Odontornithes: A Monograph on the Extinct Toothed Birds of North America, *U.S. Geol. Explor., 40th Parallel*, Washington, DC.

Martin, LD & Tate, J (1976) The skeleton of *Baptornis advenus* (Aves: Hesperornithiformes), *Smithsonian Contributions to Paleobiology*, 27: 35–66.

Rees, J & Lindgren, J (2005) Aquatic birds from the Upper Cretaceous (Lower Campanian) of Sweden and the biology and distribution of hesperornithiforms, *Palaeontology*, 48(6): 1,321–1,329.

Chapter 10

Chiappe, LM & Dyke, G (2002) The Mesozoic radiation of birds, *Annual Review of Ecology and Systematics*, 33: 91–124.

Clarke, JA (2004) Morphology, phylogenetic taxonomy, and systematics of *Ichthyornis* and *Apatornis* (Avialae: Ornithurae), *Bulletin of the American Museum of Natural History*, 286: 1–179.

Clarke, JA, Tambussi, CP, Noriega, JI, Erickson, GM, & Ketcham, RA (2005) Definitive fossil evidence for the extant avian radiation in the Cretaceous, *Nature*, 433: 305–308.

Cooper, A & Penny, D (1997) Mass survival of birds across the Cretaceous–Tertiary boundary: molecular evidence, *Science*, 275: 1,109–1,113.

Cracraft, J (2001) Avian evolution, Gondwana biogeography and the Cretaceous–Tertiary mass extinction event, *Proceedings of the Royal Society of London*, 268: 459–469.

Dyke, GJ, Dortangs, RW, Jagt, JW, Mulder, EW, Schulp, AS, & Chiappe, LM (2002) Europe's last Mesozoic bird. *Naturwissenschaften*, 89(9), 408–411.

Dyke, GJ & van Tuinen, M (2004) The evolutionary radiation of modern birds (Neornithes): reconciling molecules, morphology and the fossil record, *Zoological Journal of the Linnean Society*, 141: 153–177.

Feduccia, A (2003) 'Big bang' for tertiary birds?, *Trends in Ecology and Evolution*, 18(4): 172–176.

Marsh, OC (1880) Odontornithes: A Monograph on the Extinct Toothed Birds of North America, *U.S. Geol. Explor., 40th Parallel*, Washington, DC.

Rowe, T, Kishi, K, Merck Jr, J, & Colbert, M (1998) *The Age of Dinosaurs*, 3rd edition. Educational Interactive Multimedia on CD-ROM for Macintosh and PC Computers, W. H. Freeman & Co., New York.

van Tuinen, M & Hedges, SB (2001) Calibration of avian molecular clocks, *Molecular Biology and Evolution*, 18: 206–213.

Index

tail **25**, 25, **127**, 128, 150, 155, **212**, 228

tail feathers 70, 76, 77–78, 93, 127, 128, 140, 155–56, 166, **172–73**, 175–77, **176**

tail locomotor module **170**, 172, 182

take-off run 96–98, 140–41

Tarsitano, Samuel 37

tarsometatarsus **25**, 27, 167, 213, 225

Tate, James Jr. 224

tectonic forces 7, 153

teeth
in birds 24, 127, 155, 157, 160, 169, **189**, **192**, 199, 214, 218, 222, **223**, 226, 234
in nonavian dinosaurs 11, 65, 117, 127

temporal openings 8, 24, 127

"temporal paradox" 49–52, **51**

tendons 60

terrestrial animals, evolution 8

terrestrial birds **139**, 139, 142, 179, 219, 246–47

Tertiary Period **6**, 240

tetanurans 13, 41, 55

Tethys Sea/Ocean 5, 136, 190, 201, 220, 224

tetrapods 8, 21, 54, 170

Teviornis gobiensis 241

Teylers Museum (Netherlands) 120–21

Thailand 12

"thecodont" hypothesis 36–39, **36**, 50–51, 54, 87

therizinosaurids 65

thermoregulation 79 *see also* insulation

theropod hypothesis 32–34, **36**, 36, 37, 39–52, **89**, 142

theropods 12–16, **17**, 23, 30, 112

thoracic sac **143**, 193

thrushes 195

thrust 80, 86, **86**, 96, **97**, 99

thyreophorans 10

Tibet 6

tibia 27, 166

tibiotarsus **25**, 27, 150, **150**, 167, 171, 213

tinamous 186, 237, 238

titanosaurs 12

toes 27, 162, 213, **216**
opposable 167, **170**, 194, 198
rotation of 129–30, 228

Tokaryk, Tim 224

toothlessness, evolution of 160, 218

traction 99

trailing vane 59, 93

tree-climbing 70, 87–94, **89**, 138, **139**, 178–81, 198

trees 137, 139–40, 198

Triassic Period 3–5, **5**, **6**, 144

Triceratops horridus 11, **64**

triosseal canal **26**, **26**, 128, **128**, 192

Troodon 47, **127**
T. formosus 46

troodontids **42**, **44**, 44, 71, 72, 95–96, 101

trunk of birds 24–25

tubercles 81–83

turbulence **194**, 195

Turgai Strait 220, 224

turtles 9, 30

two-toed earless skink 57, **58**

tyrannosaurids 83

tyrannosauroids 67

Tyrannosaurus rex 3, 12, 15, **17**, 32, 41, **64**, **64**, 83, 161

Ukhaa Tolgod (Mongolia) 43, 112, **112**, 216

ulna **25**, 26, 150

uncinate processes 25, **25**

unfolding of wing 218, **218**

uniformitarianism 223

USA *see* North America; *see also* particular states

Utahraptor ostrommaysorum 68

Uzbekistan 189

Valley of Gwangi (film, 1969) 190

vaned feathers **62**, 63, 65, 67, 67, **68**, **70**, 70, **80**

vanes 58–59, **59**, 76, 140

Vega Island (Antartica) 241

Vegavis iaai 241, 241

Velociraptor 15, 22, 41, **41**, **42**, 49, **51**, 53, **68**, 74, **89**, 107, **149**
V. mongoliensis 12

vertebrae 184 *see also* saddle-shaped vertebrae joints; spine

vertebrates 8–9, 21

Viohl, Günter 123, 124

volcanism 181, 225

von Meyer, Hermann 120

von Nopcsa, Baron Franz 95, **95**, 99

von Siemens, Ernst Werner 121–22, 133

Vorona berivotrensis 154, 215

vulture, Indian white-backed 31

Wagner, Andreas 130

Walker, Alick 39

Walker, Cyril 184–86, **187**, 195

warm-bloodedness 20, 79

water-tight egg 8

webbed feet **227**, 228

Wegener, Alfred 7

weight and thrust **97**, 97

Wellnhofer, Peter 133, 137

Wellnhoferia grandis **132**, 133–34

Western Interior Seaway *see* Pierre Seaway

whales 3

Wild, Rupert 39

William of Ockham 20

Willinston, S. W. 94

wing-assisted running 97–99, **98**

wing loading 86, 99, 101, 128, 194

wing locomotor module **170**, 182, 193

wingbeat cycle **26**, 96, 128, 193

wings 172, 214
convergent evolution 20, 20, **90**
measurements 86
specialization 26, 193

wingspan 86, 140, 173

wingstroke **26**, 92, 128, 156, 192, **193**

wishbone 13, 26, 34, **34**, **40**, 41, 169, 199–200, 210 *see also* furcula

Witmer, Lawrence M. 33

Workerszell (Germany) 124

wrist 26, 42, 96, 129, 140, 156, **156**, 166

Wyoming Dinosaur Center 125

x-rays **167**

Xu Xing 68, 93

Yalden, Derek 88, 138

Yangtze River (China) 181

Yanornis martini 100, **100**, 213–16, 219, **214**, 238

Yixian Formation (China) 48, 59, 68, 161, 181, **189**

Yixianornis grabaui 213–16, **214**, 238

Yucatán Peninsula 6

Yungavolucris bretincola 199

Zhou Zhonghe 88, 161, 169

Zupaysaurus rougieri 13